The DV-Xα Molecular-Orbital Calculation Method

Tomohiko Ishii • Hisanobu Wakita
Kazuyoshi Ogasawara • Yang-Soo Kim
Editors

The DV-Xα Molecular-Orbital Calculation Method

Springer

Editors
Tomohiko Ishii
Department of Advanced Materials Science
Faculty of Engineering
Kagawa University
Takamatsu, Japan

Hisanobu Wakita
Faculty of Science
Fukuoka University
Johnan-ku, Fukuoka, Japan

Kazuyoshi Ogasawara
Department of Chemistry
School of Science and Technology
Kwansei Gakuin University
Hyogo, Japan

Yang-Soo Kim
Korea Basic Science Institute
Suncheon, Republic of Korea

ISBN 978-3-319-36437-7 ISBN 978-3-319-11185-8 (eBook)
DOI 10.1007/978-3-319-11185-8
Springer Cham Heidelberg New York Dordrecht London

Printed on acid-free paper

Springer is part of Springer Science+Business Media (www.springer.com)

Foreword

The discrete variational Xα (DV-Xα) molecular orbital calculation method is one of the most versatile methods for estimating the electronic structures of atom-aggregates or clusters in both ground and excited states. It has been used extensively for solid-state chemistry and physics; materials science; and X-ray, electron, IR, Raman, and UV–vis spectroscopy with great success.

To develop the research on this method, the Society for DV-Xα in Japan was established in 1988, and an annual meeting has been held every summer since then. The International Workshops on DV-Xα have been held five times, once each in Hungary (1996), South Korea (1998), Japan (2001), South Korea (2006) and Japan (2008). The Society for DV-Xα in Korea was established in 2010 and has greatly promoted the research activity in Korea since then.

This proceedings, entitled *The DV-Xα Molecular-Orbital Calculation Method* is published in commemoration of the international symposium of the "Sixth International Conference on the DV-Xα Method (DV-Xα 2010) and the 23rd DV-Xα Annual Meeting," held in Daejeon, Korea, at the KBSI (Korean Basic Science Research Institute) on August 4–6 in 2010. This international conference was organized by the Organizing Committee 2010, which consisted of members of the DV-Xα societies in Japan and Korea who aimed to discuss the research on molecules, particles, crystals, glasses, metals, semiconductors, surface, boundary defects from a physics perspective, organic and inorganic chemistry, materials science and many forms of spectroscopy, the theory of DV-Xα, and related experiments and theories.

The international symposium consisted of 3 plenary, 20 invited and 27 poster sessions. Plenary lecturers Prof. Kazuyuki Tatsumi (the president of the IUPAC) talked about Fe/S cluster preparation, Prof. Di Zhang (Shanghai Jio Tong Univ.) about morphogenetic preparation, and Prof. Masahiko Morinaga (Nagoya Univ.) about alloy designing. Over 80 participants studiously discussed the subject of each lecture and eagerly shared ideas and possible future developments for this excellent method.

Finally, the organizing committee of DV-Xα 2010 concluded by publishing a number of important lectures in book-style proceedings for a notable memory of our activity, a wider understanding of the DV-Xα method, and to receive evaluations from other scientific fields.

To support this international symposium and the publication of this volume, special thanks should be expressed to both DV-Xα societies in Japan and Korea. The financial supporting funds from the JSPS (Japan Society for the Promotion of Sciences) and the KBSI are also acknowledged.

Fukuoka
25 July, 2014

Hisanobu Wakita
Myung-Chul Chang
Tomohiko Ishii
Yang-Soo Kim

Preface

Theoretical and computational sciences have become one of the most exciting areas of research. For example, the Nobel Prize in chemistry in 2013 was awarded to three theoretical chemists, Professors Martin Karplus, Michael Levitt and Arieh Warshel whose works have greatly influenced many scientists and engineers. The activity at the DV-Xα annual meeting to discuss computational sciences has continued for more than 25 years after the establishment of the DV-Xα society in Japan and Korea. The DV-Xα method has been utilized in not only the fundamental sciences such as physics and chemistry but also in a broader range of engineering areas such as metallurgy, material engineering, electricity, amorphous materials, photology, universology, bio-engineering, and pharmacology. This proceedings volume, entitled *The DV-Xα Molecular-Orbital Calculation Method* is published in commemoration of the international symposium of the "Sixth International Conference on the DV-Xα Method (DV-Xα 2010) and The 23rd DV-Xα Annual Meeting" held in Daejeon, KBSI, Korea, in 2010. In this volume there are 14 most powerful and interesting papers reported as recent activities involving the DV-Xα method. Aspects of the fundamental sciences, including details of calculation methods and related programs, fundamental calculation theories and their expressions, and inter-atomic interaction potential and total energy calculations are included. In addition, a material design method which works by means of chemical bonding, discussions of optical properties of certain materials, application examples of dye-sensitized solar cells, innovative applications of water, the creation of an electrode material for lithium secondary batteries, discussions of the lifetimes of positrons, crystal structural optimization techniques, and the mechanisms of the luminescence of metal complexes are also discussed as examples of applications of the DV-Xα method. We are very proud to publish such a high-quality and fruitful proceedings volume which offers discussions and examples of applications which have progressed in 4 years since the previous DV-Xα international meeting held in 2010.

Takamatsu, Japan Tomohiko Ishii
Hyogo, Japan Kazuyoshi Ogasawara
Suncheon, Republic of Korea Yang-Soo Kim
Fukuoka, Japan Hisanobu Wakita
22 July 2014

Contents

Part III Applications

Part I
Fundamentals

Chapter 1
The DV-Xα Molecular Orbital Calculation Method and Recent Development

Yoshiyuki Kowada and Kazuyoshi Ogasawara

1.1 Introduction

Recently, theoretical calculation methods are widely applied in many kinds of the material researches according to the outstanding progress of the computational technology. Among them the DV-Xα method has been widely applied to the development and analysis of the organic and inorganic materials since this method has characteristic advantages compared with the other molecular orbital methods. In this chapter, we would like to introduce the development history and the recent progress of this method.

The DV-Xα cluster method is one of the first principle molecular orbital calculation methods based on the Xα method. The Xα method is one of the electronic state calculation methods by the Hatree-Fock-Slater method proposed by J. C. Slater on 1951. This method is applied not only to the electronic state calculation of molecules and atoms, but also to the energy band calculation of solid-state materials. All atomic orbitals in the periodic table were calculated with Hartree-Fock-Slater method by Herman and Skillman on 1963 and the table of the eigen values, wavefunctions and potentials were reported (Herman and Skillman 1963). On 1965, the calculation results of transuranium elements were reported by Liberman, Waber, and Cromer, in which the Dirac equation with Xα potential was used (Liberman et al. 1965). The first application of Xα potential to

Y. Kowada (✉)
Department of Natural Sciences, Hyogo University of Teacher Education,
942-1 Shimokume, Kato-shi, Hyogo 673-1494, Japan
e-mail: ykowada@hyogo-u.ac.jp

K. Ogasawara
Department of Chemistry, School of Science and Technology,
Kwansei Gakuin University, 2-1 Gakuen, Sanda, Hyogo 669-1337, Japan
e-mail: ogasawara@kwansei.ac.jp

T. Ishii et al. (eds.), *The DV-Xα Molecular-Orbital Calculation Method*,
DOI 10.1007/978-3-319-11185-8_1

the molecular orbital theory was the multiple scattering (MS) Xα method developed by K. H. Johnson on 1966 (Johnson 1966).

In the Xα method, the simple notation of the exchange potential, which is proportion to the third power of the electrondensity ρ with the approximation of the free electron model, is used. Furthermore, the parameter α to this equation, shown in Eq. (1.1), is adapted to expanded this approximation to the general atoms and molecules (Slater 1974),

$$\mathrm{V_{XC}}(\mathrm{r}_\uparrow) = -3\alpha\left[\frac{3}{4\pi}\rho_\uparrow(\mathrm{r})\right]^{\frac{1}{3}} \tag{1.1}$$

here $\rho_\uparrow$ is the density of the up-spin electron, which means that the exchange potential $\mathrm{V_{XC}}(\mathrm{r}_\uparrow)$ is the function of only up-spin electron , so that the exchange interaction is observed only between electrons with the same spin. The same equation is obtained for the down-spin electrons. The total electron density is described as $\rho = \rho_\uparrow + \rho_\downarrow$. Consequently, the Xα method can be considerably shortened the calculation time than those of any other MO methods.

This method is applied not only to the electronic state calculation of molecules and atoms, but also to the energy band calculation of solid-state materials. After Gaspar and Korn, the densities functional method is developed with more restrict theoretical expansions. Essentially the Xα approximation is still good agreement with the density functional method.

D.E. Ellis, A. Rosen, and H. Adachi et al. developed the DV-Xα cluster method at Northwestern University in the 1970s (Ellis et al. 1976; Rosen et al. 1976) There are several characteristic advantages on this method. That is, the discrete-variational method can be numerically solved the Schrödinger equation at the discrete micro-spatial volume on the sample points scattered in three-dimensional space. The evaluation of the elements of Hamiltonian, H_{ij}, and overlap matrices, S_{ij}, are achieved as weighted sums of integrated values at the sample points instead of the conventional Rayleigh-Rits method,

$$S_{ij} = \sum_k \omega(r_k)\phi_i^*(r_k)\phi_j(r_k) \tag{1.2}$$

$$H_{ij} = \sum_k \omega(r_k)\phi_j^*(r_k)h(r_k)\phi_j(r_k) \tag{1.3}$$

where r_k is one of the total N sample points in the three-dimensional real space and $\omega(r_k)$ is the integration weight or reciprocal of the sample point density at r_k. As the result, the DV-Xα method can calculate the eigenvalues and the eigen functions of molecular orbitals of the cluster models containing any elements in the periodic table. That is the reason why this method is widely applied to the calculation of the electronic states of the various kinds of inorganic materials.

After the initial period of the development of the DV-Xα method, the respective researchers have been modified the original program and added unique ideas.

At present, there are several versions (program sets) in the present, such as the program set named "dvem" developed by D. E. Ellis, "ADF" by Baerends, "DMOL" by Delly. In Japan, Adachi et al. have been hardly developed program set "scat" to the present (Adachi et al. 1978). The Initial version of scat was developed on the mainframe computer in Osaka University and ported to UNIX workstation in early 1990s. In that time, several programs to draw figures in the scat came to work with the graphical interface on the X-window system. The windows PC version was appeared in the middle of 1990s and this version is the most popular in Japan now.

The development of DV-Xα method has been achieved up to the present. The relativistic version, which included relativistic effect with Dirac equation, is developed at the same time with the non-relativistic version by Schrödinger equation. The results with the relativistic version of DV-Xα method were reported in early 1970s (Rosen and Ellis 1974). The relativistic version had been improved and we can use the program set named rscat (Adachi 1977). The rscat is very easy to use because of the same operation scheme with the non-relativistic scat.

By using the DV-Xα method, the eigenvalue and eigen function of each molecular orbital could be obtained accurately for all elements in the periodic table. However, there are still several problems to be solved for further calculation. One of these problems is the many electron interaction, which is necessary to obtain the optical spectra of 3d transition metal ions and 4f rare-earth ions. In 2001, the program named "gemcalc" has been developed by Ogasawara et al., which can include the many electron interactions by the configuration interaction calculations (Ogasawara et al. 2001). They have already reported many papers about the absorption spectra of the rare-earth ions and phosphors (Ishii et al. 2004b; Ogasawara et al. 2005; Watanabe et al. 2006; Toyoshima et al. 2007; Yoshida and Ogasawara 2009). This calculation method, called "DVME method", has been very attractive since it is one of few methods including both many electron interactions and the relativistic effect. Another problem is the total energy calculation of the clusters and molecules. In the DV-Xα method, the numerical eigen function is used and the integration is adopted to obtain the resonance and overlap integrals. The quite small numeric errors of the integrals at the sample points are accumulated and it affects to the total energy. However, there are two different version of total energy calculation program named TESDA and Coulomb and now we can obtain total energy of any cluster models by the DV-Xα method (Kowada et al. 2009a).

In this chapter, we would like to introduce the recent evaluation of the DV-Xα method. The first topic is the electronic state of solid-state electrolyte in which the calculation of moving ions in solid-state materials and total energy of the cluster model (Kowada et al. 2009). The second is the electronic state of the rare-earth ions in phosphate glasses with the calculation of the absorption and fluorescence spectra of the Pr^{3+} and Tb^{3+} ions (Kowada et al. 2009b).

1.2 Total Energy Calculation

Since there are several programs for the energy calculation, among them we use the program Coulomb to obtain the total energy (Kowada et al. 2009a). The calculated energy obtained in the cluster method is the total energy of the local cluster. Then, we called the total energy as local cluster energy (LCE). First we would like to note that the outline of the calculation procedure of the total energy by the program Coulomb. In the DV-Xα cluster method, the molecular orbitals are obtained by the SCC approximation (Adachi et al. 1978). For the energy calculation, the MOs are re-calculated by the Self Consistent Field (SCF) approximation. In the Coulomb, the SCF molecular orbital calculation is achieved in the following procedure. The wave functions of molecular orbitals calculated by the DV-Xα cluster method are used for the basis functions in the SCF calculation.

$$\phi_\ell = \sum_i C_{i\ell}\chi_i \tag{1.4}$$

The electron density ρ(r) is obtained from the molecular orbitals and the Coulomb potentials between two electrons, V_{ee}, are calculated.

$$\rho(r) = \sum f_\ell \phi_\ell^2(r) \tag{1.5}$$

$$V_{ee}(r_1) = \int \frac{\rho(r_2)}{r_{12}} dr_2 \tag{1.6}$$

The eigenvalues and C_{ij} are obtained by following equations.

$$\left(\widetilde{\mathbf{H}} - \widetilde{\varepsilon}\,\widetilde{\mathbf{S}}\right)\widetilde{\mathbf{C}} = 0 \tag{1.7}$$

$$(H)_{ij} = (K.E.)_{ij} + (V_{Ze})_{ij} + (V_{ee})_{ij} + (V_{ex})_{ij} \tag{1.8}$$

$$(K.E.)_{ij} = \int \chi_i(r_1)\left\{-\frac{1}{2}\nabla_1^2\right\}\chi_j(r_1)dr_1 \tag{1.9}$$

$$(V_{Ze})_{ij} = \int \chi_i(r_1)\{\sum_{n=1}^{\infty} \frac{(-Z_V)}{r_{1V}}\}\chi_j(r_1)dr_1 \tag{1.10}$$

$$(V_{ee})_{ij} = \int \chi_i(r_1)\int \frac{\rho(r_2)}{r_{12}} dr_2\,\chi_j(r_1)dr_1 \tag{1.11}$$

$$(V_{ex\uparrow})_{ij} = -3\alpha\left(\frac{3}{4\pi}\right)^{\frac{1}{3}} \int \chi_{i\uparrow}(r_1)(\rho_\uparrow(r_1))^{\frac{1}{3}}\chi_{j\uparrow}(r_1)dr_1 \tag{1.12}$$

After the SCF calculation, the energy of the cluster is calculated.

$$E = E_{K.E.} + E_{ze} + E_{ee} + E_{ex} + E_{zz} \tag{1.13}$$

where $E_{K.E.}$, E_{Ze}, E_{ee}, E_{ex} and E_{ZZ} mean the kinetic energy of the electrons, the energy of the nuclear-electron attraction, the energy of the electron–electron repulsion, the energy of the exchange term, and the energy of the nuclear-nuclear repulsion, respectively. Each energy term is calculated by the weighted sum of the each value on the sample points except E_{ZZ}. In the program Coulomb, there are two major improvements for the energy calculation. The first is the improvement of the estimation of the Coulomb potential. In the DV-Xα molecular calculation, the electron density is calculated as the spherically symmetric charge densities centered on the nuclei of the molecule. In the Coulomb, however, the Coulomb potential is estimated by the electron distribution based on each molecular orbital. The second is the SCF calculation, which makes the molecular orbitals of the lowest energy with the linear combination of the fixed atomic orbitals obtained by the SCC calculation. These improvements give enough accuracy for the total cluster energy. The number of the sample points used in the energy calculation is 2,000 per atom, which is four times larger than that in the chemical bonding analyses, for the enough precision of the wave functions of molecular orbitals.

1.3 Relativistic DVME Method

The detail of this method will be described in another chapter, so that the outline and the estimation of the theoretical fluorescence spectra are noted here. In the DVME method, Slater determinants corresponding to all the possible electronic configurations are constructed using the four-component fully relativistic molecular spinors obtained by the relativistic DV-Xα cluster calculations. Only electrons occupying the molecular spinors mainly composed of Ln-4f states in the grand state are treated explicitly. Then by diagonalization of the many-electron Hamiltonian, the multiplet energies are obtained as eigenvalues and the many-electron wave functions are obtained as linear combination of the Slater determinants. The relativistic many-electron Hamiltonian is expressed as

$$H = \sum_{i=1}^{n}\left[c\alpha\mathbf{p}_i + \beta c^2 - \sum_{\nu}\frac{Z_\nu}{|\mathbf{r}_i - \mathbf{R}_\nu|} + V_0(\mathbf{r}_i) + \sum_{\mu}\frac{Z_\mu^{eff}}{|\mathbf{r}_i - \mathbf{R}_\mu|}\right] + \sum_{i=1}^{n}\sum_{j>i}^{n}\frac{1}{|r_i - r_j|} \tag{1.14}$$

where α, β are the Dirac matrices, c the velocity of light, r_i, p_i the position and the momentum operator of the *i*th electron, Z_ν and R_ν the charge and position of the νth nucleus, $Z_{\mu eff}$ and R_μ the effective charge and position of the μth ion outside the model cluster, n the number of explicitly treated electrons. $V_0(r_i)$ is the potential from the remaining electrons (Watanabe and Kamimura 1989).

The oscillator strengths of electric dipole transitions in the absorption spectra are calculated as follows:

$$I_{if} = 2\Delta E \left| \left\langle \Psi_f \middle| \sum_k r_k \cdot e \middle| \Psi_i \right\rangle \right|^2 \tag{1.15}$$

Ψ_i and Ψ_f are the initial and final states with energies of E_i and E_f, respectively. ΔE is the energy difference between E_i and E_f. In the case of fluorescence spectra, the transition probabilities of electric dipole transitions are obtained by the following equation:

$$I_{if} = 2\Delta E^3 \left| \left\langle \Psi_f \middle| \sum_k r_k \cdot e \middle| \Psi_i \right\rangle \right|^2 \tag{1.16}$$

In this calculation, the higher energy state is adopted as the initial state of the fluorescence spectra and usually the grand states is the final state. We could obtained the transition probabilities of the fluorescence spectra in the same manner as the absorption spectra.

1.4 Recent Application of the DV-Xα Method

1.4.1 Electronic State of Mobile Li Ions in Super-Ionic Conductors

1.4.1.1 Introduction

Recent 10 years we have investigated the electronic state of the ion movement in the solid-state materials (Kowada et al. 1998, 2000, 2004, 2008a, b, 2009a, b, c; 2010; Imanaka et al. 2000; Araki et al. 2001). Usually the electronic state calculation of moving ions are very difficult to obtain since Born-Oppenheimer approximation is adapted in popular molecular orbital method. The movement of the ions in the solid sate materials, however, are much slower than that of the electrons that contribute to have any interactions with the surrounding ions in the crystal structure. This means the variation of the interactions between the moving ions and the surrounding matrices could be estimated approximately with several model clusters with the different positions of the moving ions.

In these studies, we have paid attentions on the bonding state of the moving ions from the viewpoints of the interactions named covalency and ionicity between the moving ions and the surrounding ions. Thus the covalent bonding of the moving ions should support an important part of the ion movement in the solid-state materials. We would like to introduce the study of the electronic states of the

Li_3N crystal in which the interaction of the moving ions in superionic conducting materials is analyzed. This material is a typical Li ion conductor and has high ionic conductivity. Furthermore the Li and the N ions usually have covalent interactions in the solid-state materials. We also suggest that how useful the DV-Xα method is to get the relationship between the high ionic conductivity and the covalent interactions of the moving cations in these systems.

1.4.1.2 Li_3N Crystal

There are several studies about the high Li ion conductivity of Li_3N crystal by experimental measurements and theoretical calculations (Wahl and Holland 1978; Walker and Catlow 1981; Koyama et al. 2002; Sarnthein et al. 1996). Several experimental studies reported that the Li_3N crystal contained H^+ ion impurities in interstitial positions and they make a certain concentration of the Li ion vacancies. These vacancies are yield on the planes, which contains both Li and N ions, called Li_2N plane. It is reported that the activation energy of the movement of the Li ion on Li_2N planes is 0.004 eV (Sarnthein et al. 1996). In the Li_3N crystal there is the other kind of the Li ions on the planes containing only the Li ions, which connect between the Li_2N planes. This kind of the Li ions hardly moves compared with that on the Li_2N plane. In addition to this, the Li_3N crystal has the anisotropicity in the ionic conductivity and the Li ion conductivity along the c axis is much smaller than those of the a and b axes.

Thus the Li ion can move easily in the Li_2N plane and hardly move in the Li plane. This is something strange from the viewpoint of the chemical bonding. In the Li_2N plane, the moving Li ions might have strong interaction with the N ions compared with those on the Li plane. Generally the larger interaction between the Li and the N ions seems to impose constraints on the fast movement of the Li ion. The experimental results, however, suggest that the fast movement of the Li ions in the Li_3N crystal is not dependent on the smaller interaction between the Li and the N ions. The chemical bonding of the Li ion should contribute to the fast movement of the Li ion in Li_3N crystal.

1.4.1.3 Model Cluster of Mobile Li Ion in Li_3N Crystal

In order to calculate the movement of the Li ion in the Li_3N crystal, several model clusters are constructed by the crystal structure of the Li_3N crystal (Schulz and Thiemann 1979). This crystal is assigned to the space group of $P_{6/mmm}$ and the lattice constants a and c are 3.65 and 3.88 Å, respectively. The schematic diagrams of the model clusters are shown in Fig. 1.1. The Li_3N crystal has layer structure of two kinds of plane structures, the Li_2N and the Li planes. On the Li_2N plane, there are Li and N ions that make hexagons. On the Li plane there are only Li ions that are bonding with the N ions on the Li_2N planes and are bridging between two Li_2N

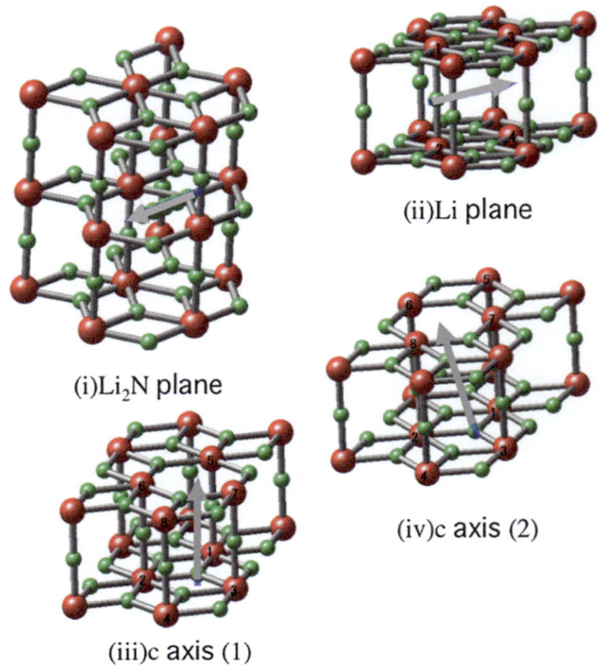

Fig. 1.1 Schematic diagram of the Li_3N model clusters. *Green sphere*: Li ion, *Red sphere*: N ion

planes. The conduction path of the Li ions reported by several experimental studies is included in the Li_2N plane. Figure 1.1 (i) shows the model clusters for the conduction path in Li_2N plane. In this model, one Li ion is removed from the Li_2N plane so that the neighboring Li ion can move to the vacancy. The movement of the Li ion is simulated by using several model clusters with the different positions of the Li ion along the estimated conduction path, since it is fundamentally very difficult to manipulate the movement of the atoms and ions in the molecular orbital calculations. Figure 1.1 (ii) shows the model cluster for the movement of the Li ions on the Li plane. In this model cluster one Li ion is moved to the neighboring vacancy of Li site. This path is not a suitable path for the Li ion in the Li_3N crystal reported by several experimental studies.

The Li_3N crystal has anisotropy of the Li ion conductivity and the conductivity along the c axis is much smaller than those in the a and b axes. Then we have used model clusters of the movement of the Li ion toward the c axis shown in Fig. 1.1 (iii) and (iv) for the comparison of the movement toward the different directions. Since there are several possibilities of the conduction path toward c axis, we have adopted two possible paths for the model clusters.

1.4.1.4 Local Cluster Energy of Mobile Li Ion

Figure 1.2a and b show the variation of the LCE with the different positions of the moving Li ion in the clusters (i) and (ii), respectively. In this figure, the ordinate

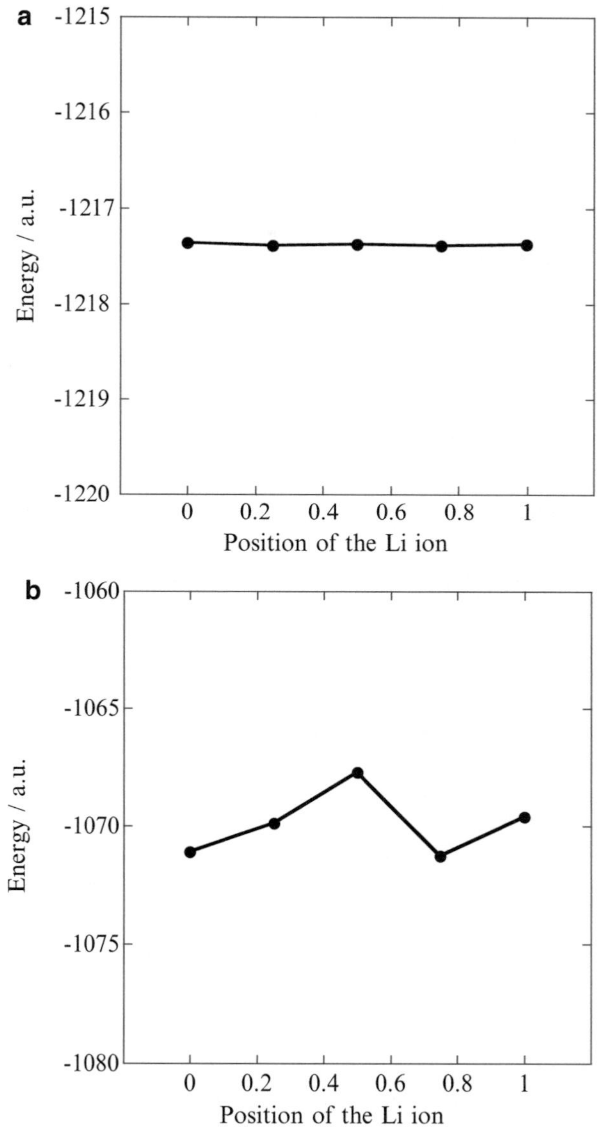

Fig. 1.2 The position dependence of total energy of the clusters, (**a**) cluster (i), (**b**) cluster (ii)

shows the LCE of the clusters in a.u. and the abscissa shows the relative position of the moving Li ion. In the case of cluster (i), the energy changes very small with the positions of the moving Li ion, that is, the difference between the minimum and the maximum is 0.031 a.u. This value is larger than the result of the band calculation (Sarnthein et al. 1996), since the LCE usually includes lager interaction with surface atoms. On the other hand, the energy change with the Li position in the cluster (ii) is 3.796 a.u. and is much larger than that in the cluster (i). Thus the LCE evaluated by the Coulomb could be useful to discuss the movement of the Li ion in the Li ion conductors, although the cluster size, the number of sample points, the basis functions and the other computational parameters should be clarified to obtain the qualitative total energy of the cluster. Thus the Li ions on the Li_2N plane could move easier than those on the Li plane.

1.4.1.5 Bonding Nature and Li Ion Movement

The bonding state of the moving Li ions is analyzed, since the smaller energy different with the movement of the Li ion might be dependent on the chemical bonding of the moving Li ion.

Figure 1.3 shows net charge of the moving Li ions in clusters (i)–(iv). In this figure, the abscissa shows the relative position of the moving Li ion. Since the length of the conduction path (i)–(iv) are different, the value of 0 means the initial position and 1 means the final position of the moving Li ion so that the position of the moving lithium ion in each path is normalized.

In the cluster (i), shown as closed circles, the net charge is 0.41 at initial position. It changes little with the movement of the Li ion and becomes 0.43 at the final position. The charge of the Li ion is almost the same and no obvious changes are observed in the cluster (ii), shown as closed squares. In the Li_3N crystal, however, there is no obvious change of the net charge of the Li ion during the movement. In the cases of the clusters (iii) and (iv), shown as closed triangles and diamonds, the net charge of the Li ion at initial position is 0.28, which is clearly smaller than those in the clusters (i) and (ii). The net charge of the Li ion is increased at $x = 0.25$

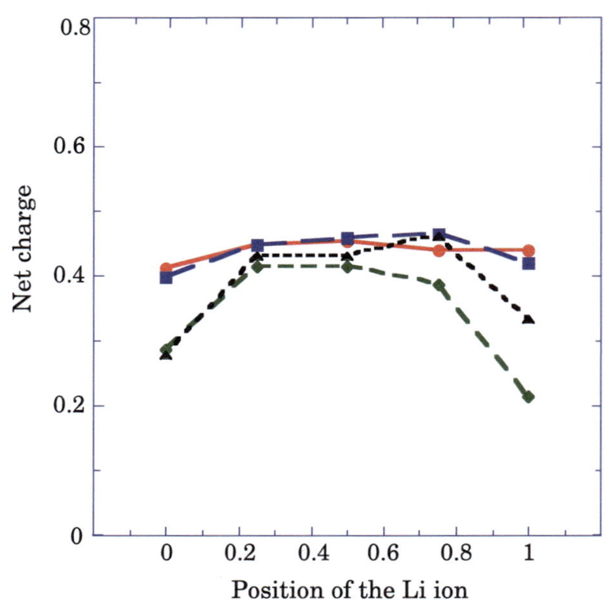

Fig. 1.3 The relationship between the net charge and the position of the moving Li ion in the model clusters (i)–(iv). ●: cluster (i), ■: cluster (ii), ▲: cluster (iii), ◆: cluster (iv)

and is almost constant until x = 0.75. At the final position the net charge becomes smaller again. This result might be caused by the structures of the model clusters. In the clusters (iii) and (iv), the initial and final positions are located on the surface of the model clusters. Though the surface atoms usually have larger electron density than the inner atoms, the net charges of the Li ions became smaller at the initial and the final positions in the clusters (iii) and (iv). There are some difficulties of the calculations of the conduction path along with the c axis because the cluster size becomes two times larger than the clusters (i) and (ii) if all the Li ions around the conduction path through the c axis are located inside of the model clusters. However, the net charge at the positions inside of the clusters, from x = 0.25 to x = 0.75, shows similar values as those observed in the clusters (i) and (ii).

Thus the net charges of the moving Li ions in the different paths are very similar each other and we can conclude that there is no obvious change of the ionic interaction of the moving Li ion in the experimentally reported conduction path compared with the other non-conduction paths.

Figure 1.4 shows the total bond overlap populations (TBOP) of the moving Li ion in the model clusters shown in Fig. 1.1. The TBOP means the summation of all the bond overlap populations between the moving Li ion and the surrounding ions, which shows the covalency of the bonding state of the moving Li ions.

In the cluster (i), which is the model of the Li_2N plane, TBOP is 0.82 at the initial position and is slightly decreased with the movement of the Li ion. There is a minimum of TBOP at x = 0.5. The TBOP is increased after x = 0.5 and is 0.79 at the final position. Basically the similar change is observed in the TBOP of the moving Li ion on the Li plane. In this plane the TBOP is 0.80 at the initial position and is decreased with the movement of the Li ion. A minimum of TBOP is observed at x = 0.5 and finally the TBOP becomes 0.78. At the initial and the final positions the

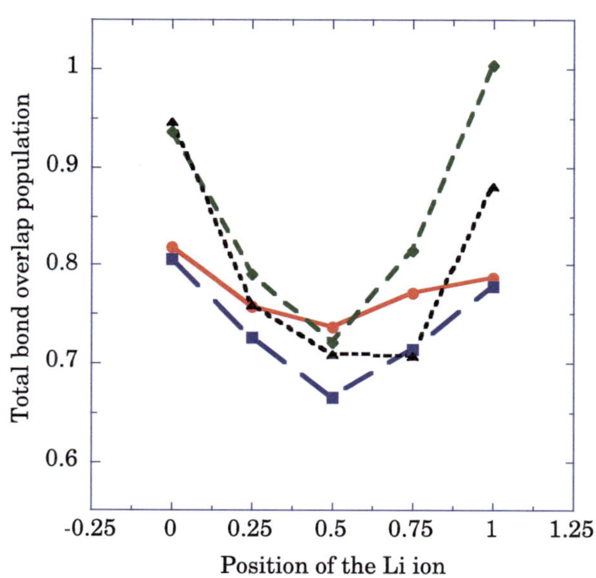

Fig. 1.4 The relationship between the total bond overlap population and the position of the moving Li ion in the model clusters (i)–(iv). ●: cluster (i), ■: cluster (ii), ▲: cluster (iii), ◆: cluster (iv)

TBOP of the moving Li ion on the Li_2N plane are almost the same as those on the Li plane. However, at the middle position, x = 0.5, the TBOP on the Li_2N plane is almost 0.1 larger than that on the Li plane. As the result, the difference between the maximum and the minimum of the TBOP of the Li ion on the Li_2N plane is 0.077, which is much smaller than that on the Li plane, 0.13. This smaller change of the TBOP through the path is a characteristic of the bonding nature of the moving Li ion on the Li_2N plane. Furthermore, this change is consistent with the results of the chemical bonding analyses.

In the case of the movement of the Li ion toward the c axis, the different changes are observed. In the model cluster (iii) for the c axis movement, the TBOP of the moving Li ion is 0.95 at the initial position. The TBOP of the moving Li ion is decreased largely with the movement and becomes minimum, 0.72, at x = 0.5. The change of the TBOP with the movement of the Li ion through the path toward the c axis is about 0.2, which is much larger than those in the clusters (i) and (ii). The same tendency is observed in the cluster (iv). The initial TBOP of the Li ion is 0.94 and the minimum is 0.70 at x = 0.75. The difference between the maximum and the minimum is 0.35. In general the surface ions have relatively large interaction with the surrounding ions. This "surface effect" is the reason of the large change of the covalency in these model clusters. However, since the net charge of the moving Li ion is almost the same as those on the Li and Li_2N plane in the positions in the range from x = 0.25 to 0.75 as shown in Fig. 1.3, the result suggests that surface effect of the bonding state of the moving Li ion could be neglect in these positions.

1.4.1.6 Conclusion

As the results, the variation of LCE with the movement of the Li ion suggests that the movement of the Li ion through the conduction path make the smaller energy change than another path. The net charge of the moving Li ion in any paths has very small change with the movement. Contrary to this, the variations of the total bond overlap populations of the moving Li ion are different from one another. Especially the TBOP of the Li ion changes smaller on the Li_2N plane, which contains the conduction path, than those on the other paths. The smaller change of TBOP of the moving Li ion is dependent on the little change of the BOP of both Li-Li and Li-N. These results suggest that the small change of the TBOP of the moving Li ion makes smaller energy difference with the movement and could be one of the origins of the fast movement of the Li ion in the Li_3N crystal.

1.4.2 Electronic States of Lanthanide Ions in Phosphate Glasses

1.4.2.1 Introduction

Rare-earth ion doped oxide glasses have been attractive for their application to white LED, the fiber-LASER amplification, and displays, etc. Many people have made effort to study about the absorption and luminescence spectra of these glassy materials (Miniscalco 1991; Wang et al. 1994; Souza et al. 2000; Reddy et al. 2000).

In these studies, theoretical analyses of the electronic state of rare-earth ions in oxide glasses are very important. Usually the absorption and fluorescence spectra of these glasses are analyzed by the empirical method (Ajithkumar et al. 2000; Tanabe et al. 2000, 2002; Jose et al. 2003; Moorthy et al. 2005). There are several empirical methods to analyze 4f-4f transition spectra of rare-earth ions. But in these methods, we need the experimental parameters to obtain the results.

Recently the relativistic DVME method, which is one of the first principles configuration interaction method is developed and there have been already many studies about electronic state of rare-earth ions in solid-state materials (Ishii et al. 2004a, b; Ogasawara et al. 2005; Watanabe et al. 2006; Toyoshima et al. 2007; Yoshida and Ogasawara 2009). However, there are few studies about fluorescence spectra of rare-earth ions in the oxide materials. We have calculated fluorescence spectra of the Pr^{3+} and Tb^{3+} ions in phosphate matrices by the relativistic DVME method. These phosphate materials have been expected as new laser materials for various applications because of their good fluorescence efficiency in the visible and infrared region.

1.4.2.2 Fluorescence Spectra of Pr^{3+} Ion in Phosphate Glass

Figure 1.5 shows model cluster, $H_{18}PrP_7O_{28}$, for the calculation of Pr^{3+} ions in the phosphate glasses used. This cluster includes seven surrounding PO_4 units terminated with H ions for the neutrality of the cluster. The atom positions are determined from $Pr(PO_3)_3$ crystal structure (Jouini et al. 2003). In this case, Madelung potential has not used for the calculation of the relativistic DV-Xα method. Though the cluster is a model cluster for $Pr(PO_3)_3$ crystal, it can be applied to the calculation of glassy materials because this model has no long-range structure.

Figure 1.6 shows the theoretical spectrum of the $H_{18}PrP_7O_{28}$ cluster. In the spectrum, the several initial states, such as $^3P_2$, $^3P_1$, $^3P_0$ and $^1I_6$, are used to obtain the theoretical fluorescence spectrum of Pr^{3+} ion in phosphate matrices. In this figure, the experimental Pr^{3+} fluorescence spectrum in the $Pr(PO_3)_3$ crystal is shown for comparison (Jouini et al. 2003). In the theoretical spectrum, there are two intense peaks around 350 and 570 nm. The peak positions shift slightly smaller

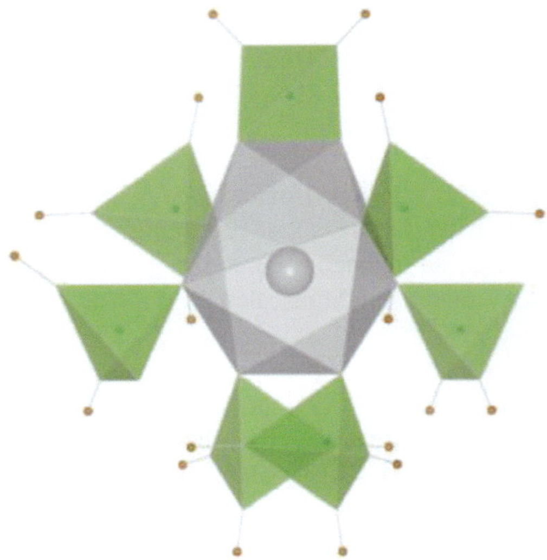

Fig. 1.5 Schematic diagram of the $H_{18}PrP_7O_{28}$ cluster

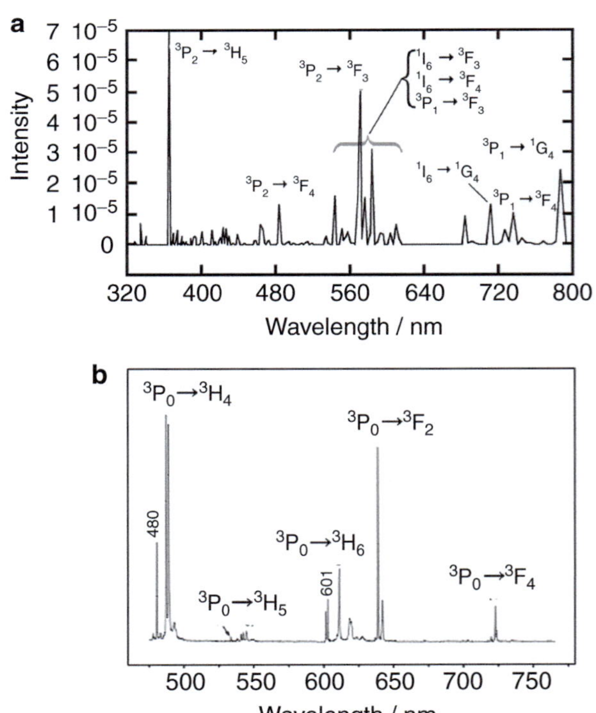

Fig. 1.6 Theoretical fluorescence spectrum of the $H_{18}PrP_7O_{28}$ cluster with several initial states. (a) Theoretical spectrum, (b) experimental spectrum reported by A. Jouini et al. 2003

wavelength but the relative positions of these major peaks are corresponding to those in the experimental spectrum. Furthermore, the relative peak intensities are also similar to those in the experimental spectrum. However, the assignments of the peaks shown in the figure are very complicated. For an example, the assignments of the intense peaks around 550–610 nm are several transitions from $^3P_2$, $^1I_6$, and $^3P_1$ to $^3F_3$, and from $^1I_6$ to $^3F_4$, though the corresponding peaks around 600–650 nm in the experimental spectrum are assigned to the transitions from $^3P_0$ to $^3H_6$ and $^3F_2$. This result suggests that there are several initial states in the fluorescence of Pr^{3+} ion in phosphate matrices, and it is very difficult to make assignment of the peaks by only the empirical method. This result also shows the relativistic DVME method is very useful to analyze fluorescence spectra of rare-earth ions in phosphate matrices and the first principles calculation is necessary to analyze the fluorescence spectra of rare-earth ions.

1.4.2.3 Fluorescence Spectra of Tb^{3+} Ion in Phosphate Glass

Figure 1.7 shows model clusters, $H_{15}TbP_6O_{24}$, used in the calculation of the electronic state of the Tb^{3+} ion. In this model cluster the atom positions are estimated by the ionic radii of Tb^{3+} ion and the structure of $Lu(PO_3)_3$ crystal, since there is no report about the crystal structure of Terbium phosphate crystals (Yuan et al. 2008). The crystal structure of $Lu(PO_3)_3$ is monoclinic, space group C_C and cell parameters a = 13.972 Å, b = 20.018 Å, c = 9.9556 Å, β = 127.351°. The model cluster includes only one lanthanide ion site. This cluster includes six surrounding PO_4 units terminated with H ions for the neutrality of the cluster, though the cluster of the Pr^{3+} ion, which is a heavy-lanthanide, has seven PO_4 units.

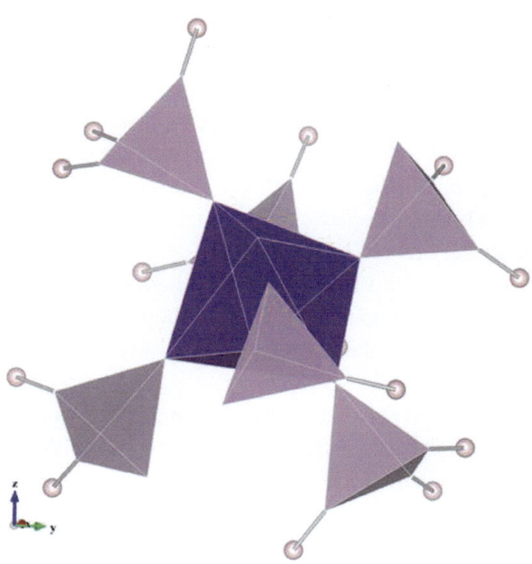

Fig. 1.7 Schematic diagram of the $H_{18}TbP_7O_{28}$ cluster (These figures were drawn with VESTA developed by K. Momma and F. Izumi)

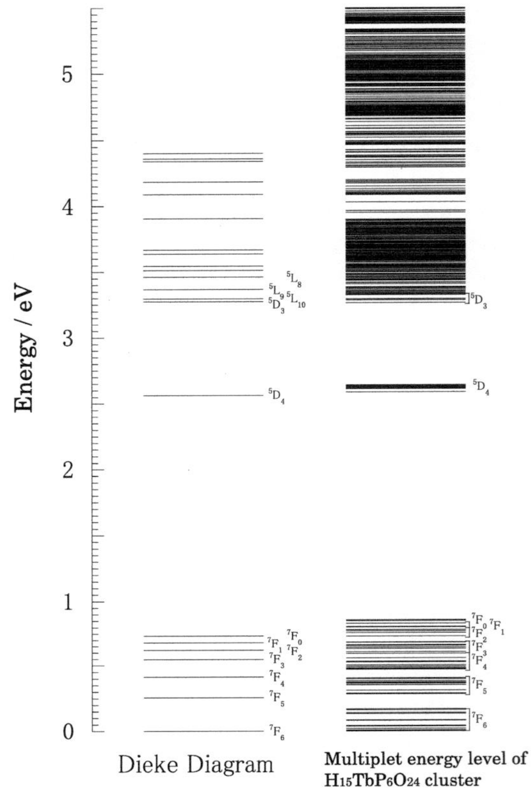

Fig. 1.8 Multiplet energy level structure of $H_{18}TbP_7O_{28}$

Figure 1.8 shows the obtained multiplet level structure of the Tb^{3+} ion in the $H_{15}TbP_6O_{24}$ cluster. Tough there are only 14 molecular orbitals mainly contributed by 4f atomic orbitals of the Tb^{3+} ion in the usual one electron approximation molecular orbital method, there are 3,003 multiplet levels after the electron configuration interaction calculation. The level structure of the multiplet states is basically corresponding to that reported by Dieke (1968).

Figure 1.9 shows the calculated and experimental absorption spectra of the Tb^{3+} ion in the $H_{15}TbP_6O_{24}$ cluster. The experimental spectrum is the result of the Tb^{3+} ion doped calcium meta-phosphate glass, which is prepared to compare with the theoretical ones.

In the relativistic DVME method, the energy of the multiplet levels in the 4f-4f transitions is usually over estimated (Brik and Ogasawara 2007). In these theoretical spectra, the energy scale is corrected with a factor, 0.75 for easiness of comparison.

In the experimental spectra, there are small peaks at about 2.51 in the visible region. The same peak is observed in the theoretical spectra and the main components of this peak is the transition from $^5D_4$ to $^7F_6$. Basically, the peak position observed in the theoretical spectrum is in good agreement with the experimental one. Thus, relativistic DVME method is useful to assign the peaks in the absorption spectra of Tb^{3+} ion in phosphate glasses.

Fig. 1.9 Absorption spectra of $H_{18}TbP_7O_{28}$. The experimental spectra were measurements of the $50CaO{\bullet}50P_2O_5$ glass with 1 mol% of Tb_2O_3. (**a**) Experimental spectrum, (**b**) theoretical spectrum

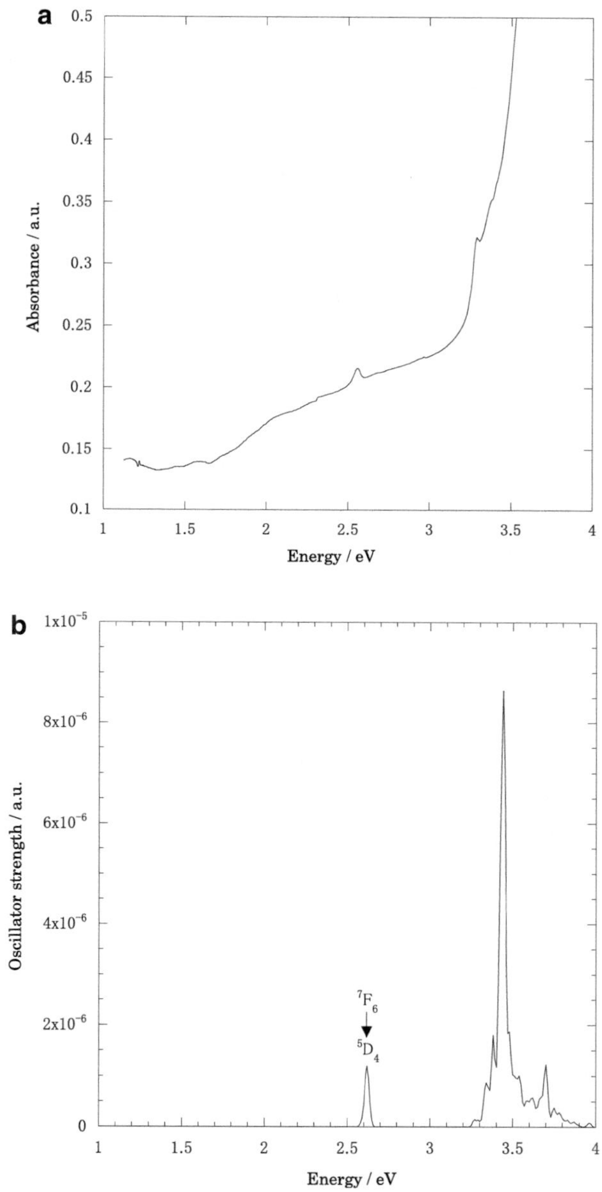

Figure 1.10a is the experimental fluorescence spectrum of Tb^{3+} doped $50P_2O_5{\bullet}50CaO$ glass under excitation with 3.29 eV. In this spectrum, two intense peaks around 1.58 and 3.32 eV are reflection of the excitation light. There is an intense peak at 2.29 eV and another at 2.54 eV. These peaks are assigned to the transition from $^5D_4$ to $^7F_5$ and from $^5D_4$ to $^7F_4$, respectively, by referring the Dieke diagram. The other small peaks are observed at 2.00, 2.12, 2.71, 2.85 and 3.00 eV,

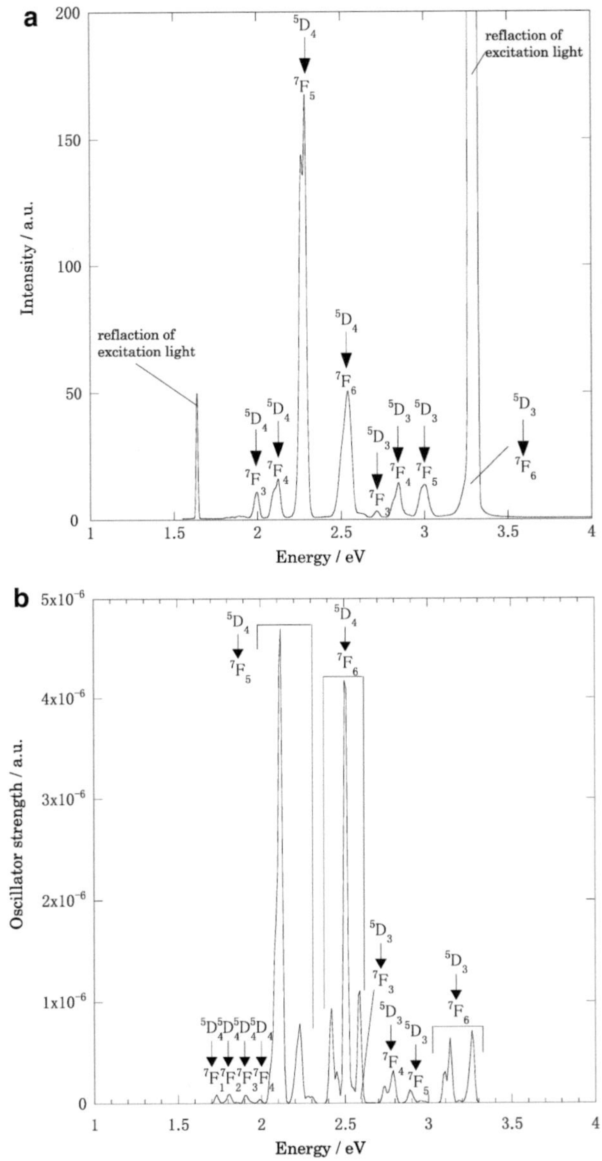

Fig. 1.10 Fluorescence spectra of $H_{18}TbP_7O_{28}$. (**a**) Experimental spectrum, (**b**) theoretical spectrum of Tb^{3+} ion doped $50CaO{\bullet}50P_2O_5$ glass

whose assignments are described in the figure. Figure 1.10b shows the theoretical spectrum of the $H_{15}TbP_6O_{24}$ cluster. In the spectrum, there are two intense peaks around 2.12 and 2.51 eV; those are corresponding to the peaks from $^5D_4$ to $^7F_5$ and from $^5D_4$ to $^7F_6$ in the experimental ones, respectively. The peaks observed in the experimental spectra are usually assigned by referring the Dieke diagram, so that the assignments are sometime incorrect especially in the spectra in the lanthanide ions with the complex multiplet energy levels. In the case of the Pr^{3+} ion, the assignments of the peaks by the empirical manner are much different from the theoretical ones. To the contrary, the assignments of the peaks observed in the theoretical spectrum in Fig. 1.10b are almost the same as the assignments in the experimental spectrum. This result still shows the relativistic DVME method is very useful to analyze fluorescence spectra of rare-earth ions in phosphate matrices and the first principles calculation is necessary to analyze the fluorescence spectra of rare-earth ions.

Conclusion

The recent development and applications of the DV-Xα cluster method is introduced. The total energy calculation, called local cluster energy (LCE), is applied to the movement of the Li ions in the Li_3N crystal, which is one of the super-ionic conducting materials. The movement of the Li ion is simulated by using several model clusters with different positions of a Li ion, which is located on the conduction path. As the results, the LCE showed that the movement of the Li ion through the conduction path makes the smaller energy change than another path. This result suggests that the energy calculation program Coulomb is very useful to analyze the energy change of the cluster models for ionic conductors.

The relativistic DVME method, which is one of the first principles CI calculation methods, is applied to the calculation of the theoretical fluorescence spectra of rare-earth ions in phosphate glasses. The model cluster with surrounding PO_4 units terminated with H ions are suitable to obtain the absorption and fluorescence spectra in the phosphate glasses. The theoretical spectra with several initial states are good agreement with the experimental one, though the spectra with only one initial state are not in the case of the Pr^{3+} ion. These results suggested that the relativistic DVME method is very useful and necessary to analyze fluorescence spectra of rare-earth ions in phosphate matrices.

References

Adachi H (1977) Relativistic molecular orbital theory in the Dirac-Slater model. Technol Rept Osaka Univ 27(1364–1393):569–576

Adachi H, Tsukada M, Satoko C (1978) Discrete variational Xα cluster calculations. I. Application to metal clusters. J Phys Soc Jpn 45(3):875–883

Ajithkumar G, Gupta PK, Jose G, Unnikrishnan NV (2000) Judd-Ofelt intensity parameters and laser analysis of Pr^{3+} doped phosphate glasses sensitized by Mn^{2+} ions. J Non-Cryst Solids 275 (1):93–106

Araki R, Hayashi A, Kowada Y, Tatsumisago M, Minami T (2001) Electronic states calculated by the DV-Xα cluster method for lithium ion conductive Li2S-SiS2-Li4SiO4 oxysulfide glasses. J Non-Cryst Solids 288(1–3):1–7

Brik MG, Ogasawara K (2007) Comparative study of the absorption spectrum of Li_2CaSiO_4:Cr^{4+}: first-principles fully relativistic and crystal field calculations. Opt Mater 30(3):399–406

Dieke GH (1968) Spectra energy levels of rare earth ions in crystals. Wiley, New York

Ellis DE, Adachi H, Averill FW (1976) Molecular cluster theory for chemisorption of first row atoms on nickel (100) surfaces. Surf Sci 58(2):497–510

Herman F, Skillman S (1963) Atomic structure calculations. Prentice-Hall, Englewood Cliffs

Imanaka N, Tamura S, Adachi G, Kowada Y (2000) Electronic state of trivalent ionic conductors with $Sc_2(WO_4)_3$-type structure. Solid State Ion 130(3–4):179–182

Ishii T, Ogasawara K, Adachi H, Burmester P, Huber G (2004a) First-principles analysis for the optical absorption spectra of metal ions in solids. Int J Quant Chem 99(4):488–494

Ishii T, Brik MG, Ogasawara K (2004b) First-principles analysis method for the multiplet structures of rare-earth ions in solids. J Alloy Compd 380(1–2 Spec Iss):136–140

Johnson KH (1966) "Multiple-scattering" model for polyatomic molecules. J Chem Phys 45(8):3085–3095

Jose G, Thomas V, Jose G, Paulose PI, Unnikrishnan NV (2003) Application of a modified Judd-Ofelt theory to Pr^{3+} doped phosphate glasses and the evaluation of radiative properties. J Non-Cryst Solids 319(1–2):89–94

Jouini A, Ferid M, Gâcon JC, Grosvalet L, Thozet A, Trabelsi-Ayadi M (2003) Crystal structure and optical study of praseodymium polyphosphate $Pr(PO_3)_3$. Mater Res Bull 38(11–12):1613–1622

Kowada Y, Adachi H, Tatsumisago M, Minami T (1998) Electronic states of Ag ions in AgI-based superionic conducting glasses. J Non-Cryst Solids 232–234:497–501

Kowada Y, Yamada Y, Tatsumisago M, Minami T, Adachi H (2000) Variation of electronic state of AgI-based superionic conductors with movement of Ag ions. Solid State Ion 136–137:393–397

Kowada Y, Okamoto M, Tanaka I, Adachi H, Tatsumisago M, Minami T (2004) Chemical bonding of Ag ions in AgI-based superionic conducting glasses. J Non-Cryst Solids 345–346:489–493

Kowada Y, Tatsumisago M, Minami T, Adachi H (2008a) Electronic state of sulfide-based lithium ion conducting glasses. J Non-Cryst Solids 354(2–9):360–364

Kowada Y, Tatsumisago M, Minami T, Adachi T (2008b) Chemical bonding of mobile cations in superionic conductors. Adv Quant Chem 54:255–270

Kowada Y, Nishitani W, Ogasawara K (2009a) Total cluster energy calculation of lithium ion conductors by the DV-Xα method. Int J Quant Chem 109(12):2658–2663

Kowada Y, Noma S, Ogasawara K (2009b) Fluorescence spectra of Pr^{3+} ions in phosphate materials calculated by the DVME method. Int J Quant Chem 109(12):2753–2757

Kowada Y, Tatsumisago M, Minami T (2009c) Chemical bonding and lithium ion conductions in Li_3N. Solid State Ion 180(6–8):462–466

Kowada Y, Hayashi A, Tatsumisago M (2010) Chemical bonding of Li ions in $Li_7P_3S_{11}$ crystal. J Phys Soc Jpn 79(Suppl A):65–68

Koyama Y, Yamada Y, Tanaka I, Nishitani SR, Adachi H, Murayama M, Kanno R (2002) Evaluation of migration energy of lithium ions in chalcogenides and halides by first principles calculation. Mater Trans 43(7):1460–1463

Liberman D, Waber JT, Cromer DT (1965) Self-consistent-field Dirac-Slater wave functions for atoms and ions. I. Comparison with previous calculations. Phys Rev 137(1A):A27–A34

Miniscalco WJ (1991) Erbium-doped glasses for fiber amplifiers at 1500 nm. J Lightwave Tech 9 (2):234–250

Moorthy LR, Jayasimhadri M, Radhapathy A, Ravikumar RVSSN (2005) Lasing properties of Pr^{3+}-doped tellurofluorophosphate glasses. Mater Chem Phys 93(2–3):455–460

Ogasawara K, Iwata T, Koyama Y, Ishii T, Tanaka I, Adachi H (2001) Relativistic cluster calculation of ligand-field multiplet effects on cation $L_{2,3}$ x-ray-absorption edges of $SrTiO_3$, NiO and CaF_2. Phys Rev B 64(11): Art no 115413, pp 1154131–1154135

Ogasawara K, Watanabe S, Toyoshima H, Ishii T, Brik M G, Ikeno H, Tanaka I (2005) Optical spectra of trivalent lanthanides in $LiYF_4$ crystal. J Solid State Chem 178(2 Spec Iss):412–418

Reddy MR, Raju SB, Veeraiah N (2000) Optical absorption and fluorescence spectra studies of Ho^{3+} ions in $PbO\text{-}Al_2O_3\text{-}B_2O_3$ glass system. J Phys Chem Solids 61(10):1567–1571

Rosen A, Ellis DE (1974) Relativistic molecular wavefunctions: XeF_2. Chem Phys Lett 27(4):595–599
Rosen A, Ellis DE, Adachi H, Averill FW (1976) Calculations of molecular ionization energies using a self-consistent-charge Hartree-Fock-Slater method. J Chem Phys 65(9):3629–3634
Sarnthein J, Schwarz K, Blöchl PE (1996) Ab initio molecular-dynamics study of diffusion and defects in solid Li_3N. Phys Rev B 53(14):9084–9091
Schulz H, Thiemann KH (1979) Defect structure of the ionic conductor lithium nitride. Acta Crystallogr Sect A A35(2):309–314
Slater JC (1974) Quantum theory of molecules and solids, vol 4. McGraw-Hill, New York
Souza Filho AG, Mendes Filho J, Melo FEA, Custódio MCC, Lebullenger R, Hernandes AC (2000) Optical properties of Sm^{3+} doped lead fluoroborate glasses. J Phys Chem Solids 61(9):1535–1542
Tanabe S, Sugimoto N, Ito S, Hanada T (2000) Broad-band 1.5 μm emission of Er3+ ions in bismuth-based oxide glasses for potential WDM amplifier. J Lumin 87:670–672
Tanabe S, Hayashi H, Hanada T, Onodera N (2002) Fluorescence properties of Er^{3+} ions in glass ceramics containing LaF_3 nanocrystals. Opt Mater 19(3):343–349
Toyoshima H, Watanabe S, Ogasawara K, Yoshida H (2007) First-principles calculations of 4f-5d optical absorption spectra in $BaMgAl_{10}O_{17}$:Eu. J Lumin 122–123(1–2):104–106
Wahl J, Holland U (1978) Local ionic motion in the superionic conductor Li_3N. Solid State Commun 27(3):237–241
Walker JR, Catlow CRA (1981) Defect structure and ionic conductivity in lithium nitride. Philosophical Magazine A 43(2):265–272
Wang JS, Vogel EM, Snitzer E (1994) Tellurite glass: a new candidate for fiber devices. Opt Mater 3(3):187–203
Watanabe S, Kamimura H (1989) First-principles calculations of multiplet structures of transition metal deep impurities in II–VI and III–V semiconductors. Mater Sci Eng B3(3):313–324
Watanabe S, Ishii T, Fujimura K, Ogasawara K (2006) First-principles relativistic calculation for f4-5d transition energy of Ce^{3+} in various fluoride hosts. J Solid State Chem 179(8):2438–2442
Yoshida H, Ogasawara K (2009) Theoretical analysis of phosphors based on first-principles cluster calculations using the relativistic DVME method. ECS Trans 16(31):35–40
Yuan J–L, Zhang H, Zhao J–L, Chen H–H, Yang X–X, Zhang G–B (2008) Synthesis, structure and luminescent properties of $Lu(PO_3)_3$. Opt Mater 30(9):1369–1374

Part II
Recent Theoretical Progress

Chapter 2
Algebraic Molecular Orbital Theory

Jun Yasui

2.1 Introduction

Huzinaga described the principle of MO theory from the beginning before the Hartree-Fock-Roothaan theory in his text book (Huzinaga 1980). We can see there a new direction of MO theory. He gave us several hints in his book, for example, effectiveness of nonlinear variational method to obtain better eigenfunctions of atomic and molecular Hamiltonian, an excellent behavior of Slater-type basis functions, and so on. He also mentioned the Hartree-Fock variety which implies a multivariable theory. Inspired by Huzinaga, we shall propose an algebraic MO equation in order to construct a multivariable theory for quantum chemistry.

2.1.1 *Multivariable Problem*

Quantum chemistry has been developing in the field of physics, chemistry, and biology. Problems of molecular science spread in a wide range of space from nanometer order to meter order, energy from rotational transition energy level to hard x-ray energy level, and time from attosecond order to day order.

In the development of quantum chemistry, molecular orbital calculation has been fairly successful as a practical numerical method. However, there are underlying problems that are not well understood in quantum chemistry. Chemistry is a science that deals with fundamentally multivariable relations existing in atoms and molecules. For example, electronic state and molecular property depend on the molecular parameters, such as atomic position, atomic number, and number of

---

J. Yasui (✉)
Frontier Research Center, Canon Inc., 30-2 Shimomaruko 3-chome, Ohta-ku,
Tokyo 146-8501, Japan
e-mail: yasui.jun@canon.co.jp

T. Ishii et al. (eds.), *The DV-Xα Molecular-Orbital Calculation Method*,
DOI 10.1007/978-3-319-11185-8_2

constituent nuclei and electrons. However, the traditional molecular orbital theory, such as the Hartree–Fock–Roothaan (HFR) theory (Roothaan 1951) and the Kohn–Sham density functional theory (Hohenberg and Kohn 1964; Kohn and Sham 1964) have been developed without considering multivariable problems in chemistry. In such theories parameters of molecular structure are treated as numerical parameters. For example, MO energy is plotted numerically along the axis of molecular structures. To express the electronic state as a function of molecular parameters is still a basic problem in quantum chemistry. In this chapter multivariable problems that have been overlooked in the development of MO theory are discussed.

2.1.2 Variational Principle

On the basis of analytical basis set expansion method, MO equation is defined by applying the method of Lagrangian multiplier. The analytical basis set expansion method has fundamentally two kinds of parameter, linear combination coefficient and nonlinear orbital exponent. The roll of these parameters is a variational parameter in defining MO equation, and an optimization parameter to obtain approximate solutions of MO equation. However, in the traditional MO theory only linear combination coefficients are adopted, and nonlinear orbital exponents of basis function are fixed at numerical value. This means that the definition of the traditional MO equation is restricted in linear variational space, so that eigenstate cannot reach the exact eigenstate of the Hamiltonian even when the basis set is extended.

In contrast to analytical basis function such as Slater-type function and Gaussian-type function, numerical atomic basis function (Averill and Ellis 1973) is used in the discrete-variational-Xα (DV-Xα) method based on the Hartree-Fock-Slater (HFS) model which is easily applicable from atoms to clusters (Adachi et al. 1978; Adachi 2006; Slater 1974, 1979). Numerical atomic basis function does not have an orbital exponent, but it is numerically generated at every SCF cycle in the MO calculation by solving the Schrödinger equation of atomic radial function for each atomic potential in molecule. In the DV-Xα method, the Fock matrix is also defined at every SCF cycle by molecular integrals of numerical atomic basis functions. Although the HFS equation is defined on the linear variational principle, the MO is obtained through a doublly optimization process, i.e., optimization of numerical atomic basis functions and diagonalization of the Fock matrix. Owing to the effect of double optimization, proper MOs can be obtained for atoms and clusters with small basis set in the DV-Xα method on the HFS model. However, the traditional MO theories are not applicable to algebraic theory as long as molecular integral is expressed in numerical value. Mukoyama proposed an analytical expression of atomic wave function which is obtained by nonlinear fitting of STFs to a numerical atomic wave function obtained in a similar way to the DV-Xα method (Mukoyama and Yasui 1992).

The traditional MO equation is defined by using the Lagrangian function which is furnished with a single constraint, the orthonormal condition of MOs. The exact eigenfunction of molecular Hamiltonian satisfies several conditions, such as Kato's cusp condition (Kato 1957; Steiner 1963; Pack and Brown 1966; Chong 1967; Nakatsuji and Izawa 1989; Chapman and Chong 1970; Poling et al. 1971; Rassolov and Chipman 1996; Nagy and Sen 2001; Janosfalvi et al. 2005), the virial theorem, and the Hellmann–Feynman theorem (Feynman 1939; Feinberg et al. 1970; Stanton 1962; Fernández Rico et al. 2007). Although these conditions are necessary conditions being obeyed by the exact solutions of the Schrödinger equation, they are not adopted in the traditional Lagrangian function to define MO equation. However, Reinhardt and Hoggan, Koga and Galek showed that better solutions of the HFR equation can be obtained by using STFs and by adopting the nuclear cusp condition and the virial theorem (Reinhardt and Hoggan 2009; Koga et al. 1999; Galek et al. 2005). These studies suggested that multiple constraint condition is effective in solving the HFR equation. It is interesting that they adopted the nuclear cusp condition and the virial theorem because the former is a local condition and the latter is a global condition.

2.1.3 Trial Functions and Molecular Integrals

Molecular orbital is a model of electronic eigenfunction of the Schrödinger equation of molecule on the mean field approximation. Usually the Hartree–Fock (HF) equation is solved by using basis set expansion method. However, it is practically impossible to obtain the exact solution of the HF equation even when basis set is extended in size; the basis set is not able to expand itself fully in the Hilbert space of the Schrödinger equation as long as it is finite.

The molecular orbital model is used to calculate the electronic structure not only on the HF one-body model but also beyond the HF level. The first group beyond the HF level uses a single determinant for molecular orbitals; density functional theory belongs to this group. In the second group, there are configuration interaction method, perturbation theory, and coupled cluster theory, which use the virtual MOs of the HF model. For this reason, good basis functions are required for better MO calculations beyond the mean field approximation.

Despite of difficulties in evaluating multicenter molecular integrals, Slater-type function (STF) is a superior basis function because it satisfies Kato's nuclear-electron cusp condition and has a good long-range behavior similar to the Schrödinger hydrogen wave function (Kotani et al. 1955; Yasui and Saika 1982). STF has been adopted in advanced MO calculation softwares, such as ADF (Baerends et al. 1973), STOP (Bouferguene et al. 1996), and SMILE (Fernández Rico et al. 2001, 2004), however, it is impossible to address algebraic MO calculation by those software. This is because there is no symbolic information of molecular parameters in molecular integrals expressed by numerical value. STF is a superior basis function but analytical molecular integral over STF is required.

2.1.4 SCF Method

It is crucial to execute MO calculation for any size of atomic system without unnecessary calculations (Dirac 1929). In the traditional MO calculation, SCF iterative technique is used in solving a nonlinear MO equation. This technique is poor at convergence for large molecular system because of quasidegenerate electronic energy structure and huge eigenvalue problem. How to solve the Schrödinger equation without SCF iterative technique is an essential problem for large molecular system.

2.1.5 Nonadiabatic Process

Fundamentally, there is a time dependent discrepancy of motion between electron and nucleus due to a large difference of mass. Electron as a particle with fast velocity and nucleus as a particle with low velocity exchange their momenta via Coulomb interaction in time-dependent process. This cooperative interaction between electrons and nuclei is called nonadiabatic interaction. This is an essential problem in molecular science, such as chemical reaction, slow atomic collision, spectroscopy for electron and nucleus, surface science, solid-state physics, and so on, however, this has been existing as a difficult problem in molecular science since the beginning of quantum mechanics of molecule (Domcke et al. 2004; Baer 2006; Köppel et al. 2009; Nakamura 2012; Longuest-Higgins 1975; Berry 1984; Baer 2002).

The Schrödinger equation consists of fast variables and slow variables in molecule, electron and nucleus. The standard approach for solving the molecular Schrödinger equation is the Born–Oppenheimer (BO) approximation. In this approximation, molecular system has a double-layered energy structure by the large mass difference in which electron motions and nucleus motions are separated as follows. Since the slow variable is a good adiabatic parameter in the BO approximation, the Schrödinger equation for the fast variable can be solved by fixing the slow variable.

After obtaining eigenstates of the fast variables, equation for the slow variables can be defined using the eigenvalues of the first variable and the momentum operator shifted by the gauge potential A defined by

$$A_{mn}(R) = i\int \varphi_m^* \nabla_R \varphi_n dV, \tag{2.1}$$

where i is the imaginary unit, φ stands for the wave function of a fast particle, and R represents the molecular structure. The gauge potential A is designated the nonadiabatic coupling (NAC) vector. The Schrödinger equation for the slow variable satisfies the hidden local gauge symmetry which keeps the phase of total wave function in time by cancelling global phase of the fast particle, whereas the motion of the slow particle brings about a phase shift to the fast particle during a time interval t in adiabatic process. This phase, designated the Longuet-Higgins/Berry phase, is defined by the path integral of NAC along a motion of the slow particle on the potential energy surface (PES) as

$$\gamma_{mn}(t) = \int_0^t A_{mn}(R(t)) \cdot dR(t). \tag{2.2}$$

In addition of dynamical phase, the time-dependent wave function of the fast particle, electronic wave function, has the Longuet-Higgins/Berry phase.

The nonadiabatic electronic transitions are found in various mechanisms in chemistry, molecular physics and molecular biology. This transition is driven by NAC in the critical region of avoided crossing or conical intersection. This transition has been studied as an electron transition between adiabatic PESs on the BO approximation (Adachi and Yasui 1987).

NAC and the Longuet-Higgins/Berry phase always coexist as a reciprocal relation in the dual-layered molecular energy system. Study of nonadiabatic molecular dynamics is a most difficult but a most interesting problem in chemistry. To disentangle problems in cooperative dynamics between electrons and nuclei is an essential and important problem in molecular science.

2.1.6 Aim of Our Study

As ignored problems which have been existing subconsciously as important problems from the beginning of quantum chemistry, we point out multivariable problems, which is concerned with functional relation between electronic state and molecular structure, introduction of nonlinear variational parameters to define molecular orbital equation, multiple constraint conditions for including electron correlation effect into the Lagrangian function, multivariable expression of molecular integrals over the STFs, cooperative dynamics of electrons and nuclei in molecular systems, method for solving the HFR equation without SCF iterative technique, and so on. Quantum chemistry has been developed without recognizing these multivariable problems.

The aim of our study is to construct a theoretical base to solve these multivariable problems beyond the traditional quantum chemistry constructed on the numerical method. As a first step to construct algebraic quantum chemistry, we shall propose an algebraic MO theory.

2.2 Theory

In this section, we propose a method for defining an algebraic MO equation by the variational principle. Two innovative methods are given, polynomial expression of molecular integrals and the Lagrangian function with extended constraint condition.

2.2.1 *Polynomial Expression of Molecular Integrals*

In order to retain an atomic picture in a multivariable MO theory, we adopt a linear combination of atomic orbitals approximation (LCAO-MO)

$$\phi_k\left(C,\vec{r},\zeta,R\right)=\sum_{\alpha}^{atom}\sum_{p}^{AO(\alpha)}C_{\alpha p}^{k}\chi_{\alpha p}\left(\vec{r}_{\alpha},\zeta_{\alpha p},R_{\alpha},\left\{n_{\alpha p},l_{\alpha p},m_{\alpha p}\right\}\right), \tag{2.3}$$

in this study. On the left hand side of Eq. (2.3), ϕ_k denotes for the k-th molecular orbital, and on the right hand side of Eq. (2.3), C the vector of linear combination coefficient of basis function, ζ the set of orbital exponent of the basis function and R the atomic configuration, α the sequence number of constituent atoms, $AO(\alpha)$ the number of basis functions centered at α, p the sequence number of the basis functions at α, $\chi_{\alpha p}$ the p-th basis function centered at α, $\{n, l, m\}$ is the set of quantum numbers of the basis function, and r_α the polar coordinates of electron with origin at α. In the following the basis function χ is written by omitting the set of quantum numbers $\{n, l, m\}$ in short.

Molecular integral is integration over coordinates of electron which is the fastest particle in molecule. On the assumption that atomic configuration R and nonlinear parameter ζ are constant during integration, molecular integral is defined as a symbolic function with respect to R and ζ. Molecular integral is a minimum symbolic element in the algebraic MO theory on the BO approximation.

In order to construct algebraic MO equation, we prepare polynomial expression of molecular integrals through two steps. In the first step, we execute molecular integral analytically remaining R and ζ in symbol. Even in case when analytical expression of molecular integral is tremendously complicated, analytical molecular integral is a minimum symbolic element that enables to express MO equation in symbol. In the second step, we transfer analytical molecular integral by the Taylor expansion method given by Eq. (2.4) from analytical expression to polynomial expression in order to give an algebraic structure to the MO theory (Yasui 2010, 2011). A method of analytical expression of molecular integrals over STFs is described in detail in the next chapter.

$$f(x)=\sum_{p=0}^{\infty}\frac{1}{p!}\sum_{k=0}^{p}\binom{p}{k}(-x_0)^{p-k}\left[\left(\frac{d}{dx}\right)^{p}f(x)\right]_{x=x_0}x^k\cong\sum_{i=0}^{N}A_i(x_0)x^i, \tag{2.4}$$

where x_0 means the center of the Taylor expansion distributed so as to keep the polynomials in the same order in any region of variables. The last term in Eq. (2.4) is the approximate expression of single variable functions. This method can be easily extended to multivariable functions. Let us show an example for the case of overlap integrals. Overlap integral between basis functions centered at A and B separated by R is given

$$S(\zeta_a, \zeta_b, R)_{AB}^{ab} = \int \chi^*\left(\vec{r}_A, \zeta_a, R_A\right)\chi\left(\vec{r}_B, \zeta_b, R_B\right)dV. \tag{2.5}$$

Applying the Taylor expansion method, overlap integral can be approximately expressed in polynomial with respect to atomic distance R, and orbital exponents ζ_a and ζ_b

$$S(\zeta_a, \zeta_b, R)_{AB}^{ab} \cong \sum_{p_a, p_b, p_R} A\left(\{n_a, l_a, m_a\}_A, \{n_b, l_b, m_b\}_B\right)_{p_a, p_b, p_R} \zeta_a^{p_a} \zeta_b^{p_b} R^{p_R}, \tag{2.6}$$

where A stands for coefficient of polynomials expressed by rational number of fields. Coefficients of polynomial molecular integrals can be calculated and stored in advance for R and ζ over full range. Equation (2.6) shows that overlap integral is a polynomial ring with respect to R, ζ_a and $\zeta_{b.}$ The same procedure is applicable for other one-electron molecular integrals and two-electron molecular integrals to change their expressions from analytical to polynomial. By introducing of polynomial expression of molecular integrals into MO equation, we can define MO equation as an algebraic equation. Molecular integral expressed in polynomial works as a polynomial ring in the algebraic MO theory.

2.2.2 *Total Electronic Energy*

Owing to the polynomial expression of molecular integrals with respect to atomic configuration parameters and nonlinear orbital exponent parameters, we can define total electronic energy as a multivariable function

$$E[N, n, Z, R, C, \zeta] = T[N, n, Z, R, C, \zeta] + V[N, n, Z, R, C, \zeta], \tag{2.7}$$

where N is the number of atoms in molecule, n the number of electrons in molecule, Z the set of nuclear charge for each atom in molecule, R stands for the set of atomic configuration parameters, C is the set of linear combination coefficients, and ζ the set of nonlinear orbital exponents of basis functions, T and V on the right hand side stand for the total kinetic energy of electrons and the total potential energy, respectively.

We write total kinetic energy of electrons in LCAO approximation

$$T[N,n,R,C,\zeta] = \sum_{\alpha,\beta}^{N} \sum_{p}^{AO(\alpha)} \sum_{q}^{AO(\beta)} D_{\alpha p,\beta q}[n,C] t_{\alpha p,\beta q}[R,\zeta], \tag{2.8}$$

where α and β run over all pairs of atoms in molecule, $AO(\alpha)$ stands for the order of basis functions on the α th atom, and a density matrix on the right hand side is defined as

$$D_{\alpha p,\beta q}[n,C] = 2 \sum_{k}^{occu.MO} C_{\alpha p}^{*k} C_{\beta q}^{k}, \tag{2.9}$$

where k runs over occupied molecular orbitals, and kinetic molecular integral on the right hand side in Eq. (2.8) is defined in atomic unit as

$$t_{\alpha p,\beta q}[R,\zeta] = \int dV \chi_{\alpha p}^{*}\left(\vec{r}_{\alpha}, \zeta_{\alpha p}, R_{\alpha}\right)\left(-\frac{1}{2}\Delta\right)\chi_{\beta q}\left(\vec{r}_{\beta}, \zeta_{\beta q}, R_{\beta}\right). \tag{2.10}$$

Total electronic potential energy is expressed as a sum of nuclear-electron attraction potential energy V^{Ne}, electron-electron repulsion potential energy V^{ee}, and nuclear- nuclear repulsion potential energy V^{NN}

$$V[N,n,Z,R,C,\zeta] = V^{Ne}[N,n,Z,R,C,\zeta] + V^{ee}[N,n,R,C,\zeta] + V^{NN}[N,Z,R]. \tag{2.11}$$

We write nuclear-electron attraction potential energy on the right hand side in Eq. (2.11) in LCAO approximation as

$$V^{Ne}[N,n,Z,R,C,\zeta] = \sum_{\alpha,\beta}^{N} \sum_{p}^{AO(\alpha)} \sum_{q}^{AO(\beta)} D_{\alpha p,\beta q}[n,C] V_{\alpha p,\beta q}^{Ne}[N,Z,R,\zeta], \tag{2.12}$$

where

$$V_{\alpha p,\beta q}^{Ne}[N,R,R,\zeta] = \sum_{\gamma}^{N} Z_{\gamma} v_{\alpha p,\beta q}^{\gamma}[R,\zeta] \tag{2.13}$$

and nuclear-electron attraction molecular integral on the right hand side in Eq. (2.13) is defined in atomic unit as

$$v_{\alpha p,\beta q}^{\gamma}[R,\zeta] = \int dV \chi_{\alpha p}^{*}\left(\vec{r}_{\alpha}, \zeta_{\alpha p}, R_{\alpha}\right)\frac{1}{R_{\gamma}}\chi_{\beta q}\left(\vec{r}_{\beta}, \zeta_{\beta q}, R_{\beta}\right), \tag{2.14}$$

where R_γ stands for the distance of γ th atom from the origin of the internal coordinates for molecule. We write total electron repulsion potential in Eq. (2.11) in LCAO approximation as

$$V^{ee}[N,n,R,C,\zeta] = \frac{1}{2}\sum_{\alpha,\beta,\gamma,\delta}^{N}\sum_{p}^{AO(\alpha)}\sum_{q}^{AO(\beta)}\sum_{r}^{AO(\gamma)}\sum_{s}^{AO(\delta)} D_{\alpha p,\beta q}[n,C]\left[I_{\alpha p,\beta q,\gamma r,\delta s}[\zeta,R] - \frac{1}{2}I_{\alpha p,\gamma r,\beta q,\delta s}[\zeta,R]\right]D_{\gamma r,\delta s}[n,C], \tag{2.15}$$

where α, β, γ and δ run over all combinations of four atoms in molecule. We define electron-electron repulsion molecular integral in Eq. (2.15) in atomic unit as

$$I_{\alpha p,\beta q,\gamma r,\delta s}[\zeta,R] = \int\int dV_1 dV_2 \chi^*_{\alpha p}\left(\vec{r}_{1\alpha},\zeta_{\alpha p},R_\alpha\right)\chi^*_{\beta q}\left(\vec{r}_{1\beta},\zeta_{\beta q},R_\beta\right) \frac{1}{r_{12}}\chi_{\gamma p}\left(\vec{r}_{2\gamma},\zeta_{\gamma r},R_\gamma\right)\chi_{\delta s}\left(\vec{r}_{2\delta},\zeta_{\delta s},R_\delta\right). \tag{2.16}$$

Nuclear repulsion potential on the right hand side in Eq. (2.11) is written in atomic unit

$$V^{NN}[N,Z,R] = \sum_{\alpha,\beta}^{N} Z_\alpha Z_\beta \frac{1}{R_{\alpha\beta}}, \tag{2.17}$$

where $R_{\alpha\beta}$ is a atomic distance between α th atom and β th atom.

The variables in total electronic energy in Eq. (2.7) are separated to two groups. In the first group, number of electrons n and linear variational parameter C belong to the density matrix. In the second group, number of atom N, atomic charge Z, atomic configuration R, and nonlinear orbital exponent ζ belong to the molecular integral. We can prepare molecular integrals in Eqs. (2.5), (2.10), (2.14), and (2.16) in advance for full range of R and ζ, and use them in algebraic MO calculations.

2.2.3 Extension in the Variational Principle

On the variational principle, MO equation is defined with the method of the Lagrangian multiplier. In this study we construct the algebraic MO equation by extending the Lagrangian function with respect to nonlinear variation and constraint condition.

By polynomial expression of molecular integrals with respect to orbital exponent of basis function, we introduce nonlinear variational parameter to the definition of the Lagrangian function. This enables the variation principle in nonlinear space. Huzinaga's book shows significant effects of nonlinear optimization to obtain better eigenstates.

The traditional MO equation, for example the HFR equation and the KS equation, is defined using the Lagrangian function with a single constraint condition. The Lagrangian function for the HFR equation consists of the total energy and a constraint condition in Eq. (2.18)

$$L_{HFR}[C] = E_{HFR}[C] + \lambda_s G_s[C], \tag{2.18}$$

where C stands for linear combinational coefficient of basis function, λ_S on the right hand side in Eq. (2.18) is the Lagrangian multiplier, and G_s means constraint condition of orthonormality of molecular orbitals. The Lagrangian function denoted by L_{HFR} in Eq. (2.18) leads the HFR MO equation by applying the variational principle

$$\frac{\partial}{\partial C} L_{HFR}[C] = 0. \tag{2.19}$$

Equation (2.19) shows that the HFR equation is defined only in the linear variational space.

While the HF equation or the KS equation is defined using the Lagrangian function with a single constraint condition, orthonormal condition of MO, we define the algebraic MO equation using the Lagrangian function with multiple constraint conditions in Eq. (2.20)

$$L_{multi}[N, n, Z, R, C, \zeta, \lambda] = E[N, n, Z, R, C, \zeta] + \sum_i \lambda_i G_i[N, n, Z, R, C, \zeta], \tag{2.20}$$

where L_{multi} on the left hand side is a Lagrangian function with multi constraint conditions, λ_i stands for the Lagrangian multiplier for the i th constraint condition, and G_i a necessary condition which is obeyed by the exact solution of the Schrödinger equation. We name the additional constraint condition in the Lagrangian function as the Schrödinger condition. We furnish the new Lagrangian function with a set of constraint conditions each of which has a different action in approaching the limit of variation. Choice of constraint condition depends on the aim. By the new Lagrangian function with nonlinear variational parameters and multiple constraint conditions, we define an algebraic MO equation on the variational principle

$$\frac{\partial}{\partial \alpha} L_{multi}[N, n, Z, R, C, \zeta, \lambda] = 0, \ (\alpha = C, \zeta). \tag{2.21}$$

Exact eigenfunction of the molecular Hamiltonian of the Schrödinger equation should obey the Schrödinger condition, such as Kato's cusp condition, the virial theorem, the Hellmann–Feynman theorem (Feynman 1939) and the Liu–Parr–Nagy identity (Liu et al. 1995). Each condition subjects the trial function to a different action as discussed below.

The orthonormal condition of molecular orbitals is generally expressed as

$$C^{*}(R)S(\zeta,R)C(R) = 1. \tag{2.22}$$

At any molecular structure R, linear combination coefficient C and nonlinear orbital exponent of basis functions ζ cooperate to satisfy the orthonormal condition. In the algebraic MO theory the orthonormal condition in Eq. (2.22) is expressed by simultaneous polynomial equation. All other Schrödinger condition is the same.

Kato's cusp condition is a condition for eigenfunctions at the singular point of Coulomb interaction or in the vicinity of atomic center. In the algebraic MO theory the nuclear-electron cusp condition is adopted mainly as a local constraint condition. Three equivalent expressions for the nuclear-electron cusp condition are given as

$$\psi(0) + \frac{\psi'(0)}{Z} = 0, \tag{2.23}$$

$$\rho(0) + \frac{\rho'(0)}{2Z} = 0, \tag{2.24}$$

$$Z = -\frac{1}{2}\frac{\rho'(0)}{\rho(0)} = -\frac{1}{2}\lim_{r\to 0} d\log_e \rho(r), \tag{2.25}$$

where ψ is the eigenfunction, Z the atomic charge, ρ (r) the electron charge density. The electron charge density at nucleus in Eq. (2.24) has a relation to the Fermi contact interaction. The energy of Fermi contact interaction is proportional to square of electron charge density at atomic center. Since every atomic center is a singular point of Coulomb interaction in molecule, the nuclear-electron cusp condition holds at every atom. The Eq. (2.25) means that the atomic number Z is proportional to the derivation of electron charge density over electron charge density at nucleus. The nuclear-electron cusp condition implies that the atomic positions are determined so as to satisfy the nuclear-electron cusp condition in molecule.

In this study the molecular electron charge density in LCAO approximation is expressed as

$$\rho[N,n,R,C,\zeta]\left(\vec{r}\right) = \sum_{\alpha}^{N} \rho_{\alpha,\alpha}[n,R,C,\zeta]\left(\vec{r}\right) + \sum_{\alpha\geq 2}^{N}\sum_{\beta(<\alpha)}^{N-1} \rho_{\alpha,\beta}[n,R,C,\zeta]\left(\vec{r}\right), \tag{2.26}$$

where electron charge distributions on atom or between atoms are defined as

$$\rho_{\alpha,\alpha}[n,R,C,\zeta]\left(\vec{r}\right) = \sum_{p}^{AO(\alpha)}\sum_{q}^{AO(\alpha)} D_{\alpha p,\alpha q}[n,C]\chi_{\alpha p}^{*}\left(\vec{r}_{\alpha},\zeta_{\alpha p},R_{\alpha}\right)\chi_{\alpha q}\left(\vec{r}_{\alpha},\zeta_{\alpha q},R_{\alpha}\right), \tag{2.27}$$

$$\rho_{\alpha,\beta}[n,R,C,\zeta]\left(\vec{r}\right)=\sum_{p}^{AO(\alpha)}\sum_{q}^{AO(\beta)}D_{\alpha p,\beta q}[n,C]\chi_{\alpha p}^{*}\left(\vec{r}_{\alpha},\zeta_{\alpha p},R_{\alpha}\right)\chi_{\beta q}\left(\vec{r}_{\beta},\zeta_{\beta q},R_{\beta}\right), \quad (2.28)$$

respectively. The sum of electron charge density at nucleus can be expressed as

$$\rho[N,n,R,C,\zeta](0)=\sum_{\alpha}^{N}\rho_{\alpha,\alpha}[n,R,C,\zeta](0)+\sum_{\alpha>2}^{N}\sum_{\beta(<\alpha)}^{N-1}\rho_{\alpha,\beta}[n,R,C,\zeta](0), \quad (2.29)$$

where

$$\rho_{\alpha,\alpha}[n,R,C,\zeta](0)=\sum_{p}^{AO(\alpha)}\sum_{q}^{AO(\alpha)}D_{\alpha p,\alpha q}[n,C]\chi_{\alpha p}^{*}(0,\zeta_{\alpha p},R_{\alpha})\chi_{\alpha q}(0,\zeta_{\alpha q},R_{\alpha}), \quad (2.30)$$

$$\begin{aligned}\rho_{\alpha,\beta}[n,R,C,\zeta](0)=&\sum_{p}^{AO(\alpha)}\sum_{q}^{AO(\beta)}\\&\left[\chi_{\alpha p}^{*}(0,\zeta_{\alpha p},R_{\alpha})\chi_{\beta q}(R_{\alpha},\zeta_{\beta q},R_{\beta})+\chi_{\alpha p}^{*}(R_{\beta},\zeta_{\alpha p},R_{\alpha})\chi_{\beta q}(0,\zeta_{\beta q},R_{\beta})\right].\end{aligned} \quad (2.31)$$

There is another relation with respect to electron charge density at nucleus. The Liu–Parr–Nagy (LPN) identity expressed as

$$\rho(0)=-\frac{1}{4\pi}\int\frac{\nabla^{2}\rho(r)}{|r|}dV, \quad (2.32)$$

has a qualification of constraint condition in the Lagrangian function. Although the LPN identity has an electron charge density at nucleus on the left hand side, it does not require the relation to the atomic charge Z on the right hand. The electron charge density at nucleus appears in Eqs. (2.24) and (2.32).

The virial theorem in the Born-Oppenheimer approximation is expressed as

$$2T[N,n,R,C,\zeta]+V[N,n,Z,R,C,\zeta]+\sum_{\alpha}^{N}\vec{R}_{\alpha}\cdot\frac{\partial E[N,n,Z,R_{\alpha},C,\zeta]}{\partial\vec{R}_{\alpha}}=0. \quad (2.33)$$

Equation (2.33) means that the kinetic energy and the potential energy are balanced with help of the third term at any molecular geometry point on the adiabatic total energy surface (Mayer 2010). Since the exact solution of the HFR equation satisfies the virial theorem, the virial theorem can be adopted as a constraint condition in the algebraic Lagrangian function.

The Hellmann–Feynman theorem

$$\frac{\partial}{\partial \alpha} E[N, n, Z, R, C, \zeta] = \left\langle \Psi(\alpha) \middle| \frac{\partial}{\partial \alpha} H[N, n, Z, R, C, \zeta] \middle| \Psi(\alpha) \right\rangle, \tag{2.34}$$

can be adopted as a constraint condition in the algebraic MO Lagrangian function, where α is a parameter included in the Hamiltonian. Equation (2.34) shows the general form of the Hellmann–Feynman theorem from which specific theorem, such as the virial theorem and the electrostatic theorem, is derived.

There are mathematical conditions which are satisfied with the exact eigenfunction of the molecular Hamiltonian, for example, closure relation for quantum mechanical operators. We can select the combination of constraint conditions according to the aim.

Although it is expected that the Lagrangian function extended with nonlinear variational parameters and multiple constraint conditions would be effective, it is necessary to keep the basis set in an adequate size.

2.3 Discussion

In the previous section we proposed a method for defining an algebraic MO equation in detail. In this section we will discuss about advantage and expectation in the algebraic MO theory.

2.3.1 Multivariable Theory for Chemistry

It is worth constructing a multivariable theory for quantum chemistry because chemistry is a science of atoms and molecules described by many quantities associated with nucleus and electron, and parameters of external fields such as electromagnetic field and boundary conditions applied to atoms and molecules. Mechanism of chemical reaction and that of electromagnetic property of molecule are fundamentally multivariable problems.

As a methodology for multivariable problems, numerical method is inadequate because much information of molecular parameters disappears in numerical operations of addition and multiplication.

The object of chemistry is not always a forward problem. The order of searching solutions depends on the problem. It is difficult to remove the distinction between a forward problem and an inverse problem on the numerical method. On the other hand, the algebraic MO theory is adequate for multivariable problems and because many variables are expressed by symbol so that their symbolic meanings are conserved during calculation process.

The algebraic MO theory is also adequate for inverse problem because the order of searching solutions is interchangeable.

The algebraic MO is an approximate equation of the molecular Schrödinger equation in affine space (Yasui 2010, 2011). Symbolic calculation on polynomials is essential in this theory.

2.3.2 Polynomial Expression of Molecular Integrals over STFs

Most innovative part in this study is an introduction of polynomial expression of molecular integrals to molecular orbital theory. This enables to construct the algebraic molecular orbital theory in affine space. Mathematically, molecular integral expressed in polynomial plays a roll of polynomial ring, and changes the form of molecular orbital equation from matrix equation to simultaneous polynomial equation. The set of solutions of polynomial MO equation is the approximate affine variety of the Schrödinger equation. This is a basic idea of the algebraic MO theory with many variables. Molecular integral expressed in polynomial is indispensable to the algebraic quantum chemistry.

Many studies of molecular integral with STO, GTO, and other basis functions have been done in the development quantum chemistry, however, the use is confined to numerical calculations. In order to construct the algebraic MO theory, it is necessary at least that molecular integral is obtained analytically as a function of atomic configuration parameters and orbital exponents of basis function. By analytical integration, the parameters of molecular integral keep themselves in symbolic expression. In addition to that, in order to define the algebraic MO equation for the second step, it is necessary to transform molecular integral from analytical expression to polynomial one. This transformation is done by the method of Taylor expansion.

The aim of quantum chemistry is various. In some studies, the precise calculation is needed. In other studies, it is important to grasp the essence of chemistry neither too much nor too little in accuracy. For these aims, polynomial molecular integral is extremely advantageous especially for large molecules. The accuracy of polynomial molecular integral is controllable in a wide range of variables by using adequate expansion centers. On the controllability of accuracy, it is possible to make a seamless approximation or an effective Hamiltonian on the first principle.

With respect to calculation error, there is no mixing of numerical errors in symbolic polynomial calculation because the coefficients of polynomial ring are expressed by rational numbers in this study.

While the algebraic MO theory is constructed in real-space in this study, we can define it in momentum-space by using the Shibuya-Wolfman integral which is molecular integral of exponential type basis function in momentum space (Shibuya and Wulfman 1965; Avery 2000).

2.3.3 Advantage of Extension of the Variational Principle

The effectiveness of variational principle depends on the set of trial functions, the set of variational parameters, and the constraint conditions. In principle, LCAO-MO theory is defined by the set of linear parameters and nonlinear parameters, however, in most traditional MO theory, nonlinear variational parameters are omitted. On the other hand, in the algebraic MO equation, orbital exponent is introduced as a nonlinear variational parameter in the definition of MO equation through polynomial expression of molecular integrals. In the algebraic MO theory compatibility of linear variational parameter and nonlinear variational parameter is attained. The action of orbital exponent as a nonlinear variational parameter is effective, which is described in Huzinaga's book. In case of atomic collision, for example, overlap of wave functions between atoms begins to contact with their tails of wave functions in the asymptotic region determined by the orbital exponent of basis functions. Contact of wave function brings about initial separation of molecular energy level in the same symmetry, and works to stabilize total electronic energy of atomic system. Using linear parameters and nonlinear parameters, trial functions fit precisely the details of variational active space of the molecular Hamiltonian so as to approach local minimum of total electronic energy under the set of Schrödinger conditions. In addition to accurate eigenenergy in the algebraic MO equation, more precise eigenfunction will be obtained by the extend variational method for degenerated state or quasidegenerated state of the Hamiltonian.

The set of the Schrödinger condition acts to extend variational function space in which each condition takes a different action to minimize total electronic energy in molecule. Thanks to polynomial expression of molecular integrals, we can extend the Lagrangian function without difficulty.

The orthonormal condition is a nonlocal constraint condition. In the algebraic MO theory the orthonormal constraint condition in Eq. (2.22) works not only in linear variational space through the density matrix but also in nonlinear variational space through the overlap matrix. On the other hand, in the traditional MO theory, nonlinear orbital exponents in overlap matrix are fixed in numerical value so that it is difficult to optimize MOs at arbitrary molecular geometrical point.

Atomic center is a singular point of Coulomb potential in molecule. By Kato's nuclear-electron cusp condition expressed in several form in Eqs. (2.23), (2.24) and (2.25), wave functions of the Schrödinger equation is able to avoid the divergence at atomic centers. As described in Eq. (2.29), atomic charge at atomic center includes contributions from the outer region. This means that cusp and tail of basis functions hold Kato's nuclear-electron cusp condition. Exact wave functions or exact electron charge densities, which obey Kato's nuclear-electron cusp condition, distribute in molecule so as to determine molecular structure, in other words, it might be said that Kato's nuclear-electron cusp condition gives a principle to determine molecular structure.

While the LPN identity in Eq. (2.32) includes electron charge density at atomic center, it does not require atomic number Z in the definition. Since electron charge density at atomic center appears both in Kato's nuclear-electron cusp condition and the LPN identity conditions, these conditions are intrinsically the same. By this reason, the LPN identity is approximately applicable to non-exponential basis functions, such as Gauss type functions.

The virial theorem in Eq. (2.33) holds at any molecular structure in the stationary state. Equation (2.33) means that kinetic energy and potential energy are balanced exactly with a help of the third term, which is an inner product of atomic position vector and gradient vector of the electronic total energy surface. It is interesting that the kinetic energy and the potential energy are global quantities but the third term is a summation of local quantities. Since the kinetic energy is always positive and the potential energy is always negative, the third term has to work as an adjuster to hold the virial theorem. This suggests that the third term needs accuracy. In the third term expressed in polynomial, first derivation of electronic total energy with respect to atomic position vector is obtained analytically. At any molecular geometrical point, the kinetic energy, the potential energy and the third term are able to take a balance among themselves by optimizing linear variational parameters and nonlinear variational parameters in this study.

It is well recognized that the fulfillment of the Hellmann–Feynman theorem is difficult by poor basis functions. This suggests that the Hellmann–Feynman theorem works as a strong constraint condition for defining a precise MO equation. It is expected that the Hellmann–Feynman theorem expressed in polynomials with respect to linear parameters and nonlinear parameters will play an effective roll as a strong constraint condition in the Lagrangian function.

Compared with a single constraint condition, the set of constraint conditions will generate more precise constraint condition as a whole. Multiple constraint condition increases a degree of freedom of trial function in variational active space.

By the cooperative effect by nonlinear variational parameter and multiple constraint condition in the Lagrangian function, the algebraic MO equation will provide more accurate eigenenergies and MOs than the traditional MO equation does.

2.3.4 Advantage in Calculation of Electron Correlation

Since the Schrödinger equation includes the electron correlation or the many body effect completely, electron correlation energy is defined by the total energy difference between the HF equation and the Schrödinger equation. Both the HF theory and the algebraic MO theory is based on the one-body model, however, the latter is superior in the variational principle as described in this chapter. This means that the total energy difference between the algebraic MO equation and the Schrödinger is expected to be smaller than in the case of the HF theory.

The eigenstate obtained by the algebraic MO calculation includes the electron correlation in some degree because the Lagrangian function adopts the Schrödinger conditions as constraint conditions. Electron correlation or electron many-body effect appears in coincident collision among electrons over two particles. Electron correlation appears not only in the stable eigenstate but also in the transition process. Electronic polarizability and multielectron transition, for example, are sensitive to the electron correlation on the ground state and during the transition process, respectively. The algebraic MO theory is useful in the calculation of electronic properties in which many-body interactions are active.

Having size consistency, many-body perturbation theory (MBPT) is valid for calculation of electronic correlations, particularly for large molecules. For a long time, various variation theory and perturbation theory of electronic correlation have been studied on the base of the traditional MO theory with orbital exponents of basis functions being fixed. For this reason, it is difficult for the traditional MO theory to achieve a high accuracy of electron correlation energy. On the other hand, it is expected that the algebraic MO theory will lead a rapid convergence in MBPT.

In a case when nonadiabatic interaction and electron correlation is close in energy, or in a critical region where a drastic reorganization of electronic configurations occurs, it may be difficult to analyze those complex interaction. For these cases, mathematics of polynomial will be useful for disentangling the complex.

2.3.5 Integer Variables in Quantum Chemistry

Number of atom, atomic number and number of electron is a basic information in chemistry. The Schrödinger equation includes atomic number Z, number of atom and number of electron as integer parameter. This means that there exists at least a relation between the integer parameter and the quantum eigenstate in molecule. However, it is impossible to know the relation from the HFR calculations because atomic number Z is dealt as a numerical value. The traditional theoretical chemistry has no awareness to solve MO equation with respect to those integers. It is interesting that Kato's nuclear-electron cusp condition connects directly atomic number Z and wave function at atomic centers in molecule.

2.3.6 Advantage of Polynomial Equation

M.P. Barnett informed us many papers about applications of polynomial equation in the field of chemistry except quantum chemistry (Barnett et al. 2004). M. Minimair and M.P. Barnett explained how to solve polynomial equations by using Gröbner bases (Minimair and Barnett 2004). Algebraic computational theory of polynomials has been developed in recent years (Cox et al. 2007), however, there has been no algebraic method in the development of quantum chemistry.

There are several mathematical advantages in solving a simultaneous algebraic equation. Firstly, it is possible to identify the existence of solutions rigorously even if it is difficult to solve the equation under a difficult condition. Secondary, identification of the number of solutions is possible. Lastly, it is possible to identify the relations among variables. In some cases, there is a remarkable possibility to find a useful relation among molecular variables. In addition to that, there is a great advantage in the algebraic method, that is, there is no distinction between forward problem and inverse problem. We can obtain solutions in any sequence of variables from polynomial equation.

The SCF iterative technique is a standard method for solving a nonlinear MO equation by regarding it as a linear equation. For small molecules SCF iterative technique is useful, however, it is inconvenient in case when a quasidegenerate electronic state exists. In some cases, quasidegenerate state cannot be solved by the SCF iterative technique. On the other hand, the algebraic MO equation needs no SCF iterative technique because nonlinear polynomial MO equation is solved algebraically. This is a great practical advantage of the algebraic MO equation.

2.3.7 Advantage in the Born-Oppenheimer Approximation

Study of nonadiabatic interaction between electron and nucleus is interesting as a fundamental interaction in chemistry. There are two interesting problems after obtaining PESs on the BO approximation. The one is to solve nuclear eigenstates on PES by using NAC. The another is to solve nonadiabatic electron transition probability between PESs by using NAC.

In the BO approximation, total molecular wave function is expressed by the sum of product of electronic wave function and nuclear wave function. The phase of electronic wave functions and that of nuclear wave functions cancel each other in the stationary state. This cancelation mechanism is a hidden nature in the nuclear Schrödinger equation. Nuclear quantum state is sensitive to NAC and PES because energy separation between nuclear quantum states is much smaller than that of electronic adiabatic PESs. In order to give a reliable description for nuclear state, it is better to express NAC and PES by continuous parameters, not by discrete numerical parameters. Owing to the method of polynomial expression of NAC, the nuclear Schrödinger equation will be solved in high accuracy. Construction of the coupled Schrödinger equation for electrons and nuclei is a fundamental problem not only for nonadiabatic process but also for molecular dynamics on the BO approximation in a wide range of time.

Within the adiabatic approximation, the time-dependent electronic wave function has the dynamical phase and the Longuet-Higgins/Berry phase which are acquired by the eigenenergy of electronic state and NAC, respectively. Acquisition of NAC and the Longuet-Higgins/Berry phase needs consistency in solving the nuclear Schrödinger equation. Acquisition of electronic eigenstate and nuclear eigenstate on the BO approximation is a multivariable problem. In symbolic calculation of NAC and the Longuet-Higgins/Berry phase, polynomial method will be useful. NAC and the Longuet-Higgins/Berry phase expressed as a multivariable function will play an important roll as a reciprocal relation in the Born-Oppenheimer molecular dynamics.

Nonadiabatic electron transition appears in fundamental process in many fields of molecular science, such as chemical reaction, especially important in photochemistry, slow atomic collision, spectroscopy for electron and nucleus, surface science, solid-state physics. The nonadiabatic electron transition probability is a time-dependent problem. In slow atom-atom collision, the transition near the avoided crossing region of adiabatic electronic states obeys a selection rule different from that for dipole transition. To discover a region where the nonadiabatic electron transition occurs between PESs is an interesting problem.

Conclusion

In this chapter we proposed a method for defining an algebraic molecular orbital equation in order to construct a multivariable symbolic quantum theory for molecular science. We defined the algebraic MO equation as a simultaneous multivariable polynomial equation on the nonlinear variational principle by introducing polynomial expression of molecular integrals over STFs with respect to atomic configuration parameters and nonlinear orbital exponents of basis functions. In this algebraic theory there is no need of SCF iterative technique and there is no distinction between forward problem and inverse problem. The algebraic MO theory is beyond the Hartree-Fock theory in spite of one-body model because our theory includes electron correlation by adoption of the Schrödinger condition as constraint condition in the Lagrangian function. Using controllability of polynomial molecular integral accuracy, a seamless effective Hamiltonian can be defined without lack of the first principle in a wide range of physical dimension. This study brings a new practical method on the first principle for large molecules with accordance to the aim of study.

Acknowledgments I would like to express my heartfelt thanks to the late Professor A. Saika who gave me the theme of molecular integrals over STFs to study the electron correlation effect in NMR chemical shifts in my graduate school; Professor Shigeru Huzinaga whose book sparked my interest in molecular orbital theory; Professor Tsutomu Watanabe who gave me an interest in slow atomic collisions including nonadiabatic problems of chemistry in abundance; Professor Takeshi Mukoyama who impregnated me with a wide interest in inner shell ionization process and atomic collision; Professor Hirohiko Adachi who introduced me the DV-Xα calculation method based on DFT theory and showed a method for the calculation of transition probability of nonadiabatic electron transition; Professor Tai-Ichi Shibuya who pointed out the relation between STF molecular integral and Shibuya-Wolfman integral in momentum space; and Dr. Akihito Kikuchi in Canon Inc. who showed his great mathematical interest in this study.

References

Adachi H (2006) DV-Xα method and molecular structure. In: Adachi H, Mukoyama T, Kawai J (eds) Hartree-Fock-Slater method for materials science. Springer, Berlin/New York

Adachi H, Yasui J (1987) Theoretical calculation of transition probability for nonadiabatic process by discrete-variational Xα method. Bul Hyogo Univ Teacher Educ 7:61–68

Adachi H, Tsukada M, Satoko C (1978) Discrete variational Xα cluster calculations. I. Application to metal clusters. J Phy Soc Jpn 45:875–883

Averill FW, Ellis DE (1973) An efficient numerical multicenter basis set for molecular orbital calculations: application to FeCl4. J Chem Phys 59:6412–6418

Avery J (2000) Hyperspherical harmonics and generalized Sturmians. Kluwer, Dordrecht

Baer R (2002) Born-Oppenheimer invariants along nuclear configuration paths. J Chem Phys 117:7405–7408

Baer M (2006) Beyond Born-Oppenheimer: electronic nonadiabatic coupling terms and conical intersections, 1st edn. Wiley-Interscience, Hoboken

Baerends EJ, Ellis DE, Ros P (1973) Self-consistent molecular Hartree-Fock-Slater calculations I. Computational procedure. Chem Phys 2:41–51

Barnett MP, Capitani JF, von zur Gathen J, Gerhard J (2004) Symbolic calculation in chemistry, selected examples. Int J Quant Chem 100:80–104

Berry MV (1984) Quantal phase factors accompanying adiabatic changes. Proc R Soc Lond A 392:45–57

Bouferguene A, Fares M, Hoggan PE (1996) STOP: Slater Type Orbital Package for general molecular electronic structure calculations. Int J Quant Chem 57:801–810

Chapman JA, Chong DP (1970) Cusp constraints for atomic wavefunctions. Can J Chem 48:2722–2726

Chong DP (1967) Coalescence conditions as constraints in open-shell SCF theory. J Chem Phys 47:4907–4909

Cox D, Little J, O'Shea D (2007) Ideals, varieties, and algorithms, 3rd edn. Springer, New York

Dirac PAM (1929) Quantum mechanics of many-electron systems. Proc Roy Soc (London) A 123:714–733

Domcke W, Yarkony DR, Köppel H (2004) Conical intersections: electronic structure, dynamics & spectroscopy. World Scientific, Singapore

Feinberg MJ, Ruedenberg K, Mehler EL (1970) The origin of bonding and antibonding in the hydrogen molecule-ion. Adv Quant Chem 5:27–98

Fernández Rico J, Lopez R, Aguado A, Ema I, Ramírez G (2001) New program for molecular calculations with Slater-type orbitals. J Quant Chem 81:148–153

Fernández Rico J, Lopez R, Ramírez G, Ema I (2004) Efficiency of the algorithms for the calculation of Slater molecular integrals in polyatomic molecules. J Comp Chem 25:1987–1994

Fernández Rico J, López R, Ema I, Ramírez G (2007) Generation of basis set with high degree of fulfillment of the Hellmann-Feynman theorem. J Comp Chem 28:748–758

Feynman RP (1939) Forces in molecules. Phys Rev A 56:34–343

Galek PTA, Handy NC, Cohen AJ, Chan GKL (2005) Hartree-Fock orbitals which obey the nuclear cusp condition. Chem Phys Lett 404:156–163

Hohenberg P, Kohn W (1964) Inhomogeneous electron gas. Phys Rev B 136:864–871

Huzinaga S (1980) Molecular orbital method (Japanese). Iwanami Book Publisher, Tokyo, Cambridge University Press (2000)

Janosfalvi Z, Sen KD, Nagy Á (2005) Cusp conditions for non-interacting kinetic energy density of the density functional theory. Phys Lett A 344:1–6

Kato T (1957) On the eigenfunctions of many-particle systems in quantum mechanics. Comm Pure Appl Math 10:151–177

Koga T, Kanayama K, Watanabe S, Thakkar AJ (1999) Analytical Hartree-Fock wave functions subject to cusp and asymptotic constraints: He to Xe, Li+ to Cs+ H- to I-. Int J Quant Chem 71:491–497

Kohn W, Sham LL (1964) Self-consistent equations including exchange and correlation effects. Phys Rev A 140:1133–1138

Köppel H, Yarkony DR, Barentzen H (2009) The Jahn-Teller-effect: fundamentals and implications for physics and chemistry, 1st edn. Springer, Heidelberg

Kotani M, Amemiya A, Ishiguro AE, Kimura T (1955) Table of molecular integrals. Maruzen, Tokyo

Liu S, Parr RG, Nagy Á (1995) Cusp relations for local strongly decaying properties in electronic systems. Phys Rev A 52:2645–2651

Longuest-Higgins HC (1975) Intersection of potential energy surfaces in polyatomic molecules. Proc Roy Soc Lond A 344:147–156

Mayer I (2010) Simple theorems, proofs and derivations in quantum chemistry. Kluwer/Plenum Publishers, New York

Minimair M, Barnett MP (2004) Solving polynomial equations for chemical problems using Gröbner bases. Mol Phys 102:2521–2535

Mukoyama T, Yasui J (1992) Analytical expression of the Hartree-Fock wave functions. Bull Inst Chem Res Kyoto Univ 70:385–391

Nagy Á, Sen KD (2001) Ground- and excited-state cusp conditions for the electron density. J Chem Phys 115:6300–6308

Nakamura H (2012) Nonadiabatic transition: concepts, basic theories and applications, 2nd edn. World Scientific, Singapore

Nakatsuji H, Izawa M (1989) Calculation of hyperfine splitting constants with Slater-type cusp basis by the symmetry adapted cluster-configuration interaction theory. J Chem Phys 91:6205–6214

Pack RT, Brown WB (1966) Cusp conditions for molecular wave functions. J Chem Phys 45:556–559

Poling SM, Davidson ER, Vincow G (1971) Calculation of the hyperfine splittings of CH. cusp constraint of a wavefunction. J Chem Phys 54:3005–3013

Rassolov VA, Chipman DM (1996) Behavior of electronic wave functions near cusps. J Chem Phys 104:9908–9912

Reinhardt P, Hoggan PE (2009) Cusps and derivatives for wave-functions expanded in Slater orbitals: a density study. Int J Quant Chem 109:3191–3198

Roothaan CCJ (1951) New developments in molecular orbital theory. Rev Mod Phys 23:69–89

Shibuya T, Wulfman CE (1965) Molecular orbitals in momentum space. Proc Roy Soc A 286:376–389

Slater JC (1974) Quantum theory of molecules and solids. McGraw-Hill, New York

Slater JC (1979) The calculation of molecular orbitals. Wiley, New York

Stanton RE (1962) Hellmann-Feynman theorem and correlation energies. J Chem Phys 36:1298–1300

Steiner E (1963) Charge densities in atoms. J Chem Phys 39:2365–2366

Yasui J (2010) Polynomial expressions of molecular integral functionals over Slater-type-orbitals and its application to the extension of Hartree-Fock-Roothaan equation. Bull Soc DV Xα 23:54–59

Yasui J (2011) Algebraic molecular orbital equation. Bull Soc DV Xα 24:47–54

Yasui J, Saika A (1982) Unified analytical evaluation of two-center, two-electron integrals over Slater-type orbitals. J Chem Phys 76:468–472

Chapter 3
Analytical Expression of Molecular Integrals over Slater-Type Functions for Generating Their Polynomial Expressions

Jun Yasui

3.1 Introduction

Electromagnetic property of molecule changes depending on the molecular structure. It is a fundamental problem to express molecular quantum system symbolically as a function of molecular structure. Molecular orbital equation is defined almost on the LCAO-MO method in which basis function is located at atomic center. The Hartree-Fock-Roothaan equation (Roothaan 1951), the Hartree-Fock-Slater equation (Slater 1951, 1974, 1979), and the Kohn-Sham equation (Kohn and Sham 1964) adopt the LCAO-MO method. In these standard equations, however, the Fock matrix consists of molecular integrals calculated numerically in advance of diagonalization of the matrix, and eigenvalue problem is solved numerically using SCF iteration scheme. As long as the MO equation employs numerical molecular integrals, we cannot deal with molecular property as an analytical function of molecular structure because symbolic information of molecular structure vanishes in the process of numerical calculations. The important role of analytical molecular integral is to connect electronic state and atomic configuration because molecular integral is a parametric function of atomic configuration.

Molecular calculation method mostly chooses Gaussian-type function (GTFs) as a basis function for the convenience to multicenter integrals, however, wave function expanded by GTFs does not obey the cusp condition between nucleus and electron,

$$\Psi(0) + \frac{\Psi'(0)}{Z} = 0.$$

J. Yasui (✉)
Frontier Research Center, Canon Inc., 30-2 Shimomaruko 3-chome,
Ohta-ku, Tokyo 146-8501, Japan
e-mail: yasui.jun@canon.co.jp

T. Ishii et al. (eds.), *The DV-Xα Molecular-Orbital Calculation Method*,
DOI 10.1007/978-3-319-11185-8_3

This means that GTF is not superior basis function to expand wave functions of the Schrödinger equation since the cusp condition is a necessary condition which should be obeyed by the exact wave function of the Schrödinger equation (Kato 1957).

Since wave function expanded by Slater-type functions (STFs) with exponential decay obeys the cusp condition, many studies of molecular integral over STFs have been studying in spite of difficulty of multicenter integration (Kotani et al. (1955), Silver (1971), Jones (1980, 1988, 1994), Huzinaga (1980, 2000), Yasui and Saika (1982), Wojnecki and Modrak (1993), Hierse and Pooeneer (1994), Fernández Rico et al. (1994), Yasui et al. (1996), Avery (2000), Barnett et al. (2004), Guseinov (2007), Yasui (2010, 2011)). In more general, exponential basis functions (ETFs) is intrinsically better as a basis function to approximate exact solutions of the Schrödinger equation. Molecular integrals over ETFs has been discussed and reviewed by Reinhardt and Hoggan (2009). We can also see the usefulness of STFs in the works of Mukoyama who employed STFs for the bound states and continuum states (Mukoyama and Yasui 1992; Mukoyama and Hock 1994). However, these molecular integrals over ETFs or STFs are used in numerical MO calculations at present, even in cases when the formulations of molecular integral are executed analytically or symbolically. On the other hand, we need molecular integrals not only in analytical form but also in the form which keeps its symbolic expression in the MO equation.

Our MO theory is constructed on the algebra. The aim of this study is to formulate analytical molecular integral over STFs which is adequate to generate polynomial expression of molecular integrals with respect to molecular structure and orbital exponent of STF. In addition to that, we are going to obtain polynomial molecular integrals by which efficient MO calculations are executed. By applying the Taylor expansion method to analytical molecular integrals, for example, two-center overlap integral between STFs centered at A and B with atomic separation of R is expressed approximately as

$$S(\zeta_a, \zeta_b, R)_{AB}^{ab} \cong \sum_{p_a, p_b, p_R} A\left(\{n_a, l_a, m_a\}_A, \{n_b, l_b, m_b\}_B\right)_{p_a, p_b, p_R} \zeta_a^{p_a} \zeta_b^{p_b} R^{p_R}$$

where q_a and q_b mean a set of quantum numbers, $\{n_a,l_a,m_a\}$ and $\{n_b,l_b,m_b\}$ of STFs, respectively, and coefficient A in the summation can be expressed in rational numbers. In the previous chapter we proposed the algebraic MO equation based on the polynomial expression of molecular integrals. Polynomial molecular integral makes it possible to define the effective hamiltonian without lack of the first principle by controlling molecular integral accuracy in a wide range of physical dimension. By the algebraic MO equation with the effective hamiltonian based on polynomial molecular integrals, we are going to answer to the Dirac's worry on the difficulty in studying chemistry using the Schrödinger equation (Dirac 1929).

For the first step to construct algebraic quantum chemistry, it is worth developing molecular integrals over STFs for diatomic system because diatomic molecule is a minimal molecular system with chemical bond. In diatomic system, motion of electron and nucleus is coupled nonadiabatically. It is a fundamental problem to solve nuclear motion and electron motion simultaneously beyond the Born-Oppenheimer approximation. In order to solve nuclear Schrodinger equation, it is better to express potential energy surface as an analytical function of atomic configuration.

In this study we deal with molecular integral in real-space, while it is possible to define molecular integral in momentum space (Shibuya and Wulfman 1965; Mukoyama and Kagawa 1983). As was mentioned above, analytical molecular integral is essential in quantum chemistry. We have developed the method to formulate analytical molecular integral over STFs in real-space in order to lead polynomial expression of molecular integral at first for monoatomic system and diatomic system.

3.2 General Formulation

3.2.1 Preliminaries

For the sake of convenience, we begin with a brief explanation of the notation and definitions necessary in what follows. In this article, atomic units are used.

3.2.1.1 The Coordinate System

We shall adopt the coordinate system as shown in Fig. 3.1 in this article.

A and *B* are the two fixed centers with separation *R*. *A* is the origin of this coordinate system. The STFs are centered at *A* and *B*. P_1 and P_2 are the positions of electron 1 and 2, respectively. *C* is the position of another nucleus which interacts with electron charge density expressed by the product of STFs.

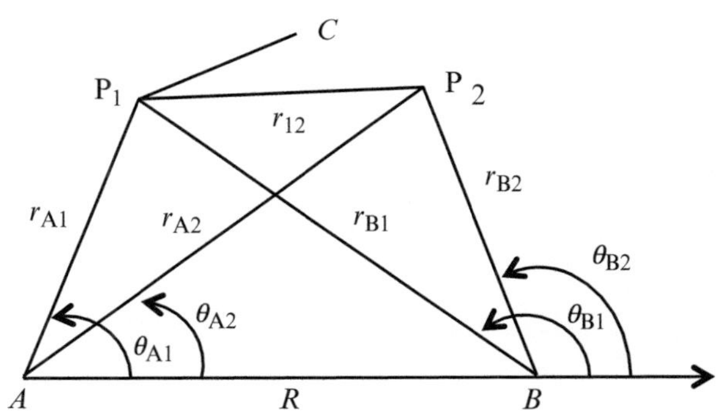

Fig. 3.1 Two-center coordinate system

3.2.1.2 Slater-Type Function

Normalized STF assigned by quantum numbers, $\{n_a, l_a, m_a\}$, centered at A is written in the form

$$\chi\left(\vec{r}\right)_A^a = R_{Aa}(r_A) Y_{l_a}^{m_a}(\theta_A, \phi_A) \tag{3.1}$$

where R_{Aa} is a normalized radial function

$$R_{Aa}(r_A) = N_r(n_a, \zeta_a) r_A^{n_a - 1} \exp(-\zeta_a r_A) \tag{3.2}$$

N_r is a normalization constant

$$N_r(n, \zeta) = (2\zeta)^{n+\frac{1}{2}} / [\Gamma(2n+1)]^{\frac{1}{2}} \tag{3.3}$$

where $\Gamma(x)$ is the gamma function,

Y_l^m is a real, normalized spherical harmonic function

$$Y_l^m(\theta_A, \phi_A) = p_l^m(\cos\theta_A) f_m(\phi) \tag{3.4}$$

p_l^m is a normalized associated Legendre function

$$p_l^m(x) = N_x(l, m) P_l^m(x) \tag{3.5}$$

and f_m is a normalized function with respect to ϕ

$$f_m(\phi) = N_\phi(m) \left\{ \begin{array}{c} \cos m\phi \\ \sin|m|\phi \end{array} \right\} \begin{array}{l} \mathit{for}\ \ m \geq 0 \\ \mathit{for}\ \ m < 0. \end{array} \tag{3.6}$$

Useful formulas with respect to the real spherical harmonic function are summarized in Appendix A.

3.2.1.3 Molecular Integrals Discussed in This Article

We formulate the following one-center and two-center one electron integrals,

Overlap one-center integrals:

$$S_{AA}^{ab} = \int dV \chi_A^a\left(\vec{r}\right) \chi_A^b\left(\vec{r}\right) \tag{3.7}$$

Overlap two-center integrals:

$$S_{AB}^{ab} = \int dV_A \chi_A^a\left(\vec{r}_A\right) \chi_B^b\left(\vec{r}_B\right) \tag{3.8}$$

Kinetic energy one-center integrals:

$$T_{AA}^{ab} = \int dV \chi_A^a(\vec{r}) \left[-\frac{1}{2} \Delta \chi_A^b(\vec{r}) \right] \tag{3.9}$$

Kinetic energy two-center integrals:

$$T_{AB}^{ab} = \int dV_A \chi_A^a(\vec{r}_A) \left[-\frac{1}{2} \Delta \chi_B^b(\vec{r}_B) \right] \tag{3.10}$$

Nuclear attraction energy one-center integrals:

$$V_{AA/A}^{ab} \equiv \int dV \chi_A^a(\vec{r}) \frac{1}{r} \chi_A^b(\vec{r}) \tag{3.11}$$

$$V_{AA/B}^{ab} \equiv \int dV \chi_A^a(\vec{r}) \frac{1}{\left| \vec{r} - \vec{R}_B \right|} \chi_A^b(\vec{r}) \tag{3.12}$$

Nuclear attraction energy two-center integrals:

$$V_{AB/A}^{ab} \equiv \int dV_A \chi_A^a(\vec{r}_A) \frac{1}{r_A} \chi_B^b(\vec{r}_B) \tag{3.13}$$

$$V_{AB/C}^{ab} \equiv \int dV_A \chi_A^a(\vec{r}_A) \frac{1}{\left| \vec{r}_A - \vec{R}_C \right|} \chi_B^b(\vec{r}_B) \tag{3.14}$$

We formulate the following one-center and two-center electron repulsion integrals,

One-center electron repulsion integrals:

$$V_{AAAA}^{abcd} \equiv \int dV_1 \int dV_2 \chi_A^a(\vec{r}_1) \chi_A^b(\vec{r}_1) \frac{1}{r_{12}} \chi_A^c(\vec{r}_2) \chi_A^d(\vec{r}_2) \tag{3.15}$$

Coulomb-type integrals:

$$V_{AABB}^{abcd} \equiv \int dV_1 \int dV_2 \chi_A^a(\vec{r}_{A1}) \chi_A^b(\vec{r}_{A1}) \frac{1}{r_{12}} \chi_B^c(\vec{r}_{B2}) \chi_B^d(\vec{r}_{B2}) \tag{3.16}$$

Hybrid-type integrals:

$$V_{AAAB}^{abcd} \equiv \int dV_1 \int dV_2 \chi_A^a(\vec{r}_{A1}) \chi_A^b(\vec{r}_{A1}) \frac{1}{r_{12}} \chi_A^c(\vec{r}_{A2}) \chi_B^d(\vec{r}_{B2}) \tag{3.17}$$

Exchange-type integrals:

$$V_{ABAB}^{abcd} \equiv \int dV_1 \int dV_2 \chi_A^a(\vec{r}_{A1}) \chi_B^b(\vec{r}_{B1}) \frac{1}{r_{12}} \chi_A^c(\vec{r}_{A2}) \chi_B^d(\vec{r}_{B2}). \tag{3.18}$$

3.2.1.4 Coordinate System of Integration

For one-center integral, integration is executed on the coordinate system with origin at A in Fig. 3.1 as follows

$$I \to \int_0^\infty r_A^2 dr_A \int_0^\pi \sin\theta_A d\theta_A \int_0^{2\pi} d\phi_A \tag{3.19}$$

where subscript of ϕ is omitted for simplicity because $\phi_\square$ and ϕ_B are defined in common.

For two-center integral, we introduce several transformations in the coordinate system shown in Fig. 3.1.

By the cosine rule, we express cosθ_A using R, r_A and r_B of triangle.

$$r_B^2 = R^2 + r_A^2 - 2Rr_A \cos\theta_A \tag{3.20}$$

$$\cos\theta_A = (2Rr_A)^{-1}\left(R^2 + r_A^2 - r_B^2\right). \tag{3.21}$$

Using the binomial expansion method,

$$\cos^n\theta_A = \sum_i^n \binom{n}{i} \sum_j^{n-i} \binom{n-i}{j} (-1)^i 2^{-n} R^{-n+2j} r_A^{n-2(i+j)} r_B^{2i}. \tag{3.22}$$

Introducing x, we can express radial distance r_B by R, r_A and x

$$r_B = \begin{cases} R + r_A x & \text{for } r_A \le R \\ r_A + Rx & \text{for } r_A \ge R \end{cases} \tag{3.23}$$

where $-1 \le x \le 1$.

3.2.1.5 Change of Variable for Two Center Integration

Using Eqs. (3.21) and (3.23), we can change variable for integration from θ_A to x, and obtain the Jacobians for two regions, respectively, as follows.

$$\sin\theta_A d\theta_A = (Rr_A)^{-1} r_B dr_B \tag{3.24}$$

$$\begin{aligned} J(r_A, R | r_A \le R) &\equiv R^{-1} r_B \\ J(r_A, R | r_A \ge R) &\equiv r_A^{-1} r_B. \end{aligned} \tag{3.25}$$

Using the new Jacobian, we can integrate two-center functions as follows

$$\begin{aligned} \int f(r_B) dV \to\ & R^{-1} \int_0^R r_A^2 dr_A \int_{-1}^1 f(r_B) r_B dx \int_0^{2\pi} d\phi \\ & + \int_R^\infty r_A dr_A \int_{-1}^1 f(r_B) r_B dx \int_0^{2\pi} d\phi \end{aligned} \tag{3.26}$$

where subscript of ϕ is omitted because ϕ_A and ϕ_B is the same (Milleur et al. 1966; Christoffersen and Ruedenberg 1968).

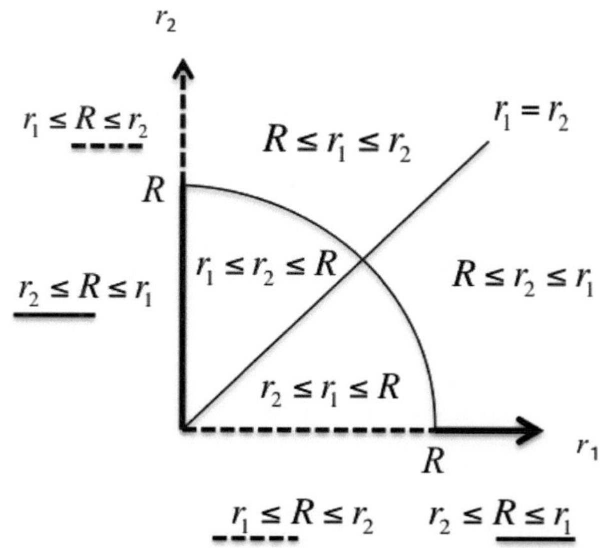

Fig. 3.2 Partition of integral domain

3.2.1.6 Domain of Integration

According large/small relation among atomic distance R, radial distance of the first electron $r_{1,}$ and radial distance of the second electron $r_{2,}$ domain of integration is separated symmetrically along the line $r_1 = r_2$ into six regions as shown in Fig. 3.2.

Six regions of integration are labeled in the following,

Region 1–1 for $r_2 \leq r_1 \leq R$, Region 1–2 for $r_2 \leq R \leq r_1$, Region 1–3 for $R \leq r_2 \leq r_1$, Region 2–1 for $r_1 \leq r_2 \leq R$, Region 2–2 for $r_1 \leq R \leq r_2$, Region 2–3 for $R \leq r_1 \leq r_2$.

3.2.1.7 One-Center Charge Density Centered on *A*

For the convenience of formulation, using a useful expression of product of spherical harmonic functions given in Eqs. (3.27, 3.A22)

$$Y_{l_a}^{m_a}(\theta, \phi) Y_{l_b}^{m_b}(\theta, \phi) = \sum_{m_{ab}} \Phi(m_a, m_b, m_{ab}) \sum_{l_{ab}} \Theta \begin{pmatrix} m_a & m_b & m_{ab} \\ l_a & l_b & l_{ab} \end{pmatrix} Y_{l_{ab}}^{m_{ab}}(\theta, \phi). \tag{3.27}$$

Charge density by the product of STFs on the same center is expressed

$$\begin{aligned} \rho_{AA}^{ab}\left(\vec{r}_A\right) &\equiv \chi_A^a\left(\vec{r}_A\right)\chi_A^b\left(\vec{r}_A\right) \\ &= R_A^a(r_A) Y_{l_a}^{m_a}(\theta_A, \phi) R_A^b(r_A) Y_{l_b}^{m_b}(\theta_A, \phi) \\ &= N_r(n_a, \zeta_a) N_r(n_b, \zeta_b) {r_A}^{n_a+n_b-2} \exp(-(\zeta_a + \zeta_b) r_A). \\ &\quad \times \sum_{m_{ab}} \Phi(m_a, m_b, m_{ab}) \sum_{l_{ab}} \Theta \begin{pmatrix} m_a & m_b & m_{ab} \\ l_a & l_b & l_{ab} \end{pmatrix} Y_{l_{ab}}^{m_{ab}}(\theta, \phi) \end{aligned} \tag{3.28}$$

3.2.1.8 Transfer of Origin of Spherical Harmonics from *B* to *A*

The following expansion is used to transfer a spherical harmonic function from center *B* to center *A* (Hobson 1931; Barnett and Coulson 1951; Buehler and Hirschfelder 1951; Löwdin 1956; Sharma 1976)

$$Y_l^m(\theta_B, \phi) = \sum_{k=|m|}^{l} T_k^{lm} R^{l-k} r_A^k r_B^{-l} Y_k^m(\theta_A, \phi) \tag{3.29}$$

where the constant T_k^{lm} is defined by

$$T_k^{lm} = (-1)^{k+m} \frac{N_x(l,m)}{N_x(k,m)} \frac{(l+|m|)!}{(l-k)!(k+|m|)!}. \tag{3.30}$$

3.2.1.9 Two-Center Charge Density Centered on *A* and *B*

Making use of Eq. (3.A29),

$$\begin{aligned}
\rho_{AB}^{ab}\left(\vec{r}_A, R\right) &\equiv \chi_A^a\left(\vec{r}_A\right)\chi_B^b\left(\vec{r}_B\right) \\
&= R_A^a(r) Y_{l_a}^{m_a}(\theta_A, \phi) \cdot R_B^b(r_B) Y_{l_b}^{m_b}(\theta_B, \phi) \\
&= N_r(n_a, \zeta_a) N_r(n_b, \zeta_b) r_A^{n_a-1} \exp(-\zeta_a r_A) r_B^{n_b-1} \exp(-\zeta_b r_B) \\
&\quad \times \sum_{k_b=|m_b|}^{l_b} T_{k_b}^{l_b m_b} \sum_{m_{ab}} \Phi(m_a, m_b, m_{ab}) \sum_{l_{ab}} \Theta\begin{pmatrix} m_a & m_b & m_{ab} \\ l_a & k_b & l_{ab} \end{pmatrix} \\
&\quad \times R^{l_b-k_b} r_A^{k_b} r_B^{-l_b} Y_{l_{ab}}^{m_{ab}}(\theta_A, \phi)
\end{aligned} \tag{3.31}$$

where r_B is kept in formula without changing to x because r_B appears from other part, for example, the Jacobian in Eq. (3.26).

3.2.1.10 One-Center Charge Density Centered on *B*

Making use of Eq. (3.A34), one-center charge density centered on *B* is expressed as

$$\begin{aligned}
\rho_{BB}^{ab}\left(\vec{r}_A, R\right) &\equiv \chi_B^a\left(\vec{r}_B\right)\chi_B^b\left(\vec{r}_B\right) \\
&= R_B^a(r_B) R_B^b(r_B) \cdot Y_{l_a}^{m_a}(\theta_B, \phi) Y_{l_b}^{m_b}(\theta_B, \phi) \\
&= N_r(n_a, \zeta_a) N_r(n_b, \zeta_b) r_B^{n_a+n_b-2} \exp(-\{\zeta_a + \zeta_b\} r_B) \\
&\quad \times \sum_{m_{ab}} \Phi(m_a, m_b, m_{ab}) \sum_{l_{ab}} \Theta\begin{pmatrix} m_a & m_b & m_{ab} \\ l_a & l_b & l_{ab} \end{pmatrix} \\
&\quad \times \sum_{k_{ab}=|m_{ab}|}^{l_{ab}} T_{k_{ab}}^{l_{ab} m_{ab}} R^{l_{ab}-k_{ab}} r_A^{k_{ab}} r_B^{-l_{ab}} Y_{k_{ab}}^{m_{ab}}(\theta_A, \phi).
\end{aligned} \tag{3.32}$$

3.2.1.11 Two-Center Integration

As a nonvanishing condition for the integration with respect to ϕ, m in spherical harmonic function Y_l^m is restricted to naught. By inserting cosθ in Eq. (3.21) to Legendre function in Eq. (3.A11), we obtain a key expression of Legendre function in which variable cosθ_A is expressed by the three distances, R, r_A and r_B

$$p_l^0(\cos\theta_A) = \sum_{v=0}^{\left[\frac{l}{2}\right]} \sum_{i=0}^{l-2v} \sum_{j=0}^{l-2v-i} \mathrm{P}(l, v, i, j) R^{-l+2(v+j)} r_A^{l-2(v+i+j)} r_B^{2i} \tag{3.33}$$

where

$$\mathrm{P}(l, v, i, j) = N_x(l, 0)\omega_v^{l0} 2^{-(l-2v)} (-1)^i \binom{l-2v}{i} \binom{l-2v-i}{j}. \tag{3.34}$$

For the spherical harmonic function Y_l^m,

$$Y_l^0(\theta_A, \phi) = \sum_{v=0}^{\left[\frac{l}{2}\right]} \sum_{i=0}^{l-2v} \sum_{j=0}^{l-2v-i} Q(l, \nu, i, j) R^{-l+2(v+j)} r_A^{l-2(v+i+j)} r_B^{2i} \tag{3.35}$$

where

$$Q(l, \nu, i, j) = N_\phi(0)\mathrm{P}(l, \nu, i, j). \tag{3.36}$$

Legendre function p_l is originally a polynomial function with respect to cosθ, but in this article it is expressed as a polynomial with respect to R, r_A, and r_B as shown on the right hand side of Eq. (3.33). Variable r_B can be transferred to x as defined by Eq. (3.23). Thanks to this expression of Legendre function given by Eq. (3.33), all molecular integrals over STFs can be expressed as analytical functions of diatomic distance, R, and orbital exponent of STF, ζ.

3.2.1.12 Two-Center Charge Density with Jacobian

Using the Jacobian in Eqs. (3.25) and (3.26), products of the Jacobian and two-center charge density ρ_{AB}^{ab} in Eq. (3.31) are expressed for $r_A \le R$ and $r_A \ge R$, respectively, as follows

$$
\begin{aligned}
&J\left(r_A, R\middle|r_A \le R\right)\rho_{AB}^{ab}\left(\vec{r}_A\middle|r_A \le R\right)_\phi \\
&\quad = 2\pi N_\phi(0) N_r(n_a, \zeta_a) N_r(n_b, \zeta_b) \\
&\qquad \times \sum_{k_b=\left|m_b\right|}^{l_b} T_{k_b}^{l_b m_b} \sum_{m_{ab}} \Phi\left(m_a, m_b, m_{ab}\right) \sum_{l_{ab}} \Theta\begin{pmatrix} m_a & m_b & m_{ab} \\ l_a & k_b & l_{ab} \end{pmatrix} \\
&\qquad \times \sum_{\nu=0}^{\left[\frac{l_{ab}}{2}\right]} \sum_{i=0}^{l_{ab}-2\nu} \sum_{j=0}^{l_{ab}-2\nu-i} \mathrm{P}(l_{ab}, \nu, i, j) \\
&\qquad \times R^{l_b-k_b-l_{ab}+2(\nu+i)-1} r_A^{n_a-1+k_b+l_{ab}-2(\nu+i+j)} \exp\left(-\zeta_a r_A\right) r_B^{n_b-l_b+2i} \exp\left(-\zeta_b r_B\right)
\end{aligned}
\tag{3.37}
$$

$$
\begin{aligned}
&\quad = 2\pi N_\phi(0) N_r(n_a, \zeta_a) N_r(n_b, \zeta_b) \\
&\qquad \times \sum_{k_b=\left|m_b\right|}^{l_b} T_{k_b}^{l_b m_b} \sum_{m_{ab}} \Phi\left(m_a, m_b, m_{ab}\right) \sum_{l_{ab}} \Theta\begin{pmatrix} m_a & m_b & m_{ab} \\ l_a & k_b & l_{ab} \end{pmatrix} \\
&\qquad \times \sum_{\nu=0}^{\left[\frac{l_{ab}}{2}\right]} \sum_{i=0}^{l_{ab}-2\nu} \sum_{j=0}^{l_{ab}-2\nu-i} \mathrm{P}(l_{ab}, \nu, i, j) \sum_{s=0}^{n_b-l_b+2i} \begin{pmatrix} n_b - l_b + 2i \\ s \end{pmatrix} \\
&\qquad \times R^{l_b-k_b-l_{ab}+2(\nu+i)-1+s} \exp\left(-\zeta_b R\right) \cdot r_A^{n_a-1+k_b+l_{ab}-2(\nu+i+j)+n_b-l_b+2i-s} \\
&\qquad \times \exp\left(-\zeta_a r_A\right) x^{n_b-l_b+2i-s} \exp\left(-\zeta_b R x\right),
\end{aligned}
\tag{3.38}
$$

$$
\begin{aligned}
&J\left(r_A, R\middle|r_A \ge R\right)\rho_{AB}^{ab}\left(\vec{r}_A\middle|r_A \ge R\right)_\phi \\
&\quad = 2\pi N_\phi(0) N_r(n_a, \zeta_a) N_r(n_b, \zeta_b) \\
&\qquad \times \sum_{k_b=\left|m_b\right|}^{l_b} T_{k_b}^{l_b m_b} \sum_{m_{ab}} \Phi\left(m_a, m_b, m_{ab}\right) \sum_{l_{ab}} \Theta\begin{pmatrix} m_a & m_b & m_{ab} \\ l_a & k_b & l_{ab} \end{pmatrix} \\
&\qquad \times \sum_{\nu=0}^{\left[\frac{l_{ab}}{2}\right]} \sum_{i=0}^{l_{ab}-2\nu} \sum_{j=0}^{l_{ab}-2\nu-i} \mathrm{P}(l_{ab}, \nu, i, j) \\
&\qquad \times R^{l_b-k_b-l_{ab}+2(\nu+i)} r_A^{n_a-2+k_b+l_{ab}-2(\nu+i+j)} \exp\left(-\zeta_a r_A\right) r_B^{n_b-l_b+2i} \exp\left(-\zeta_b r_B\right)
\end{aligned}
\tag{3.39}
$$

$$
\begin{aligned}
&\quad = 2\pi N_\phi(0) N_r(n_a, \zeta_a) N_r(n_b, \zeta_b) \\
&\qquad \times \sum_{k_b=\left|m_b\right|}^{l_b} T_{k_b}^{l_b m_b} \sum_{m_{ab}} \Phi\left(m_a, m_b, m_{ab}\right) \sum_{l_{ab}} \Theta\begin{pmatrix} m_a & m_b & m_{ab} \\ l_a & k_b & l_{ab} \end{pmatrix} \\
&\qquad \times \sum_{\nu=0}^{\left[\frac{l_{ab}}{2}\right]} \sum_{i=0}^{l_{ab}-2\nu} \sum_{j=0}^{l_{ab}-2\nu-i} \mathrm{P}(l_{ab}, \nu, i, j) \sum_{s=0}^{n_b-l_b+2i} \begin{pmatrix} n_b - l_b + 2i \\ s \end{pmatrix} \\
&\qquad \times R^{-k_b-l_b+2(\nu+i)+n_b+2i-s} \cdot r_A^{n_a-2+k_b+l_{ab}-2(\nu+i+j)+s} \exp\left(-\{\zeta_a + \zeta_b\} r_A\right) \\
&\qquad \times x^{n_b-l_b+2i-s} \exp\left(-\zeta_b R x\right).
\end{aligned}
\tag{3.40}
$$

For one-center charge density centered on B, it is formulated in the same way.

3.2.1.13 Short Summary

Two-center integrals over STF can be formulated by the use of three relations with respect to r_B. The first one is cosine rule in Eq. (3.20). The second one is the introduction of x by which r_B is expressed by R, r_A and x in Eq. (3.23). The third one is transformation of origin of spherical harmonic function from center B to center A in Eq. (3.29).

3.2.1.14 Formulas Frequently Used for the Calculation of Molecular Integrals

Let us define a half definite integral as

$$H[X, n, \zeta] \equiv \left[\int dr\, r^n \exp(-\zeta r)\right]|_{x=X}. \tag{3.41}$$

Using this, a definite integral is written by

$$\int_{X_b}^{X_a} drr^n \exp(-\zeta r) = H[X_a, n, \zeta] - H[X_b, n, \zeta]. \tag{3.42}$$

We shall line up the auxiliary functions using half definite integrals frequently used in the reduction of the molecular integrals. All molecular integrals will be deduced into five types of half definite integrals and B function

$$H[X, n, \zeta] \equiv \int^X drr^n \exp(-\zeta r) \tag{3.B1}$$

$$Hh[X, n_1, \zeta_a, n_2, \zeta_b] \equiv \int^X dr_{A_1} r_{A_1}^{n_1} \exp(-\zeta_a r_{A_1}) \int^{r_{A_1}} dr_{A_2} r_{A_2}^{n_2} \exp(-\zeta_b r_{A_2}) \tag{3.B14}$$

$$Hb[X, n_1, \zeta_a, n_2, \zeta_b] \equiv \int^X dr_A r_A^{n_1} \exp(-\zeta_a r_A) B(n_2, \zeta_b, R, r_A) \tag{3.D1}$$

where

$$B(n, \zeta, R, r_A) \equiv \int_{-1}^{1} J(r_A, R)\, dx r_B^n \exp(-\zeta r_B) \tag{3.C1}$$

$$\begin{aligned} &Hhb[X, n_1, \zeta_a, n_2, \zeta_b, n_3, \zeta_c] \\ &\quad \equiv \int^X dr_{A_1} r_{A_1}^{n_1} \exp(-\zeta_a r_{A_1}) \int^{r_{A_1}} dr_{A_2} r_{A_2}^{n_2} \exp(-\zeta_a r_{A_2}) \\ &\qquad \times \int_0^\pi \sin\theta_{A_2} d\theta_{A_2} r_{B_2}^{n_3} \exp(-\zeta_c r_{B_2}) \\ &\quad = \int^X dr_A r_A^n \exp(-\zeta_a r_A) Hb[r_A, n_2, \zeta_b, n_3, \zeta_c] \end{aligned} \tag{3.D12}$$

$$\begin{aligned} &Hbhb[X, n_1, \zeta_a, n_2, \zeta_b, n_3, \zeta_c, n_4, \zeta_d] \\ &\quad \equiv \int^X dr_{A_1} r_{A_1}^{n_1} \exp(-\zeta_a r_{A_1}) B(n_2, \zeta_b, R, r_{A_1}) Hb[r_{A_1}, n_3, \zeta_c, n_4, \zeta_d]. \end{aligned} \tag{3.D13}$$

These integrals are described in detail in the Appendices.

3.2.2 One-Electron Integral

In this section, we present the formulation of overlap integral, kinetic energy integral, nuclear attraction integral as one-center integrals.

3.2.2.1 One-Center Integral

One-center integration is carried out by using Eq. (3.19)

$$I \to \int_0^\infty r_A^2 dr_A \int_0^\pi \sin\theta_A d\theta_A \int_0^{2\pi} d\phi_A \tag{3.19}$$

Overlap Integral

Formulation of one-center overlap integral over STFs is

$$\begin{aligned} S_{AA}^{ab} &\equiv \int dV \chi_A^a\left(\vec{r}\right) \chi_A^b\left(\vec{r}\right) = \int dV \rho_{AA}^{ab}\left(\vec{r}\right) \\ &= N_r(n_a, \zeta_a) N_r(n_b, \zeta_b) \int_0^\infty r^{n_a+n_b} \exp(-(\zeta_a + \zeta_b)) dr \\ &\quad \times \int d\Omega Y_{l_a}^{m_a}(\theta, \phi) Y_{l_b}^{m_b}(\theta, \phi) \\ &= N_r(n_a, \zeta_a) N_r(n_b, \zeta_b) \ \{-H[0, n_a + n_b, \zeta_a + \zeta_b]\} \delta_{l_a, l_b} \delta_{m_a, m_b} \\ &= N_r(n_a, \zeta_a) N_r(n_b, \zeta_b)(n_a + n_b)!(\zeta_a + \zeta_b)^{-(n_a+n_b+1)} \ \delta_{l_a, l_b} \delta_{m_a, m_b} \end{aligned} \tag{3.43}$$

$$
\begin{aligned}
&= N_r(n_a, \zeta_a) N_r(n_b, \zeta_b) \\
&\quad \left\{ \frac{1}{2}\zeta_b^2 H\left[0,\, n_a + n_b,\, \zeta_a + \zeta_b\right] - n_b \zeta_b H\left[0,\, n_a + n_b - 1,\, \zeta_a + \zeta_b\right] \right. \\
&\quad \left. + \frac{1}{2}\left[n_b(n_b+1) - l_b(l_b+1)\right] H\left[0,\, n_a + n_b - 2,\, \zeta_a + \zeta_b\right] \right\} \delta_{l_a, l_b} \delta_{m_a, m_b}
\end{aligned}
\tag{3.44}
$$

where H is a half definite integral

$$
H\left[X,\, n,\, a \middle| a \neq 0\right] \equiv \int^X x^n \exp(-ax)\, dx \tag{3.B1}
$$

If X equals naught,

$$
H\left[0,\, n,\, a \middle| n \geq 0,\, a \neq 0\right] = -n! a^{-(n+1)} \tag{3.B6}
$$

For typical X, H is summarized in Appendix B.

Kinetic Energy Integral

Kinetic energy operator acting to STF is expressed as

$$
\begin{aligned}
-\frac{1}{2}\Delta \chi_A^a\left(\vec{r}\right) &= -\frac{1}{2}\zeta_a^2 \chi_A^a\left(\vec{r}\right) + n_a \zeta_a \frac{1}{r_A} \chi_A^a\left(\vec{r}\right) \\
&\quad - \frac{1}{2}\left[n_a(n_a+1) - l_a(l_a+1)\right] \frac{1}{r_A^2} \chi_A^a\left(\vec{r}\right).
\end{aligned}
\tag{3.45}
$$

One-center kinetic energy integral over STFs is defined by

$$
T_{AA}^{ab} = \int dV \chi_A^a\left(\vec{r}\right) \left[-\frac{1}{2}\Delta \chi_A^b\left(\vec{r}\right)\right]. \tag{3.46}
$$

Inserting Eq. (3.45) to Eq. (3.46), one-center kinetic energy integral is obtained

$$
\begin{aligned}
T_{AA}^{ab} &= -\frac{1}{2}\zeta_b^2 \int dV \chi_A^a\left(\vec{r}\right) \chi_A^b\left(\vec{r}\right) + n_b \zeta_b \int dV \chi_A^a\left(\vec{r}\right) \frac{1}{r_A} \chi_A^b\left(\vec{r}\right) \\
&\quad - \frac{1}{2}\left[n_b(n_b+1) - l_b(l_b+1)\right] \int dV \chi_A^a\left(\vec{r}\right) \frac{1}{r_A^2} \chi_A^b\left(\vec{r}\right) \\
&= N_r(n_a, \zeta_a) N_r(n_b, \zeta_b) \\
&\quad \left\{ \frac{1}{2}\zeta_b^2 H\left[0,\, n_a + n_b,\, \zeta_a + \zeta_b\right] - n_b \zeta_b H\left[0,\, n_a + n_b - 1,\, \zeta_a + \zeta_b\right] \right. \\
&\quad \left. + \frac{1}{2}\left[n_b(n_b+1) - l_b(l_b+1)\right] H\left[0,\, n_a + n_b - 2,\, \zeta_a + \zeta_b\right] \right\} \delta_{l_a, l_b} \delta_{m_a, m_b}.
\end{aligned}
\tag{3.47}
$$

Nuclear Attraction Energy Integral

One-center nuclear attraction energy integral is defined

$$\begin{aligned} V^{ab}_{AA/A} &\equiv \int dV \frac{\chi^a_A\left(\vec{r}_A\right)\chi^b_A\left(\vec{r}_A\right)}{r_A} \\ &= N_r(n_a,\zeta_a)N_r(n_b,\zeta_b)\int_0^\infty r_A^{n_a+n_b-1}\exp(-(\zeta_a+\zeta_b))dr_A \\ &\quad\times \int d\Omega Y^{m_a}_{l_a}(\theta_A,\phi)Y^{m_b}_{l_b}(\theta_A,\phi) \\ &= N_r(n_a,\zeta_a)N_r(n_b,\zeta_b)(-H[0,n_a+n_b-1,\zeta_a+\zeta_b])\delta_{l_a,l_b}\delta_{m_a,m_b} \\ &= N_r(n_a,\zeta_a)N_r(n_b,\zeta_b)(n_a+n_b-1)!(\zeta_a+\zeta_b)^{-(n_a+n_b)}\delta_{l_a,l_b}\delta_{m_a,m_b}\,. \end{aligned} \tag{3.48}$$

In the case center of nucleus is different from that of electron charge, nuclear attraction integral is defined

$$V^{ab}_{AA/B} \equiv \int dV\chi^a_A\left(\vec{r}\right)\frac{1}{r_B}\chi^b_A\left(\vec{r}\right) \tag{3.49}$$

$$r_B = \left|\vec{r} - \vec{R}_B\right| \ and \ R_B = \left|\vec{R}_B - \vec{R}_A\right|.$$

The inverse of r_B in Eq. (3.49) is given

$$\frac{1}{r_B} = 4\pi\sum_{l=0}^{\infty}\frac{1}{2l+1}\frac{R_B^l}{r^{l+1}}\sum_{m=-l}^{l}Y_l^m(\theta_B,\phi_B)Y_l^m(\theta,\phi) \quad for \ r \geq R_B \tag{3.50}$$

$$\frac{1}{r_B} = 4\pi\sum_{l=0}^{\infty}\frac{1}{2l+1}\frac{r^l}{R_B^{l+1}}\sum_{m=-l}^{l}Y_l^m(\theta_B,\phi_B)Y_l^m(\theta,\phi) \quad for \ r \leq R_B. \tag{3.51}$$

Using these expansion of ${r_B}^{-1}$, nuclear attraction integral in Eq. (3.49) is expressed

$$\begin{aligned} V^{ab}_{AA/B} &= 4\pi\sum_{l=0}^{\infty}\frac{1}{2l+1} \\ &\quad\times\left[R_B^l\int_{R_B}^{\infty} r^2drR^a_A(r)\frac{1}{r^{l+1}}R^b_A(r) + \frac{1}{R_B^{l+1}}\int_0^{R_B} r^2drR^a_A(r)r^lR^b_B(r)\right] \\ &\quad\times\sum_{m=-l}^{l}Y_l^m(\theta_B,\phi_B)\int d\Omega Y^{m_a}_{l_a}(\theta,\phi)Y^{m_b}_{l_b}(\theta,\phi)Y^m_l(\theta,\phi) \end{aligned} \tag{3.52}$$

Using Gaunt formula in Eq. (3.A25)

$$G\begin{pmatrix} m_a & m_b & m_c \\ l_a & l_b & l_c \end{pmatrix} \equiv \int d\Omega Y^{m_a}_{l_a}(\theta,\phi)Y^{m_b}_{l_b}(\theta,\phi)Y^{m_c}_{l_c}(\theta,\phi) \tag{3.A19}$$

and half definite integrals H in Eqs. (3.B7) and (3.B8)

$$\left\langle H\left[0, n, \alpha \middle| n \geq 0, \alpha \neq 0\right]\right\rangle = -n!\alpha^{-(n+1)} \tag{3.B6}$$

$$\left\langle H\left[R, n, \alpha \middle| n \geq 0, \alpha \neq 0\right]\right\rangle = -\sum_{k=0}^{n} A(n,k)\,\alpha^{-(k+1)} R^{n-k} \exp\left(-\alpha R\right). \tag{3.B8}$$

we can express one-center nuclear attraction integral as

$$\begin{aligned} V^{ab}_{AA/B} &= 4\pi N_r(n_a, \zeta_a) N_r(n_b, \zeta_b) \sum_{l=0}^{\infty} \frac{1}{2l+1} \\ &\times \left[R_B^l \int_{R_B}^{\infty} dr r^{n_a+n_b-(l+1)} \exp(-(\zeta_a+\zeta_b)r) \right. \\ &\left. + \frac{1}{R_B^{l+1}} \int_0^{R_B} dr r^{n_a+n_b+l} \exp(-(\zeta_a+\zeta_b)r) \right] \\ &\times \sum_{m=-l}^{l} Y_l^m(\theta_B, \phi_B) G\begin{pmatrix} m_a & m_b & m \\ l_a & l_b & l \end{pmatrix} \end{aligned} \tag{3.53}$$

$$\begin{aligned} V^{ab}_{AA/B} &= 4\pi N_r(n_a, \zeta_a) N_r(n_b, \zeta_b) \sum_{l=0}^{\infty} \frac{1}{2l+1} \\ &\times \{ -R_B^l H[R_B, n_a+n_b-(l+1), \zeta_a+\zeta_b] \\ &+ \frac{1}{R_B^{l+1}} (H[R_B, n_a+n_b+l, \zeta_a+\zeta_b] \\ &- H[0, n_a+n_b+l, \zeta_a+\zeta_b]) \} \\ &\times \sum_{m=-l}^{l} Y_l^m(\theta_B, \phi_B) G\begin{pmatrix} m_a & m_b & m \\ l_a & l_b & l \end{pmatrix} \end{aligned} \tag{3.54}$$

In this part, formulation of one electron one-center integrals; Eq. (3.44) for overlap integral, Eq. (3.47) for kinetic energy integral and Eqs. (3.48) and (3.54) for nuclear attraction energy integrals, are obtained.

3.2.2.2 Two-Center Integral

Two-center molecular integrals are derived by making use of Eqs. (3.23), (3.26), (3.29), and (3.33). Integration of two electron repulsion integrals is carried out in the six domains shown in Fig. 3.2.

Overlap Integral

Using two-center charge density in Eq. (3.31), two-center overlap integral is formulated as follows

$$
\begin{aligned}
S_{AB}^{ab} &\equiv \int dV \chi_A^a\left(\vec{r}_A\right)\chi_B^b\left(\vec{r}_B\right) = \int dV \rho_{AB}^{ab}\left(\vec{r}_A\right) \\
&= N_r(n_a,\zeta_a)N_r(n_b,\zeta_b) \\
&\quad\times \int_0^\infty r_A^2 dr_A r_A^{n_a-1}\exp\left(-\zeta_a r_A\right) r_B^{n_b-1}\exp\left(-\zeta_b r_B\right) \\
&\quad\times \sum_{k_b=\left|m_b\right|}^{l_b} T_{k_b}^{l_b m_b} \sum_{m_{ab}} \Phi(m_a,m_b,m_{ab}) \sum_{l_{ab}} \Theta\begin{pmatrix} m_a & m_b & m_{ab} \\ l_a & k_b & l_{ab} \end{pmatrix} \\
&\quad\times R^{l_b-k_b} r_A^{k_b} r_B^{-l_b} \int d\Omega_A Y_{l_{ab}}^{m_{ab}}(\theta_A,\phi)
\end{aligned}
\tag{3.55}
$$

$$
\begin{aligned}
&= 2\pi N_\phi(0)N_r(n_a,\zeta_a)N_r(n_b,\zeta_b) \\
&\quad\times \sum_{k_b=\left|m_b\right|}^{l_b} T_{k_b}^{l_b m_b} R^{l_b-k_b} \sum_{m_{ab}} \Phi(m_a,m_b,m_{ab}) \sum_{l_{ab}} \Theta\begin{pmatrix} m_a & m_b & m_{ab} \\ l_a & k_b & l_{ab} \end{pmatrix} \\
&\quad\times \int_0^\infty dr_A r_A^{n_a+k_b+1}\exp(-\zeta_a r_A) r_B^{n_b-l_b-1}\exp(-\zeta_b r_B) \\
&\quad\times \int_0^\pi \sin\theta_A d\theta_A p_l^0(\cos\theta_A)
\end{aligned}
\tag{3.56}
$$

Using the associated Legendre function p_l^0 as a function of R, r_A, and r_B given by Eq. (3.33),

$$
\begin{aligned}
S_{AB}^{ab} &= 2\pi N_\phi(0)N_r(n_a,\zeta_a)N_r(n_b,\zeta_b) \\
&\quad\times \sum_{k_b=\left|m_b\right|}^{l_b} T_{k_b}^{l_b m_b} \sum_{m_{ab}} \Phi(m_a,m_b,m_{ab}) \sum_{l_{ab}} \Theta\begin{pmatrix} m_a & m_b & m_{ab} \\ l_a & k_b & l_{ab} \end{pmatrix} \\
&\quad\times \int_0^\infty dr_A r_A^{n_a+k_b+1}\exp(-\zeta_a r_A) r_B^{n_b-l_b-1}\exp(-\zeta_b r_B) \\
&\quad\times \int_0^\pi \sin\theta_A d\theta_A \sum_{\nu=0}^{\left[\frac{l_{ab}}{2}\right]} \sum_{i=0}^{l_{ab}-2\nu} \sum_{j=0}^{l_{ab}-2\nu-i} \mathrm{P}(l_{ab},\nu,i,j) R^{l_b-k_b-l_{ab}+2(\nu+i)} r_A^{l_{ab}-2(\nu+i+j)} r_B^{2i}
\end{aligned}
\tag{3.57}
$$

$$
\begin{aligned}
&= 2\pi N_\phi(0)N_r(n_a,\zeta_a)N_r(n_b,\zeta_b) \\
&\quad\times \sum_{k_b=\left|m_b\right|}^{l_b} T_{k_b}^{l_b m_b} \sum_{m_{ab}} \Phi(m_a,m_b,m_{ab}) \sum_{l_{ab}} \Theta\begin{pmatrix} m_a & m_b & m_{ab} \\ l_a & k_b & l_{ab} \end{pmatrix} \\
&\quad\times \sum_{\nu=0}^{\left[\frac{l_{ab}}{2}\right]} \sum_{i=0}^{l_{ab}-2\nu} \sum_{j=0}^{l_{ab}-2\nu-i} \mathrm{P}(l_{ab},\nu,i,j) R^{l_b-k_b-l_{ab}+2(\nu+i)} \\
&\quad\times \int_0^\infty dr_A r_A^{n_a+k_b+l_{ab}-2(\nu+i+j)+1}\exp(-\zeta_a r_A) \\
&\quad\times \int_0^\pi \sin\theta_A d\theta_A r_B^{n_b-l_b+2i-1}\exp(-\zeta_b r_B).
\end{aligned}
\tag{3.58}
$$

Using auxiliary function *Hb* in Eq. (3.D1),

$$
\begin{aligned}
&Hb[X, n_1, \zeta_a, n_2, \zeta_b] \\
&\quad \equiv \int^X dr_A r_A^{n_1} \exp(-\zeta_a r_A) \int_0^\pi \sin\theta_A d\theta_A r_B^n \exp(-\zeta_b r_B) \\
&\quad \equiv \int^X dr_A r_A^{n_1} \exp(-\zeta_a r_A) \int_{-1}^1 J(r_A, R) dx r_B^n \exp(-\zeta_b r_B) \\
&\quad = \int^X dr_A r_A^{n_1} \exp(-\zeta_a r_A) B(n_2, \zeta_b, R, r_A)
\end{aligned}
\tag{3.D1}
$$

which is described in detail in Appendix D, the two-center overlap integral is found to be

$$
\begin{aligned}
S_{AB}^{ab} &= 2\pi N_\phi(0) N_r(n_a, \zeta_a) N_r(n_b, \zeta_b) \\
&\times \sum_{k_b=|m_b|}^{l_b} T_{k_b}^{l_b m_b} \sum_{m_{ab}} \Phi(m_a, m_b, m_{ab}) \sum_{l_{ab}} \Theta\begin{pmatrix} m_a & m_b & m_{ab} \\ l_a & k_b & l_{ab} \end{pmatrix} \\
&\times \sum_{\nu=0}^{\left[\frac{l_{ab}}{2}\right]} \sum_{i=0}^{l_{ab}-2\nu} \sum_{j=0}^{l_{ab}-2\nu-i} \mathrm{P}(l_{ab}, \nu, i, j) R^{l_b - k_b - l_{ab} + 2(\nu+i)} \\
&\times \big[Hb\big[R, n_a + k_b + l_{ab} - 2(\nu+i+j) + 1, \zeta_a, n_b - l_b + 2i - 1, \zeta_b \big| X \le R\big] \\
&\quad - Hb\big[0, n_a + k_b + l_{ab} - 2(\nu+i+j) + 1, \zeta_a, n_b - l_b + 2i - 1, \zeta_b \big| X \le R\big] \\
&\quad - Hb\big[R, n_a + k_b + l_{ab} - 2(\nu+i+j) + 1, \zeta_a, n_b - l_b + 2i - 1, \zeta_b \big| X \ge R\big]\big].
\end{aligned}
\tag{3.59}
$$

Symbolic calculation system makes it easy to formulate two-center overlap integral expressed in Eq. (3.59) for any combination of quantum number $\{n, l, m\}$.

Kinetic Energy Integral

Kinetic energy integral is defined as

$$
T_{AB}^{ab} = \int dV \chi_A^a(\vec{r}) \left[-\frac{1}{2} \Delta \chi_B^b(\vec{r}) \right] \tag{3.60}
$$

$$
\begin{aligned}
T_{AB}^{ab} = &-\frac{1}{2}\zeta_b^2 \int dV \chi_A^a(\vec{r}) \chi_B^b(\vec{r}) + n_b \zeta_b \int dV \chi_A^a(\vec{r}) \frac{1}{r_A} \chi_B^b(\vec{r}) \\
&- \frac{1}{2}[n_b(n_b+1) - l_b(l_b+1)] \int dV \chi_A^a(\vec{r}) \frac{1}{r_A^2} \chi_B^b(\vec{r})
\end{aligned}
\tag{3.61}
$$

Kinetic energy integral is similar to the overlap integral. Making use of formula of two-center overlap integral, S_{AB}^{ab}, in Eqs. (3.58) and (3.59), two-center kinetic energy integral is obtained as follows:

$$\begin{aligned}
T_{AB}^{ab} &= 2\pi N_\phi(0) N_r(n_a, \zeta_a) N_r(n_b, \zeta_b) \\
&\times \sum_{k_b=|m_b|}^{l_b} T_{k_b}^{l_b m_b} \sum_{m_{ab}} \Phi(m_a, m_b, m_{ab}) \sum_{l_{ab}} \Theta\begin{pmatrix} m_a & m_b & m_{ab} \\ l_a & k_b & l_{ab} \end{pmatrix} \\
&\times \sum_{\nu=0}^{\left[\frac{l_{ab}}{2}\right]} \sum_{i=0}^{l_{ab}-2\nu} \sum_{j=0}^{l_{ab}-2\nu-i} \mathrm{P}(l_{ab}, \nu, i, j) R^{l_b-k_b-l_{ab}+2(\nu+i)}
\end{aligned}$$

$$\begin{aligned}
&\Big\{ -\frac{1}{2}\zeta_b^2 \big[Hb\big[R, n_a + k_b + l_{ab} - 2(\nu+i+j) + 1, \zeta_a, n_b - l_b + 2i - 1, \zeta_b \big| X \le R\big] \\
&\quad - Hb\big[0, n_a + k_b + l_{ab} - 2(\nu+i+j) + 1, \zeta_a, n_b - l_b + 2i - 1, \zeta_b \big| X \le R\big] \\
&\quad - Hb\big[R, n_a + k_b + l_{ab} - 2(\nu+i+j) + 1, \zeta_a, n_b - l_b + 2i - 1, \zeta_b \big| X \ge R\big]\big] \\
&\quad + n_b\zeta_b \big[Hb\big[R, n_a + k_b + l_{ab} - 2(\nu+i+j), \zeta_a, n_b - l_b + 2i - 1, \zeta_b \big| X \le R\big] \\
&\quad - Hb\big[0, n_a + k_b + l_{ab} - 2(\nu+i+j), \zeta_a, n_b - l_b + 2i - 1, \zeta_b \big| X \le R\big] \\
&\quad - Hb\big[R, n_a + k_b + l_{ab} - 2(\nu+i+j), \zeta_a, n_b - l_b + 2i - 1, \zeta_b \big| X \ge R\big]\big] \\
&\quad - \frac{1}{2}[n_b(n_b+1) - l_b(l_b+1)] \\
&\quad \big[Hb\big[R, n_a + k_b + l_{ab} - 2(\nu+i+j) - 1, \zeta_a, n_b - l_b + 2i - 1, \zeta_b \big| X \le R\big] \\
&\quad - Hb\big[0, n_a + k_b + l_{ab} - 2(\nu+i+j) - 1, \zeta_a, n_b - l_b + 2i - 1, \zeta_b \big| X \le R\big] \\
&\quad - Hb\big[R, n_a + k_b + l_{ab} - 2(\nu+i+j) - 1, \zeta_a, n_b - l_b + 2i - 1, \zeta_b \big| X \ge R\big]\big]\Big\}.
\end{aligned} \tag{3.62}$$

Nuclear Attraction Energy Integral

Making use of two-center overlap integral in Eq. (3.59), nuclear attraction integral is defined in the case Coulomb interaction between two-center electron charge centered at A and B and a nuclear charge located at A as

$$\begin{aligned}
V_{AB/A}^{ab} &\equiv \int dV \chi_A^a\left(\vec{r}_A\right) \left[\frac{1}{r_A}\right] \chi_B^b\left(\vec{r}_B\right) \qquad (3.63) \\
&= 2\pi N_\phi(0) N_r(n_a, \zeta_a) N_r(n_b, \zeta_b) \\
&\times \sum_{k_b=|m_b|}^{l_b} T_{k_b}^{l_b m_b} \sum_{m_{ab}} \Phi(m_a, m_b, m_{ab}) \sum_{l_{ab}} \Theta\begin{pmatrix} m_a & m_b & m_{ab} \\ l_a & k_b & l_{ab} \end{pmatrix} \\
&\times \sum_{\nu=0}^{\left[\frac{l_{ab}}{2}\right]} \sum_{i=0}^{l_{ab}-2\nu} \sum_{j=0}^{l_{ab}-2\nu-i} \mathrm{P}(l_{ab}, \nu, i, j) R^{l_b-k_b-l_{ab}+2(\nu+i)}
\end{aligned}$$

$$\begin{aligned}&\left[Hb\left[R, n_a + k_b + l_{ab} - 2(\nu + i + j), \zeta_a, n_b - l_b + 2i - 1, \zeta_b \middle| X \leq R\right]\right.\\ &-Hb\left[0, n_a + k_b + l_{ab} - 2(\nu + i + j), \zeta_a, n_b - l_b + 2i - 1, \zeta_b \middle| X \leq R\right]\\ &\left.-Hb\left[R, n_a + k_b + l_{ab} - 2(\nu + i + j), \zeta_a, n_b - l_b + 2i - 1, \zeta_b \middle| X \geq R\right]\right].\end{aligned} \tag{3.64}$$

In the case Coulomb interaction between a nuclear charge located at C and two-center electron charge by the STF product centered at A and B shown in Fig. 3.1, nuclear attraction integral is defined as

$$V_{AB/C}^{ab} \equiv \int dV \chi_A^a\left(\vec{r}_A\right)\left[\frac{1}{r_C}\right]\chi_B^b\left(\vec{r}_B\right) \tag{3.65}$$

where

$$r_C = \left|\vec{r}_A - \vec{R}_C\right| \tag{3.66}$$

$$R_C = \left|\vec{R}_C - \vec{R}_A\right|. \tag{3.67}$$

The inverse of r_C in Eq. (3.65) is given by Eq. (3.68) for $r_A \geq R_C$ and Eq. (3.69) for $r_A \leq R_C$, respectively.

$$\frac{1}{r_C} = 4\pi\sum_{l=0}^{\infty}\frac{1}{2l+1}\frac{R_C^l}{r_A^{l+1}}\sum_{m=-l}^{l} Y_l^m(\theta_C,\phi_C)Y_l^m(\theta_A,\phi_A) \text{ for } r_A \geq R_C, \tag{3.68}$$

$$\frac{1}{r_C} = 4\pi\sum_{l=0}^{\infty}\frac{1}{2l+1}\frac{r_A^l}{R_C^{l+1}}\sum_{m=-l}^{l} Y_l^m(\theta_C,\phi_C)Y_l^m(\theta_A,\phi_A) \text{ for } r_A \leq R_C. \tag{3.69}$$

Inserting the series expression of the inverse of r_C given by Eqs. (3.68) and (3.69), three center nuclear attraction integral is formulated as follows

$$\begin{aligned}V_{AB/C}^{ab} &= 4\pi\sum_{l=0}^{\infty}\frac{1}{2l+1}\\ &\times\left[R_C^l\int_{R_C}^{\infty} r_A^2 dr_A\int_0^{\pi}\sin\theta_A d\theta_A\int_0^{2\pi} d\phi R_A^a(r_A)\frac{1}{r_A^{l+1}}R_B^b(r_B)\right.\\ &\left.+\frac{1}{R_C^{l+1}}\int_0^{R_C} r_A^2 dr_A\int_0^{\pi}\sin\theta_A d\theta_A\int_0^{2\pi} d\phi R_A^a(r_A) r_A^l R_B^b(r_B)\right]\\ &\times\sum_{m=-l}^{l} Y_l^m(\theta_C,\phi_C)Y_{l_a}^{m_a}(\theta_A,\phi)Y_{l_b}^{m_b}(\theta_B,\phi)Y_l^m(\theta_A,\phi).\end{aligned} \tag{3.70}$$

Using Eq. (3.A25) which is an integration of the triple product of spherical harmonic functions

$$\begin{aligned}
&\int_0^{2\pi} d\phi Y_{l_a}^{m_a}(\theta_A,\phi)Y_{l_b}^{m_b}(\theta_B,\phi)Y_l^m(\theta_A,\phi)\\
&\quad = 2\pi N_\phi(0)\sum_{k_b=|m_b|}^{l_b} T_{k_b}^{l_b m_b} R^{l_b-k_b} r_A^{k_b} r_B^{-l_b}\\
&\quad \times \sum_{m_{ab}} \Phi(m_a,m_b,m_{ab}) \sum_{l_{ab}} \Theta\begin{pmatrix} m_a & m_b & m_{ab}\\ l_a & k_b & l_{ab}\end{pmatrix},\\
&\quad \times \sum_{m_{abm}} \Phi(\, m_{ab} \quad m \quad m_{abm}\,) \sum_{l_{abl}} \Theta\begin{pmatrix} m_{ab} & m & m_{abm}\\ l_{ab} & l & l_{abl}\end{pmatrix}\\
&\quad \times p_{l_{abl}}^0(\cos\theta_A)
\end{aligned} \tag{3.A25}$$

$$\begin{aligned}
V_{AB/C}^{ab} &= 8\pi^2 N_\phi(0) N_r(n_a,\zeta_a) N_r(n_b,\zeta_b) \sum_{l=0}^{\infty} \frac{1}{2l+1}\\
&\times \left[R_C^l \int_{R_C}^{\infty} r_A^{n_a-l} \exp(-\zeta_a r_A) dr_A \int_0^{\pi} \sin\theta_A d\theta_A r_B^{n_b-1} \exp(-\zeta_b r_B) \right.\\
&\quad \left. + \frac{1}{R_C^{l+1}} \int_0^{R_C} r_A^{n_a+l+1} \exp(-\zeta_a r_A) dr_A \int_0^{\pi} \sin\theta_A d\theta_A r_B^{n_b-1} \exp(-\zeta_b r_B) \right]\\
&\times \sum_{m=-l}^{l} Y_l^m(\theta_C,\phi_C) \sum_{k_b=|m_b|}^{l_b} T_{k_b}^{l_b m_b} R^{l_b-k_b} r_A^{k_b} r_B^{-l_b}\\
&\times \sum_{m_{ab}} \Phi(m_a,m_b,m_{ab}) \sum_{l_{ab}} \Theta\begin{pmatrix} m_a & m_b & m_{ab}\\ l_a & k_b & l_{ab}\end{pmatrix}\\
&\times \sum_{m_{abm}} \Phi(\, m_{ab} \quad m_c \quad m_{abm}\,) \sum_{l_{abl}} \Theta\begin{pmatrix} m_{ab} & m & m_{abm}\\ l_{ab} & l & l_{abl}\end{pmatrix}\\
&\times p_{l_{abl}}^0(\cos\theta_A)\,.
\end{aligned} \tag{3.71}$$

Making use of Eqs. (3.D1) and (3.33)

$$Hb[X,n_1,\zeta_a,n_2,\zeta_b] \equiv \int^X dr_A r_A^{n_1} \exp(-\zeta_a r_A) \int_0^{\pi} \sin\theta_A d\theta_A r_B^{n_2} \exp(-\zeta_b r_B) \tag{3.D1}$$

$$p_l^0(\cos\theta_A) = \sum_{\nu=0}^{\left[\frac{l}{2}\right]} \sum_{i=0}^{l-2\nu} \sum_{j=0}^{l-2\nu-i} \mathrm{P}(l,\nu,i,j) R^{-l+2(\nu+i)} r_A^{l-2(\nu+i+j)} r_B^{2i}, \tag{3.33}$$

three center nuclear attraction integral is obtained

$$
\begin{aligned}
V_{AB/C}^{ab} &= 8\pi^2 N_\phi(0) N_r(n_a,\zeta_a) N_r(n_b,\zeta_b) \sum_{l=0}^{\infty} \frac{1}{2l+1} \\
&\times \{ -R_C^l Hb[R_C, n_a+k_b-l, \zeta_a, n_b-(l_b+1)+2i, \zeta_b | X \geq R_C] \\
&\quad + \frac{1}{R_C^{l+1}} (Hb[R_C, n_a+k_b+l+1, \zeta_a, n_b-(l_b+1)+2i, \zeta_b | X \leq R_C] \\
&\quad - Hb[0, n_a+k_b+l+1, \zeta_a, n_b-(l_b+1)+2i, \zeta_b | X \leq R_C]) \} \\
&\times \sum_{m=-l}^{l} Y_l^m(\theta_C, \phi_C) \sum_{k_b=|m_b|}^{l_b} T_{k_b}^{l_b m_b} R^{l_b-k_b} \\
&\times \sum_{m_{ab}} \Phi(m_a, m_b, m_{ab}) \sum_{l_{ab}} \Theta \begin{pmatrix} m_a & m_b & m_{ab} \\ l_a & k_b & l_{ab} \end{pmatrix} \\
&\times \sum_{m_{abm}} \Phi(\, m_{ab} \quad m_c \quad m_{abm}\,) \sum_{l_{abl}} \Theta \begin{pmatrix} m_{ab} & m & m_{abm} \\ l_{ab} & l & l_{abl} \end{pmatrix}.
\end{aligned}
\tag{3.72}
$$

It is necessary for symbolic formulation of three center nuclear attraction integral to choose an appropriate *Hb* function in Appendix D.

Electron Repulsion Integral

Considering general cases of multicenter integrals, we may write the definition of electron repulsion integrals as follows

$$
V_{ABCD}^{abcd} = \int dV_1 \int dV_2 \chi_A^a(\vec{r}_1) \chi_B^b(\vec{r}_1) \frac{1}{r_{12}} \chi_C^c(\vec{r}_2) \chi_D^d(\vec{r}_2) \tag{3.73}
$$

$$
= \int dV_1 \chi_A^a(\vec{r}_1) \chi_B^b(\vec{r}_1) U_{CD}^{cd}(\vec{r}_1) \tag{3.74}
$$

where

$$
U_{CD}^{cd}(\vec{r}_1) \equiv \int dV_2 \frac{1}{r_{12}} \chi_C^c(\vec{r}_2) \chi_D^d(\vec{r}_2) \tag{3.75}
$$

is a potential by the second electron. The inverse distance of electrons on the right hand side is given

$$
r_{12}^{-1} = 4\pi \sum_{l=0}^{\infty} \frac{1}{2l+1} \frac{r_<^l}{r_>^{l+1}} \sum_{m=-l}^{l} Y_l^m(\theta_1, \phi_1) Y_l^m(\theta_2, \phi_2). \tag{3.76}
$$

$$
\begin{aligned}
V_{ABCD}^{abcd} = 4\pi \sum_{l}^{\infty} \frac{1}{2l+1} \sum_{m=-l}^{l} \\
& \times \int_0^{\infty} r_1^2 dr_1 R_A^a(r_{A1}) R_B^b(r_{B1}) \\
& \times \int d\Omega_1 Y_{l_a}^{m_a}(\theta_{A1}, \phi_1) Y_{l_b}^{m_b}(\theta_{B1}, \phi_1) Y_{l_a}^{m_a}(\theta_{A1}, \phi_1) \\
& \times \left[\frac{1}{r_1^{l+1}} \int_0^{r_1} r_2^{2+l} dr_2 R_C^c(r_{C2}) R_D^b(r_{D2}) \int_0^{\infty} R_A^a(r_{A1}) \right. \\
& \left. + r_1^l \int_{r_1}^{\infty_1} r_2^{2-(l+1)} dr_2 R_C^c(r_{C2}) R_D^b(r_{D2}) \int_0^{\infty} R_A^a(r_{A1}) \right. \\
& \times \int d\Omega_2 Y_{l_c}^{m_c}(\theta_{C2}, \phi_2) Y_{l_d}^{m_d}(\theta_{D2}, \phi_2) Y_l^m(\theta_{A2}, \phi_2).
\end{aligned}
\tag{3.77}
$$

Potential by the Second Electron

In general, potential by the second electron is divided into two regions

$$
U_{CD}^{cd}\left(\vec{r}_1\right) = U_{CD}^{cd}\left(\vec{r}_1 \middle| r_1 \geq r_2\right) + U_{CD}^{cd}\left(\vec{r}_1 \middle| r_1 \leq r_2\right). \tag{3.78}
$$

Each potential on the right hand side is expressed by the sum of partial potentials with respect to the order l and m.

$$
U_{CD}^{cd}\left(\vec{r}_1\right) = 4\pi \sum_l \sum_{m=-l}^{l} \left[u_{CD}^{cd}\left(\vec{r}_1 \middle| r_1 \geq r_2\right)_l^m + u_{CD}^{cd}\left(\vec{r}_1 \middle| r_1 \leq r_2\right) \right] \tag{3.79}
$$

where

$$
U_{CD}^{cd}\left(\vec{r}_1 \middle| r_1 \geq r_2\right) = 4\pi \sum_l \sum_{m=-l}^{l} u_{CD}^{cd}\left(\vec{r}_1 \middle| r_1 \geq r_2\right)_l^m \tag{3.80}
$$

$$
U_{CD}^{cd}\left(\vec{r}_1 \middle| r_1 \leq r_2\right) = 4\pi \sum_l \sum_{m=-l}^{l} u_{CD}^{cd}\left(\vec{r}_1 \middle| r_1 \leq r_2\right)_l^m \tag{3.81}
$$

$$
\begin{aligned}
u_{CD}^{cd}\left(\vec{r}_1 \middle| r_1 \geq r_2\right)_l^m = & \frac{1}{2l+1} \frac{1}{r_1^{l+1}} Y_l^m(\theta_1, \phi_1) \int_0^{r_1} r_2^{2+l} dr_2 R_C^c(r_{C2}) R_D^b(r_{D2}) \\
& \times \int d\Omega_2 Y_{l_c}^{m_c}(\theta_{C2}, \phi_2) Y_{l_d}^{m_d}(\theta_{D2}, \phi_2) Y_l^m(\theta_{A2}, \phi_2)
\end{aligned}
\tag{3.82}
$$

$$u_{CD}^{cd}\left(\vec{r}_1 | r_1 \le r_2\right)_l^m = \frac{1}{2l+1} r^l Y_l^m(\theta_1, \phi_1) \int_{r_1}^{\infty} r_2^{2-(l+1)} dr_2 R_C^c(r_{C2}) R_D^b(r_{D2}) \times \int d\Omega_2 Y_{l_c}^{m_c}(\theta_{C2}, \phi_2) Y_{l_d}^{m_d}(\theta_{D2}, \phi_2) Y_l^m(\theta_{A2}, \phi_2) \tag{3.83}$$

Using the expression of potential by the second electron given by Eqs. (3.80) and (3.81), multicenter two electron repulsion integral is generally expressed

$$V_{ABCD}^{abcd} = 4\pi \sum_l \sum_{m=-l}^{l} \int dV_1 \chi_A^a\left(\vec{r}_1\right) \chi_B^b\left(\vec{r}_1\right) \times \left[u_{CD}^{cd}\left(\vec{r}_1 | r_1 \ge r_2\right)_l^m + u_{CD}^{cd}\left(\vec{r}_1 | r_1 \le r_2\right)_l^m\right] \tag{3.84}$$

One-Center Electron Repulsion Integral

Using general expression of multicenter electron repulsion integral, we can express one-center electron repulsion integral as

$$V_{AAAA}^{abcd} \equiv \int dV_1 \int dV_2 \chi_A^a\left(\vec{r}_1\right) \chi_A^b\left(\vec{r}_1\right) \frac{1}{r_{12}} \chi_A^c\left(\vec{r}_2\right) \chi_A^d\left(\vec{r}_2\right) \tag{3.85}$$

$$\equiv \int dV_1 \chi_A^a\left(\vec{r}_1\right) \chi_A^b\left(\vec{r}_1\right) U_{AA}^{cd}\left(\vec{r}_1\right) \tag{3.86}$$

$$U_{AA}^{cd}\left(\vec{r}_1\right) \equiv \int dV_2 \frac{1}{r_{12}} \chi_A^c\left(\vec{r}_2\right) \chi_A^d\left(\vec{r}_2\right) \tag{3.87}$$

$$V_{AAAA}^{abcd} = 4\pi \sum_l \sum_{m=-l}^{l} \int dV_1 \chi_A^a\left(\vec{r}_1\right) \chi_A^b\left(\vec{r}_1\right) \times \left[u_{AA}^{cd}\left(\vec{r}_1 | r_1 \ge r_2\right)_l^m + u_{AA}^{cd}\left(\vec{r}_1 | r_1 \le r_2\right)_l^m\right] \tag{3.88}$$

$$u_{AA}^{cd}\left(\vec{r}_1 | r_1 \ge r_2\right)_l^m = \frac{1}{2l+1} \frac{1}{r_1^{l+1}} Y_l^m(\theta_1, \phi_1) \times \int_0^{r_1} r_2^2 dr_2 R_A^c(r_2) R_A^d(r_2) r_2^l \, G\begin{pmatrix} m_c & m_d & m \\ l_c & l_d & l \end{pmatrix} \tag{3.89}$$

$$u_{AA}^{cd}\left(\vec{r}_1 | r_1 \le r_2\right)_l^m = \frac{1}{2l+1} r_1^l Y_l^m(\theta_1, \phi_1) \times \int_{r_1}^{\infty} r_2^2 dr_2 R_A^c(r_2) R_A^d(r_2) \frac{1}{r_2^{l+1}} G\begin{pmatrix} m_c & m_d & m \\ l_c & l_d & l \end{pmatrix} \tag{3.90}$$

In the case of one-center two electron integrals, three variables appearing in the radial part and the angular part are separated in one-center polar coordinate system. With Gaunt coefficients for the first electron and the second electron, we express one-center electron repulsion integrals as follows

$$V_{AAAA}^{abcd} = 4\pi\sum_l \frac{1}{2l+1}\sum_{m=-l}^{l} G\begin{pmatrix} m_a & m_b & m \\ l_a & l_b & l \end{pmatrix} G\begin{pmatrix} m_c & m_d & m \\ l_c & l_d & l \end{pmatrix}$$
$$\times \int_0^\infty r_1^2 dr_1 R_A^a(r_1) R_A^b(r_1)$$
$$\times \left[\frac{1}{r_1^{l+1}}\int_0^{r_1} r_2^2 dr_2 R_A^c(r_2) R_A^d(r_2) r_2^l + r_1^l \int_{r_1}^\infty r_2^2 dr_2 R_A^c(r_2) R_A^d(r_2) \frac{1}{r_2^{l+1}}\right] \quad (3.91)$$

$$V_{AAAA}^{abcd} = 4\pi N_r(n_a,\zeta_a) N_r(n_b,\zeta_b) N_r(n_c,\zeta_c) N_r(n_d,\zeta_d)$$
$$\times \sum_l \frac{1}{2l+1}\sum_{m=-l}^{l} G\begin{pmatrix} m_a & m_b & m \\ l_a & l_b & l \end{pmatrix} G\begin{pmatrix} m_c & m_d & m \\ l_c & l_d & l \end{pmatrix}$$
$$\times \left[\int_0^\infty dr_1 r_1^{n_a+n_b-(l+1)} \exp(-(\zeta_a+\zeta_b) r_1) \int_0^{r_1} dr_2 r_2^{n_c+n_d+l} \exp(-(\zeta_c+\zeta_d) r_2)\right.$$
$$\left. + \int_0^\infty dr_1 r_1^{n_a+n_b+l} \exp(-(\zeta_a+\zeta_b) r_1) \int_{r_1}^\infty dr_2 r_2^{n_c+n_d-(l+1)} \exp(-(\zeta_c+\zeta_d) r_2)\right] \quad (3.92)$$

For the convenience of symbolic formulation of molecular integrals for any quantum numbers of STF, using the half definite integrals, *Hh* and *H*, defined by Eqs. (3.B14) and (3.B1)

$$Hh[X, n_1, \alpha_1, n_2, \alpha_2] = \int^X dr_1 r_1^{n_1} \exp(-\alpha_1 r_1) \int^{r_1} dr_2 r_2^{n_2} \exp(-\alpha_2 r_2)$$
$$= \int^X dr_1 r_1^{n_1} \exp(-\alpha_1 r_1)\ H[r_1, n_2, \alpha_2] \quad (3.B14)$$

$$H[X, n, \alpha] \equiv \int^X x^n \exp(-\alpha x) dx, \quad (3.B1)$$

$$V_{AAAA}^{abcd} = 4\pi N_r(n_a,\zeta_a) N_r(n_b,\zeta_b) N_r(n_c,\zeta_c) N_r(n_d,\zeta_d)$$
$$\times \sum_l \frac{1}{2l+1}\sum_{m=-l}^{l} G\begin{pmatrix} m_a & m_b & m \\ l_a & l_b & l \end{pmatrix} G\begin{pmatrix} m_c & m_d & m \\ l_c & l_d & l \end{pmatrix}$$
$$\times \{-Hh[0, n_a+n_b-(l+1), \zeta_a+\zeta_b, n_c+n_d+l, \zeta_c+\zeta_d]$$
$$+ H[0, n_a+n_b-(l+1), \zeta_a+\zeta_b] H[0, n_c+n_d+l,\ \zeta_c+\zeta_d]$$
$$+ Hh[0, n_a+n_b+l, \zeta_a+\zeta_b, n_c+n_d-(l+1), \zeta_c+\zeta_d]\}. \quad (3.93)$$

For one-center electron repulsion integral, symbolic formulation is carried out with keeping four orbital exponents in symbol for all combinations of integer quantum numbers $\{n, l, m\}$.

Two-Center Electron Repulsion Integral

This integral is classified into, what is called, Coulomb-, hybrid-, and exchange-type integrals.

Coulomb-type

$$\begin{aligned} V_{AABB}^{abcd} &= \int dV_1 \int dV_2 \chi_A^a\left(\vec{r}_1\right)\chi_A^b\left(\vec{r}_1\right)\frac{1}{r_{12}}\chi_B^c\left(\vec{r}_2\right)\chi_B^d\left(\vec{r}_2\right) \\ &= \int dV_1 \rho_{AA}^{ab}\left(\vec{r}_1\right)U_{BB}^{cd}\left(\vec{r}_1\right) \\ &= 4\pi\sum_l \sum_{m=-l}^{l} \int dV_1 \rho_{AA}^{ab}\left(\vec{r}_1\right)\left[u_{BB}^{cd}\left(\vec{r}_1 | r_1 \geq r_2\right)_l^m + u_{BB}^{cd}\left(\vec{r}_1 | r_1 \leq r_2\right)_l^m\right] \end{aligned} \tag{3.94}$$

Hybrid-type

$$\begin{aligned} V_{AAAB}^{abcd} &= \int dV_1 \int dV_2 \chi_A^a\left(\vec{r}_1\right)\chi_A^b\left(\vec{r}_1\right)\frac{1}{r_{12}}\chi_A^c\left(\vec{r}_2\right)\chi_B^d\left(\vec{r}_2\right) \\ &= \int dV_1 \rho_{AA}^{ab}\left(\vec{r}_1\right)U_{AB}^{cd}\left(\vec{r}_1\right) \\ &= 4\pi\sum_l \sum_{m=-l}^{l} \int dV_1 \rho_{AA}^{ab}\left(\vec{r}_1\right)\left[u_{AB}^{cd}\left(\vec{r}_1 | r_1 \geq r_2\right)_l^m + u_{AB}^{cd}\left(\vec{r}_1 | r_1 \leq r_2\right)_l^m\right] \end{aligned} \tag{3.95}$$

Exchange-type

$$\begin{aligned} V_{ABAB}^{abcd} &= \int dV_1 \int dV_2 \chi_A^a\left(\vec{r}_1\right)\chi_B^b\left(\vec{r}_1\right)\frac{1}{r_{12}}\chi_A^c\left(\vec{r}_2\right)\chi_B^d\left(\vec{r}_2\right) \\ &= \int dV_1 \rho_{AB}^{ab}\left(\vec{r}_1\right)U_{AB}^{cd}\left(\vec{r}_1\right) \\ &= 4\pi\sum_l \sum_{m=-l}^{l} \int dV_1 \rho_{AB}^{ab}\left(\vec{r}_1\right)\left[u_{AB}^{cd}\left(\vec{r}_1 \,|\, r_1 \geq r_2\right)_l^m + u_{AB}^{cd}\left(\vec{r}_1 \,|\, r_1 \leq r_2\right)_l^m\right] \end{aligned} \tag{3.96}$$

where

$$\begin{aligned} & u_{BB}^{cd}\left(\vec{r}_{A1} \,|\, r_{A1} \geq r_{A2}\right)_l^m \\ &= N_r(n_c,\zeta_c)\, N_r(n_d,\zeta_d)\frac{1}{2l+1} r_{A1}^{-(l+1)}\, Y_l^m(\theta_{A1},\phi_1)\int_0^{r_{A1}} dr_{A2} r_{A2}^{2+l} \\ &\times \int d\Omega_{A2} r_{B2}^{n_c+n_d-2} \exp\left(-(\zeta_c+\zeta_d)r_{B2}\right) Y_{l_c}^{m_c}(\theta_{B2},\phi_2)\, Y_{l_d}^{m_d}(\theta_{B2},\phi_2)\, Y_l^m(\theta_{A2},\phi_2) \end{aligned} \tag{3.97}$$

$$u_{BB}^{cd}\left(\vec{r}_{A1}\,|\,r_{A1} \le r_{A2}\right)_l^m$$
$$= N_r(n_c,\zeta_c)\,N_r(n_d,\zeta_d)\frac{1}{2l+1}r_{A1}^{l}\,Y_l^m(\theta_{A1},\phi_1)\int_{r_{A1}}^{\infty} dr_{A2}r_{A2}^{2-(l+1)}$$
$$\times \int d\Omega_{A2}r_{B2}^{n_c+n_d-2}\exp\left(-(\zeta_c+\zeta_d)r_{B2}\right)Y_{l_c}^{m_c}(\theta_{B2},\phi_2)Y_{l_d}^{m_d}(\theta_{B2},\phi_2)Y_l^m(\theta_{A2},\phi_2) \tag{3.98}$$

$$u_{AB}^{cd}\left(\vec{r}_{A1}\,|\,r_{A1} \ge r_{A2}\right)_l^m$$
$$= N_r(n_c,\zeta_c)\,N_r(n_d,\zeta_d)\frac{1}{2l+1}r_{A1}^{-(l+1)}\,Y_l^m(\theta_{A1},\phi_1)$$
$$\times \int_0^{r_{A1}} dr_{A2}r_{A2}^{n_C+l+1}\exp\left(-\zeta_c r_{A2}\right)$$
$$\times \int d\Omega_{A2}r_{B2}^{n_d-1}\exp\left(-\zeta_d r_{B2}\right)Y_{l_c}^{m_c}(\theta_{B2},\phi_2)Y_{l_d}^{m_d}(\theta_{B2},\phi_2)Y_l^m(\theta_{A2},\phi_2) \tag{3.99}$$

$$u_{AB}^{cd}\left(\vec{r}_{A1}\,|\,r_{A1} \le r_{A2}\right)_l^m$$
$$= N_r(n_c,\zeta_c)\,N_r(n_d,\zeta_d)\frac{1}{2l+1}r_{A1}^{l}\,Y_l^m(\theta_{A1},\phi_1)\int_{r_{A1}}^{\infty} dr_{A2}r_{A2}^{2-(l+1)}\exp(-\zeta_c r_{A2})$$
$$\times \int d\Omega_2 r_{B2}^{n_d-1}\exp\left(-\zeta_d r_{B2}\right)Y_{l_c}^{m_c}(\theta_{A2},\phi_2)Y_{l_d}^{m_d}(\theta_{B2},\phi_2)Y_l^m(\theta_{A2},\phi_2). \tag{3.100}$$

Partial Potential Integral of Order *l* and *m*

Each for six domains of integration shown in Fig. 3.2, partial potential integral by the product of STFs located at the same center is expressed as follows

$$u_{BB}^{cd}\left(\vec{r}_{A1}\,|\,r_{A2} \le r_{A1} \le R\right)_l^m$$
$$= C(n_c,\zeta_c)\,C(n_d,\zeta_d)\frac{1}{2l+1}r_{A1}^{-(l+1)}\,Y_l^m(\theta_{A1},\phi_{A1})\int_0^{r_{A1}} dr_{A2}r_{A2}^{2+l}$$
$$\times \int d\Omega_{A2}r_{B2}^{n_c+n_d-2}\exp\left(-(\zeta_c+\zeta_d)r_{B2}\right)Y_{l_c}^{m_c}(\theta_{B2},\phi_{B2})Y_{l_d}^{m_d}(\theta_{B2},\phi_{B2})Y_l^m(\theta_{A2},\phi_{A2}) \tag{3.101}$$

$$u_{BB}^{cd}\left(\vec{r}_{A1}\,|\,r_{A2} \le R \le r_{A1}\right)_l^m$$
$$= N_r(n_c,\zeta_c)\,N_r(n_d,\zeta_d)\frac{1}{2l+1}r_{A1}^{-(l+1)}\,Y_l^m(\theta_{A1},\phi_1)\int_0^{R} dr_{A2}r_{A2}^{2+l}$$
$$\times \int d\Omega_2 r_{B2}^{n_c+n_d-2}\exp\left(-(\zeta_c+\zeta_d)r_{B2}\right)Y_{l_c}^{m_c}(\theta_{B2},\phi_2)Y_{l_d}^{m_d}(\theta_{B2},\phi_2)Y_l^m(\theta_{A2},\phi_2) \tag{3.102}$$

$$u_{BB}^{cd}\left(\vec{r}_{A1}\left|R \le r_{A2} \le r_{A1}\right.\right)_l^m = N_r(n_c,\zeta_c)\,N_r(n_d,\zeta_d)\frac{1}{2l+1}r_{A1}^{-(l+1)}Y_l^m(\theta_{A1},\phi_1)\int_R^{r_{A1}} dr_{A2}r_{A2}^{2+l} \times \int d\Omega_2 r_{B2}^{n_c+n_d-2}\exp\left(-(\zeta_c+\zeta_d)r_{B2}\right)Y_{l_c}^{m_c}(\theta_{B2},\phi_2)\,Y_{l_d}^{m_d}(\theta_{B2},\phi_2)\,Y_l^m(\theta_{A2},\phi_2) \tag{3.103}$$

$$u_{BB}^{cd}\left(\vec{r}_{A1}\left|r_{A1} \le r_{A2} \le R\right.\right)_l^m = N_r(n_c,\zeta_c)\,N_r(n_d,\zeta_d)\frac{1}{2l+1}r_{A1}^{l}Y_l^m(\theta_{A1},\phi_1)\int_{r_{A1}}^{R} dr_{A2}r_{A2}^{2-(l+1)} \times \int d\Omega_2 r_{B2}^{n_c+n_d-2}\exp\left(-(\zeta_c+\zeta_d)r_{B2}\right)Y_{l_c}^{m_c}(\theta_{B2},\phi_2)\,Y_{l_d}^{m_d}(\theta_{B2},\phi_2)\,Y_l^m(\theta_{A2},\phi_2) \tag{3.104}$$

$$u_{BB}^{cd}\left(\vec{r}_{A1}\left|r_{A1} \le R \le r_{A2}\right.\right)_l^m = N_r(n_c,\zeta_c)\,N_r(n_d,\zeta_d)\frac{1}{2l+1}r_{A1}^{l}Y_l^m(\theta_{A1},\phi_1)\int_{R}^{\infty} dr_{A2}r_{A2}^{2-(l+1)} \times \int d\Omega_2 r_{B2}^{n_c+n_d-2}\exp\left(-(\zeta_c+\zeta_d)r_{B2}\right)Y_{l_c}^{m_c}(\theta_{B2},\phi_2)\,Y_{l_d}^{m_d}(\theta_{B2},\phi_2)\,Y_l^m(\theta_{A2},\phi_2) \tag{3.105}$$

$$u_{BB}^{cd}\left(\vec{r}_{A1}\left|R \le r_{A1} \le r_{A2}\right.\right)_l^m = N_r(n_c,\zeta_c)\,N_r(n_d,\zeta_d)\frac{1}{2l+1}r_{A1}^{l}Y_l^m(\theta_{A1},\phi_1)\int_{r_{A1}}^{\infty} dr_{A2}r_{A2}^{2-(l+1)} \times \int d\Omega_{A2} r_{B2}^{n_c+n_d-2}\exp\left(-(\zeta_c+\zeta_d)r_{B2}\right)Y_{l_c}^{m_c}(\theta_{B2},\phi_2)\,Y_{l_d}^{m_d}(\theta_{B2},\phi_2)\,Y_l^m(\theta_{A2},\phi_2) \tag{3.106}$$

For each of six domains of integration shown in Fig. 3.2, two-center partial potential integral is expressed as follows

$$u_{AB}^{cd}\left(\vec{r}_{A1}\left|r_{A2} \le r_{A1} \le R\right.\right)_l^m = N_r(n_c,\zeta_c)\,N_r(n_d,\zeta_d)\frac{1}{2l+1}r_{A1}^{-(l+1)}Y_l^m(\theta_{A1},\phi_1) \times \int_0^{r_{A1}} dr_{A2}r_{A2}^{n_c+l+1}\exp(-\zeta_c r_{A2}) \times \int d\Omega_2 r_{B2}^{n_d-1}\exp\left(-\zeta_d r_{B2}\right)Y_{l_c}^{m_c}(\theta_{A2},\phi_2)\,Y_{l_d}^{m_d}(\theta_{B2},\phi_2)\,Y_l^m(\theta_{A2},\phi_2) \tag{3.107}$$

$$
\begin{aligned}
&u_{AB}^{cd}\left(\vec{r}_{A1}\mid r_{A2}\le R\le r_{A1}\right)_l^m\\
&\quad = N_r(n_c,\zeta_c)\,N_r(n_d,\zeta_d)\frac{1}{2l+1}r_{A1}^{-(l+1)}\,Y_l^m(\theta_{A1},\phi_1)\\
&\quad\times\int_0^R dr_{A2}r_{A2}^{n_c+l+1}\exp(-\zeta_c r_{A2})\\
&\quad\times\int d\Omega_2 r_{B2}^{n_d-1}\exp(-\zeta_d r_{B2})\,Y_{l_c}^{m_c}(\theta_{A2},\phi_2)\,Y_{l_d}^{m_d}(\theta_{B2},\phi_2)\,Y_l^m(\theta_{A2},\phi_2)
\end{aligned}
\tag{3.108}
$$

$$
\begin{aligned}
&u_{AB}^{cd}\left(\vec{r}_{A1}\,|R\le r_{A2}\le r_{A1}\right)_l^m\\
&\quad = N_r(n_c,\zeta_c)\,N_r(n_d,\zeta_d)\frac{1}{2l+1}r_{A1}^{-(l+1)}\,Y_l^m(\theta_{A1},\phi_1)\\
&\quad\times\int_R^{r_{A1}} dr_{A2}r_{A2}^{n_c+l+1}\exp(-\zeta_c r_{A2})\\
&\quad\times\int d\Omega_2 r_{B2}^{n_d-1}\exp(-\zeta_d r_{B2})\,Y_{l_c}^{m_c}(\theta_{A2},\phi_2)\,Y_{l_d}^{m_d}(\theta_{B2},\phi_2)\,Y_l^m(\theta_{A2},\phi_2)
\end{aligned}
\tag{3.109}
$$

$$
\begin{aligned}
&u_{AB}^{cd}\left(\vec{r}_{A1}\mid r_{A1}\le r_{A2}\le R\right)_l^m\\
&\quad = N_r(n_c,\zeta_c)\,N_r(n_d,\zeta_d)\frac{1}{2l+1}r_{A1}^{l}\,Y_l^m(\theta_{A1},\phi_1)\int_{r_{A1}}^R dr_{A2}r_{A2}^{n_c-l}\exp(-\zeta_c r_{A2})\\
&\quad\times\int d\Omega_2 r_{B2}^{n_d-1}\exp(-\zeta_d r_{B2})\,Y_{l_c}^{m_c}(\theta_{A2},\phi_2)\,Y_{l_d}^{m_d}(\theta_{B2},\phi_2)\,Y_l^m(\theta_{A2},\phi_2)
\end{aligned}
\tag{3.110}
$$

$$
\begin{aligned}
&u_{AB}^{cd}\left(\vec{r}_{A1}\mid r_{A1}\le R\le r_{A2}\right)_l^m\\
&\quad = N_r(n_c,\zeta_c)\,N_r(n_d,\zeta_d)\frac{1}{2l+1}r_{A1}^{l}\,Y_l^m(\theta_{A1},\phi_1)\int_R^\infty dr_{A2}r_{A2}^{n_c-l}\exp(-\zeta_c r_{A2})\\
&\quad\times\int d\Omega_2 r_{B2}^{n_d-1}\exp(-\zeta_d r_{B2})\,Y_{l_c}^{m_c}(\theta_{A2},\phi_2)\,Y_{l_d}^{m_d}(\theta_{B2},\phi_2)\,Y_l^m(\theta_{A2},\phi_2)
\end{aligned}
\tag{3.111}
$$

$$
\begin{aligned}
&u_{AB}^{cd}\left(\vec{r}_{A1}\,|R\le r_{A1}\le r_{A2}\right)_l^m\\
&\quad = N_r(n_c,\zeta_c)N_r(n_d,\zeta_d)\frac{1}{2l+1}r_{A1}^{l}\,Y_l^m(\theta_{A1},\phi_1)\int_{r_{A1}}^\infty dr_{A2}r_{A2}^{n_c-l}\exp(-\zeta_c r_{A2})\\
&\quad\times\int d\Omega_2 r_{B2}^{n_d-1}\exp(-\zeta_d r_{B2})\,Y_{l_c}^{m_c}(\theta_{A2},\phi_2)\,Y_{l_d}^{m_d}(\theta_{B2},\phi_2)\,Y_l^m(\theta_{A2},\phi_2)
\end{aligned}
\tag{3.112}
$$

Integration over ϕ

Integration over ϕ_1 and ϕ_2, in Eq. (3.94) for Coulomb-type integral, Eq. (3.95) for hybrid-type integral, and Eq. (3.96) for exchange-type integral, respectively, vanishes unless M_{ab}^{+} or M_{ab}^{-} equals M_{cd}^{+} or M_{cd}^{-} where

$$
\begin{aligned}
M_{ab}^{+} &= |m_a| + |m_b|,\\
M_{ab}^{-} &= \max\left(|m_a|, |m_b|\right) - \min\left(|m_a|, |m_b|\right).
\end{aligned}
\tag{3.113}
$$

The summation index m in Eqs. (3.94), (3.95), and (3.96) is accordingly limited to these nonvanishing cases, and by an interchange in the order of the double summation in Eqs. (3.94), (3.95), and (3.96) one finds

$$\sum_{l=0}^{\infty}\sum_{m=-l}^{l} \rightarrow \sum_{m=M^{\pm}}\sum_{l=m}^{\infty} \tag{3.114}$$

Then the integrals over ϕ_1 and ϕ_2 are straightforward to calculate, and will be omitted in the following.

Demonstration of Partial Integral of Coulomb-Type Integral in Region 2–3

Coulomb-type integral consists of six parts according to the integral domains from Region 1–1 to Region 2–3.

$$\begin{aligned}
V_{AABB}^{abcd} &= \int dV_1 \int dV_2 \chi_A^a\left(\vec{r}_1\right)\chi_A^b\left(\vec{r}_1\right)\frac{1}{r_{12}}\chi_B^c\left(\vec{r}_2\right)\chi_B^d\left(\vec{r}_2\right) \\
&= 4\pi\sum_l\sum_{m=-l}^{l}\int dV_1\rho_{AA}^{ab}\left(\vec{r}_1\right) \\
&\quad\times\left[u_{BB}^{cd}\left(\vec{r}_1|r_2\le r_1\le R\right)_l^m + u_{BB}^{cd}\left(\vec{r}_1|r_2\le R\le r_1\right)_l^m\right. \\
&\quad + u_{BB}^{cd}\left(\vec{r}_1|R\le r_2\le r_1\right)_l^m + u_{BB}^{cd}\left(\vec{r}_1|r_1\le r_2\le R\right)_l^m \\
&\quad\left. + u_{BB}^{cd}\left(\vec{r}_1|r_1\le R\le r_2\right)_l^m + u_{BB}^{cd}\left(\vec{r}_1|R\le r_1\le r_2\right)_l^m\right] \\
&= \sum_l\sum_{m=-l}^{l}\left[V_{AABB}^{abcd}\left(\vec{r}_1|r_2\le r_1\le R\right)_l^m + V_{AABB}^{abcd}\left(\vec{r}_1|r_2\le R\le r_1\right)_l^m\right. \\
&\quad + V_{AABB}^{abcd}\left(\vec{r}_1|R\le r_2\le r_1\right)_l^m + V_{AABB}^{abcd}\left(\vec{r}_1|r_1\le r_2\le R\right)_l^m \\
&\quad\left. + V_{AABB}^{abcd}\left(\vec{r}_1|r_1\le R\le r_2\right)_l^m + V_{AABB}^{abcd}\left(\vec{r}_1|R\le r_1\le r_2\right)_l^m\right]
\end{aligned} \tag{3.115}$$

We shall demonstrate the formulation of the partial Coulomb repulsion integral in the Region 2–3 ($R\le r_1\le r_2$) included in the second part on the right hand side of Eq. (3.94).

From Eq. (3.115), we find

$$V_{AABB}^{abcd}\left(\vec{r}_1\,|R\le r_1\le r_2\right)_l^m = 4\pi\int dV_1\rho_{AA}^{ab}\left(\vec{r}_1\right)\,u_{BB}^{cd}\left(\vec{r}_1\,|R\le r_1\le r_2\right)_l^m. \tag{3.116}$$

We start with a partial potential integral given by Eq. (3.106)

$$u_{BB}^{cd}\left(\vec{r}_{A1}\,|R\le r_{A1}\le r_{A2}\right)_l^m$$
$$= N_r(n_c,\zeta_c)\,N_r(n_d,\zeta_d)\frac{1}{2l+1}r_{A1}^l\,Y_l^m(\theta_{A1},\phi_1)\int_{r_{A1}}^{\infty}dr_{A2}r_{A2}^{2-(l+1)}$$
$$\times\int d\Omega_{A2}r_{B2}^{n_c+n_d-2}\exp(-(\zeta_c+\zeta_d)r_{B2})\,Y_{l_c}^{m_c}(\theta_{B2},\phi_2)\,Y_{l_d}^{m_d}(\theta_{B2},\phi_2)\,Y_l^m(\theta_{A2},\phi_2). \tag{3.106}$$

Using the integration of triple product of spherical harmonic functions in Eq. (3.A31)

$$\int d\Omega Y_{l_a}^{m_a}(\theta_B,\phi)Y_{l_b}^{m_b}(\theta_B,\phi)Y_{l_c}^{m_c}(\theta_A,\phi)$$
$$=2\pi N_\phi(0)\sum_{m_{ad}}\Phi(m_a,m_b,m_{ab})\sum_{l_{cd}}\Theta\begin{pmatrix} m_a & m_b & m_{ab}\\ l_a & l_b & l_{ab}\end{pmatrix}$$
$$\times\sum_{k_{ab}=|m_{ab}|}^{l_{ab}}T_{k_c}^{l_{ab}m_{ab}}\sum_{m_{abc}}\Phi(m_{ab},m_c,m_{abc})\sum_{l_{abc}}\Theta\begin{pmatrix} m_{ab} & m_c & m_{abc}\\ k_{ab} & l_c & l_{abc}\end{pmatrix}$$
$$\times\int_0^{\pi}\sin\theta_A d\theta_A$$
$$\times\sum_{\nu=0}^{\left[\frac{l_{abc}}{2}\right]}\sum_{i=0}^{l_{abc}-2\nu}\sum_{j=0}^{l_{abc}-2\nu-i}(l_{abc},\nu,i,j)R^{l_{ab}-k_{ab}-l_{abc}+2(\nu+i)}r_A^{k_{ab}+l_{abc}-2(\nu+i+j)}r_B^{2i-l_{ab}}, \tag{3.A31}$$

The expression of $u_{BB}^{cd}\left(\vec{r}_{A1}|R\le r_{A1}\le r_{A2}\right)_l^m$ is found to be

$$u_{BB}^{cd}\left(\vec{r}_{A1}|R\le r_{A1}\le r_{A2}\right)_l^m$$
$$=2\pi N_\phi(0)\,N_r(n_c,\zeta_c)\,N_r(n_d,\zeta_d)\frac{1}{2l+1}r_{A1}^lY_l^m(\theta_{A1},\phi_1)$$
$$\times\sum_{m_{cd}}\Phi(m_c,m_d,m_{cd})\sum_{l_{cd}}\Theta\begin{pmatrix} m_c & m_d & m_{cd}\\ l_c & l_d & l_{cd}\end{pmatrix}$$
$$\times\sum_{k_{cd}=\left|m_{cd}\right|}^{l_{cd}}T_{k_{cd}}^{l_{cd}m_{cd}}\sum_{m_{cdm}}\Phi(m_{cd},m,m_{cdm})\sum_{l_{cdl}}\Theta\begin{pmatrix} m_{cd} & m & m_{cdm}\\ k_{cd} & l & l_{cdl}\end{pmatrix} \tag{3.117}$$
$$\times\sum_{v=0}^{\left[\frac{l_{cdl}}{2}\right]}\sum_{i=0}^{l_{cdl}-2v}\sum_{j=0}^{l_{cdl}-2v-i}\mathrm{P}(l_{cdl},v,i,j)R^{l_{ab}-k_{ab}-l_{cdl}+2(v+i)}$$
$$\times\int_{rA1}^{\infty}dr_{A2}r_{A2}^{2-(l+1)+k_{ab}+l_{cdl}-2(v+i+j)}$$
$$\times\int_0^{\pi}\sin\theta_{A2}d\theta_{A2}r_{B2}^{n_c+n_d-2-l_{cd}+2i}\exp(-(\zeta_c+\zeta_d)r_{B2}).$$

Using the definition of B function in Eq. (3.C1) and in Eq. (3.C9)

$$B\left(n,\zeta,R,r_A\right) \equiv \int_0^{\pi} \sin\theta_A \, d\theta_A r_B^n \exp\left(-\zeta r_B\right) \tag{3.C1}$$

$$\begin{aligned} & B\left(n,\zeta,R,r_A \middle| r_A \geq R\right) \\ & = \exp(-\zeta r_A) \sum_{s=0}^{n+1} \binom{n+1}{s} R^{n+1-s} r_A^{s-1} \left\{ \int_{-1}^{1} dx x^{n+1-s} \exp(-\zeta R x) \right\} \end{aligned} \tag{3.C9}$$

or

$$\begin{aligned} & B\left(n,\zeta,R,r_A \middle| r_A \geq R\right) \\ & = \sum_{s=0}^{n+1} \binom{n+1}{s} R^{n+1-s} r_A^{s-1} \exp(-\zeta r_A) \\ & \{\langle H[1,n+1-s,\zeta R]\rangle - \langle H[-1,n+1-s,\zeta R]\rangle\} \end{aligned} \tag{3.C9}$$

$$\begin{aligned} & u_{BB}^{cd}\left(\vec{r}_{A1} \middle| R \leq r_{A1} \leq r_{A2}\right)_l^m \\ & = 2\pi N_\phi(0) N_r(n_c,\zeta_c) N_r(n_d,\zeta_d) \frac{1}{2l+1} r_{A1}^l Y_l^m(\theta_{A1},\phi_1) \\ & \times \sum_{m_{cd}} \Phi(m_c,m_d,m_{cd}) \sum_{l_{cd}} \Theta\begin{pmatrix} m_c & m_d & m_{cd} \\ l_c & l_d & l_{cd} \end{pmatrix} \\ & \times \sum_{k_{cd}=\left|m_{cd}\right|}^{l_{cd}} T_{k_{cd}}^{l_{cd} m_{cd}} \sum_{m_{cdm}} \Phi(m_{cd},m,m_{cdm}) \sum_{l_{cdl}} \Theta\begin{pmatrix} m_{cd} & m & m_{cdm} \\ k_{cd} & l & l_{cdl} \end{pmatrix} \\ & \times \sum_{\nu=0}^{\left[\frac{l_{cdl}}{2}\right]} \sum_{i=0}^{l_{cdl}-2\nu} \sum_{j=0}^{l_{cdl}-2\nu-i} \mathrm{P}(l_{cdl},\nu,i,j) \\ & \times \sum_{s=0}^{n_c+n_d-1-l_{cd}+2i} \binom{n_c+n_d-1-l_{cd}+2i}{s} R^{l_{cd}-k_{cd}-l_{cdl}+2(\nu+i)+n_c+n_d-1-l_{cd}+2i-s} \\ & \times \int_{r_{A1}}^{\infty} dr_{A2} r_{A2}^{2-(l+1)+k_{cd}+l_{cdl}-2(\nu+i+j)-s-1} \exp\left(-(\zeta_c+\zeta_d) r_{A2}\right) \\ & \times \int_{-1}^{1} dx x^{n_c+n_d-1-l_{cd}+2i-s} \exp\left(-(\zeta_c+\zeta_d) R x\right) \end{aligned} \tag{3.118}$$

$$
\begin{aligned}
& u_{BB}^{cd}\left(\vec{r}_{A1} \middle| R \le r_{A1} \le r_{A2}\right)_l^m \\
&= 2\pi N_\phi(0) N_r(n_c,\zeta_c) N_r(n_d,\zeta_d) \frac{1}{2l+1} r_{A1}^l Y_l^m(\theta_{A1},\phi_1) \\
&\times \sum_{m_{cd}} \Phi(m_c, m_d, m_{cd}) \sum_{l_{cd}} \Theta\begin{pmatrix} m_c & m_d & m_{cd} \\ l_c & l_d & l_{cd} \end{pmatrix} \\
&\times \sum_{k_{cd}=|m_{cd}|}^{l_{ab}} T_{k_{cd}}^{l_{cd} m_{cd}} \sum_{m_{cdm}} \Phi(m_{cd}, m, m_{cdm}) \sum_{l_{cdl}} \Theta\begin{pmatrix} m_{cd} & m & m_{cdm} \\ k_{cd} & l & l_{cdl} \end{pmatrix} \\
&\times \sum_{\nu=0}^{\left[\frac{l_{cdl}}{2}\right]} \sum_{i=0}^{l_{cdl}-2\nu} \sum_{j=0}^{l_{cdl}-2\nu-i} \mathrm{P}(l_{cdl},\nu,i,j) \\
&\times \sum_{s=0}^{n_c+n_d-1-l_{cd}+2i} \binom{n_c+n_d-1-l_{cd}+2i}{s} R^{l_{ab}-k_{ab}-l_{cdl}+2(\nu+i)+n_c+n_d-1-l_{cd}+2i-s} \\
&\times H[r_{A1}, 2-(l+1)+k_{ab}+l_{cdl}-2(\nu+i+j)-s-1, \zeta_d+\zeta_c] \\
&\times \langle H[1, n_c+n_d-1-l_{cd}+2i-s, (\zeta_d+\zeta_c)R] \\
&\quad - H[1, n_c+n_d-1-l_{cd}+2i-s, (\zeta_d+\zeta_c)R] \rangle
\end{aligned}
\tag{3.119}
$$

where H is an auxiliary function defined by Eq. (3.B1) in Appendix B.

$$
H[X,n,\alpha] \equiv \int^X x^n \exp(-\alpha x)\, dx \tag{3.B1}
$$

Total potential integral by the product of STFs centered at B in the Region 2–3 ($R \le r_{A1} \le r_{A2}$) is given at $\vec{r}_{A1}$

$$
U_{BB}^{cd}\left(\vec{r}_1 \middle| R \le r_1 \le r_2\right) = 4\pi \sum_l \sum_{m=-1}^{l} u_{BB}^{cd}\left(\vec{r}_1 \middle| R \le r_1 \le r_2\right)_l^m \tag{3.81}
$$

Partial Coulomb Repulsion Integrals

Partial Coulomb repulsion integrals

$$
V_{AABB}^{abcd}\left(\vec{r}_1 \middle| R \le r_1 \le r_2\right)_l^m = 4\pi \int dV_1 \rho_{AA}^{ab}\left(\vec{r}_1\right) u_{BB}^{cd}\left(\vec{r}_1 \middle| R \le r_1 \le r_2\right)_l^m \tag{3.116}
$$

can be formulated with charge density by the STF product of on the same atomic center

$$
\begin{aligned}
\rho_{AA}^{ab}\left(\vec{r}_A\right) &\equiv \chi_A^a\left(\vec{r}_A\right) \chi_A^b\left(\vec{r}_A\right) \\
&= R_A^a(r_A) Y_{l_a}^{m_a}(\theta_A,\phi) R_A^b(r_A) Y_{l_b}^{m_b}(\theta_A,\phi) \\
&= N_r(n_a,\zeta_a) N_r(n_b,\zeta_b) r^{n_a+n_b-2} \exp(-(\zeta_a+\zeta_b)) \\
&\quad \times \sum_m \Phi(m_a,m_b,m) \sum_l \Theta\begin{pmatrix} m_a & m_b & m \\ l_a & l_b & l \end{pmatrix} Y_l^m(\theta_A,\phi)
\end{aligned}
\tag{3.28}
$$

Partial Coulomb repulsion integrals $V_{AABB}^{abcd}\left(\vec{r}_1 \middle| r_1 \le r_2\right)_l^m$ is then expressed using the partial potential integral in Eq. (3.118), as follows

$$
\begin{aligned}
& V_{AABB}^{abcd}\left(\vec{r}_{A1} \middle| R \le r_1 \le r_2\right)_l^m \\
&= 4\pi \int dV_1 \rho_{AA}^{ab}\left(\vec{r}_{A1}\right) u_{BB}^{cd}\left(\vec{r}_{A1} \middle| R \le r_{A1} \le r_{A2}\right)_l^m \\
&= 8\pi^2 N_\phi(0) N_r(n_a,\zeta_a) N_r(n_b,\zeta_b) N_r(n_c,\zeta_c) N_r(n_d,\zeta_d) \\
&\times \frac{1}{2l+1} \sum_{m_{cd}} \Phi(m_c, m_d, m_{cd}) \sum_{l_{cd}} \Theta\begin{pmatrix} m_c & m_d & m_{cd} \\ l_c & l_d & l_{cd} \end{pmatrix} \\
&\times \sum_{k_{cd}=\left|m_{cd}\right|}^{l_{cd}} T_{k_{cd}}^{l_{cd} m_{cd}} \sum_{m_{cdm}} \Phi(m_{cd}, m, m_{cdm}) \sum_{l_{cdl}} \Theta\begin{pmatrix} m_{cd} & m & m_{cdm} \\ k_{cd} & l & l_{cdl} \end{pmatrix} \\
&\times \sum_{\nu=0}^{\left[\frac{l_{cdl}}{2}\right]} \sum_{i=0}^{l_{cdl}-2\nu} \sum_{j=0}^{l_{cdl}-2\nu-i} \mathrm{P}(l_{cdl}, \nu, i, j) \\
&\times \sum_{s=0}^{n_c+n_d-1-l_{cd}+2i} \binom{n_c+n_d-1-l_{cd}+2i}{s} R^{l_{cd}-k_{cd}-l_{cdl}+2(\nu+i)+n_c+n_d-1-l_{cd}+2i-s} \\
&\times \int_R^\infty dr_{A1} r_{A1}^{n_a+n_b+l} \exp\left(-(\zeta_a+\zeta_b) r_{A1}\right) \\
&\times \int d\Omega_{A1} Y_{l_a}^{m_a}(\theta_{A1}, \phi_1)\, Y_{l_b}^{m_b}(\theta_{A1}, \phi_1) Y_l^m(\theta_{A1}, \phi_{A1}) \\
&\times \int_{r_{A1}}^\infty dr_{A2} r_{A2}^{2-(l+1)+k_{cd}+l_{cdl}-2(\nu+i+j)+s-1} \exp\left(-(\zeta_c+\zeta_d) r_{A2}\right) \\
&\times \int_{-1}^1 dx x^{n_c+n_d-1-l_{cd}+2i-s} \exp\left(-(\zeta_c+\zeta_d) Rx\right)
\end{aligned}
\tag{3.120}
$$

Four integrals in the last part are calculated as follows

$$
\begin{aligned}
& \int_R^\infty dr_{A1} r_{A1}^{n_a+n_b+l} \exp\left(-(\zeta_a+\zeta_b) r_{A1}\right) \\
&\times \int d\Omega_{A1} Y_{l_a}^{m_a}(\theta_{A1}, \phi_1)\, Y_{l_b}^{m_b}(\theta_{A1}, \phi_1)\, Y_l^m(\theta_{A1}, \phi_{A1}) \\
&\times \int_{r_{A1}}^\infty dr_{A2} r_{A2}^{2-(l+1)+k_{cd}+l_{cdl}-2(\nu+i+j)+s-1} \exp\left(-(\zeta_c+\zeta_d) r_{A2}\right) \\
&\times \int_{-1}^1 dx x^{n_c+n_d-1-l_{cd}+2i-s} \exp(-(\zeta_c+\zeta_d) Rx) \\
&= G\begin{pmatrix} m_a & m_b & m \\ l_a & l_b & l \end{pmatrix} \\
&\times \int_R^\infty dr_{A1} r_{A1}^{n_1+l} \exp\left(-(\zeta_a+\zeta_b) r_{A1}\right) \\
&\times \int_{r_{A1}}^\infty dr_{A2} r_{A2}^{n_3-(l+1)} \exp\left(-(\zeta_c+\zeta_d) r_{A2}\right) \\
&\times \int_{-1}^1 dx x^{n_4} \exp\left(-(\zeta_c+\zeta_d) Rx\right)
\end{aligned}
\tag{3.121}
$$

where

$n_1 = n_a + n_b, n_3 = 1 + k_{cd} + l_{cdl} - 2(\nu + i + j) + s, n_4 = n_c + n_d - 1 - l_{cd} + 2i - s.$

For the convenience of symbolic calculation, three integrals in Eq. (3.121) are expressed as follows

$$\begin{aligned}
&\int_R^\infty dr_{A1} r_{A1}^{n_1+l} \exp\left(-(\zeta_a+\zeta_b) r_{A1}\right) \int_R^\infty dr_{A2} r_{A2}^{n3-(l+1)} \exp\left(-(\zeta_c+\zeta_d) r_{A2}\right) \\
&\quad \times \int_{-1}^{1} dx_2 x_2^{n_4} \exp\left(-(\zeta_c+\zeta_d) R x_2\right) \\
&\quad = \left\langle \int^R dr_1 r_1^{n_1+l} \exp\left(-(\zeta_a+\zeta_b) r_1\right) \int^{r_1} dr_2 r_2^{n_3-(l+1)} \exp\left(-(\zeta_c+\zeta_d) r_2\right) \right\rangle \\
&\quad \left\langle \int_{-1}^{1} dx_2 x_2^{n_4} \exp\left(-(\zeta_c+\zeta_d) R x_2\right) \right\rangle \\
&\quad = Hh\left[R, n_1+l, \zeta_a+\zeta_b, n_3-(l+1), \zeta_c+\zeta_d\right] \\
&\quad \times \left(H\left[1, n_4, (\zeta_c+\zeta_d)R\right] - H\left[-1, n_4, (\zeta_c+\zeta_d)R\right]\right)
\end{aligned} \tag{3.122}$$

where Hh and H are auxiliary half definite integrals defined by Eqs. (3.B1) and (3.B14).

It is easy to formulate all multiple integrals included in the partial potential for Coulomb-type integral given by from Eq. (3.101) to Eq. (3.106) for six regions. All multiple integrals included in the partial potential not only for Coulomb-type integral but also hybrid-type integral and exchange-type integral are summarized from Eq. (3.124) to Eq. (3.141).

By inserting the right hand side of Eq. (3.122) to Eq. (3.121), we obtain the expression of partial Coulomb repulsion integrals for the Region 2–3,

$$\begin{aligned}
&V_{AABB}^{abcd}\left(\vec{r}_1 \middle| R \le r_1 \le r_2\right)_l^m \\
&\quad = 4\pi \int dV_1 \rho_{AA}^{ab}\left(\vec{r}_{A1}\right) u_{BB}^{cd}\left(\vec{r}_{A1} \middle| R \le r_{A1} \le r_{A2}\right)_l^m \\
&\quad = 8\pi^2 N_\phi(0) N_r(n_a,\zeta_a) N_r(n_b,\zeta_b) N_r(n_c,\zeta_c) N_r(n_d,\zeta_d) \\
&\qquad \times \frac{1}{2l+1} \sum_{m_{cd}} \Phi(m_c, m_d, m_{cd}) \sum_{l_{cd}} \Theta\begin{pmatrix} m_c & m_d & m_{cd} \\ l_c & l_d & l_{cd} \end{pmatrix} \\
&\qquad \times \sum_{k_{cd}=|m_{cd}|}^{l_{cd}} T_{k_{cd}}^{l_{cd} m_{cd}} \sum_{m_{cdm}} \Phi(m_{cd}, m, m_{cdm}) \sum_{l_{cdl}} \Theta\begin{pmatrix} m_{cd} & m & m_{cdm} \\ k_{cd} & l & l_{cdl} \end{pmatrix} \\
&\qquad \times \sum_{v=0}^{\left[\frac{l_{cdl}}{2}\right]} \sum_{i=0}^{l_{cdl}-2v} \sum_{j=0}^{l_{cdl}-2v-i} \mathrm{P}(l_{cdl}, v, i, j) \; R^{l_{cd}-k_{cd}-l_{cdl}+2(v+i)} \\
&\qquad \times \sum_{s=0}^{n_c+n_d-1-l_{cd}+2i} \binom{n_c+n_d-1-l_{cd}+2i}{s} R^{n_c+n_d-1-l_{cd}+2i-s} \\
&\qquad \times G\begin{pmatrix} m_a & m_b & m \\ l_a & l_b & l \end{pmatrix} \\
&\qquad \times Hh\left[R, n_1+l, \zeta_a+\zeta_b, n_3-(l+1), \zeta_c+\zeta_d\right] \\
&\qquad \times \left(H\left[1, n_4, (\zeta_c+\zeta_d)R\right] - H\left[-1, n_4, (\zeta_c+\zeta_d)R\right]\right)
\end{aligned} \tag{3.123}$$

By inserting all partial repulsion integrals obtained in the same way into Eq. (3.114), we arrive at the final expression of Coulomb-type repulsion integral. For reminding it, we show it again.

$$\begin{aligned} V_{AABB}^{abcd} = \sum_{l} \sum_{m=-l}^{l} \Big[& V_{AABB}^{abcd}\left(\vec{r}_1 | r_2 \le r_1 \le R\right)_l^m + V_{AABB}^{abcd}\left(\vec{r}_1 | r_2 \le R \le r_1\right)_l^m \\ & + V_{AABB}^{abcd}\left(\vec{r}_1 | R \le r_2 \le r_1\right)_l^m + V_{AABB}^{abcd}\left(\vec{r}_1 | r_1 \le r_2 \le R\right)_l^m \\ & + V_{AABB}^{abcd}\left(\vec{r}_1 | r_1 \le R \le r_2\right)_l^m + V_{AABB}^{abcd}\left(\vec{r}_1 | R \le r_1 \le r_2\right)_l^m \Big] \end{aligned} \tag{3.124}$$

Formulation of hybrid-type repulsion integrals and that of exchange-type repulsion integrals are obtained almost in the same way. Symbolic calculation system makes it easy to formulate all molecular integrals over STFs for all combinations of quantum numbers of STF.

As is the same as the formulation in case of Coulomb-type integral in Eq. (3.108), multiple integrals appear in the most inner part of formula of electron repulsion integrals. Multiple integrals for all partial potentials in the six regions are presented from Eq. (3.125) to Eq. (3.142) in the following. For Coulomb-type integral and hybrid-type integral, integrations with respect to θ_1 are omitted for simplicity. All electron repulsion integrals are deduced into five types of half definite integrals. In the following, <....> means a definite integral.

Coulomb-type integral 1–1 ($r_2 \le r_1 \le R$)

$$\begin{aligned} & \int_0^R dr_1 r_1^{n_1-(l+1)} \exp(-(\zeta_a+\zeta_b)r_1) \int_0^{r_1} dr_2 r_2^{n_3+l} \exp(\pm(\zeta_c+\zeta_d)r_2) \\ & \quad \times \int_{-1}^{1} dx_2 x_2^{n_4} \exp(-(\zeta_c+\zeta_d)r_2 x_2) \\ & = \left\langle \int^R dr_1 r_1^{n_1-(l+1)} \exp(-(\zeta_a+\zeta_b)r_1) \int^{r_1} dr_2 r_2^{n_3+l} \exp(\pm(\zeta_c+\zeta_d)r_2) \right. \\ & \quad \left. \times \int_{-1}^{1} dx_2 x_2^{n_4} \exp(-(\zeta_c+\zeta_d)r_2 x_2) \right\rangle - \left\langle \int^R dr_1 r_1^{n_1-(l+1)} \exp(-(\zeta_a+\zeta_b)r_1) \right\rangle \\ & \quad \left\langle \int^0 dr_2 r_2^{n_3+l} \exp(\pm(\zeta_c+\zeta_d)r_2) \int_{-1}^{1} dx_2 x_2^{n_4} \exp(-(\zeta_c+\zeta_d)r_2 x_2) \right\rangle \\ & \quad - \left\langle \int^0 dr_1 r_1^{n_1-(l+1)} \exp(-(\zeta_a+\zeta_b)r_1) \int^{r_1} dr_2 r_2^{n_3+l} \exp(\pm(\zeta_c+\zeta_d)r_2) \right. \\ & \quad \left. \times \int_{-1}^{1} dx_2 x_2^{n_4} \exp(-(\zeta_c+\zeta_d)r_2 x_2) \right\rangle + \left\langle \int^0 dr_1 r_1^{n_1-(l+1)} \exp(-(\zeta_a+\zeta_b)r_1) \right\rangle \\ & \quad \left\langle \int^0 dr_2 r_2^{n_3+l} \exp(\pm(\zeta_c+\zeta_d)r_2) \int_{-1}^{1} dx_2 x_2^{n_4} \exp(-(\zeta_c+\zeta_d)r_2 x_2) \right\rangle \end{aligned} \tag{3.125}$$

Coulomb-type integral 1–2 ($r_2 \le R \le r_1$)

$$\begin{aligned}
&\int_R^\infty dr_1 r_1^{n_1-(l+1)} \exp(-(\zeta_a+\zeta_b)r_1) \int_0^R dr_2 r_2^{n_3+l} \exp(\pm(\zeta_c+\zeta_d)r_2) \\
&\quad \times \int_{-1}^{1} dx_2 x_2^{n_4} \exp(-(\zeta_c+\zeta_d)r_2 x_2) \\
&\quad = -\left\langle \int^R dr_1 r_1^{n_1-(l+1)} \exp(-(\zeta_a+\zeta_b)r_1) \right\rangle \\
&\quad \left\langle \int^R dr_2 r_2^{n_3+l} \exp(\pm(\zeta_c+\zeta_d)r_2) \int_{-1}^{1} dx_2 x_2^{n_4} \exp(-(\zeta_c+\zeta_d)r_2 x_2) \right\rangle \\
&\quad + \left\langle \int^R dr_1 r_1^{n_1-(l+1)} \exp(-(\zeta_a+\zeta_b)r_1) \right\rangle \\
&\quad \left\langle \int^0 dr_2 r_2^{n_3+l} \exp(\pm(\zeta_c+\zeta_d)r_2) \int_{-1}^{1} dx_2 x_2^{n_4} \exp(-(\zeta_c+\zeta_d)r_2 x_2) \right\rangle
\end{aligned} \tag{3.126}$$

Coulomb-type integral 1–3 ($R \le r_2 \le r_1$)

$$\begin{aligned}
&\int_R^\infty dr_1 r_1^{n_1-(l+1)} \exp(-(\zeta_a+\zeta_b)r_1) \int_R^{r_1} dr_2 r_2^{n_3+l} \exp(-(\zeta_c+\zeta_d)r_2) \\
&\quad \times \int_{-1}^{1} dx_2 x_2^{n_4} \exp(-(\zeta_c+\zeta_d)R x_2) \\
&= -\left\langle \int^R dr_1 r_1^{n_1-(l+1)} \exp(-(\zeta_a+\zeta_b)r_1) \int^{r_1} dr_2 r_2^{n_3+l} \exp(-(\zeta_c+\zeta_d)r_2) \right\rangle \\
&\left\langle \int_{-1}^{1} dx_2 x_2^{n_4} \exp(-(\zeta_c+\zeta_d)R x_2) \right\rangle + \left\langle \int^R dr_1 r_1^{n_1-(l+1)} \exp(-(\zeta_a+\zeta_b)r_1) \right\rangle \\
&\left\langle \int^R dr_2 r_2^{n_3+l} \exp(-(\zeta_c+\zeta_d)r_2) \right\rangle \left\langle \int_{-1}^{1} dx_2 x_2^{n_4} \exp(-(\zeta_c+\zeta_d)R x_2) \right\rangle
\end{aligned} \tag{3.127}$$

Coulomb type integral 2–1 ($r_2 \le R \le r_1$)

$$\begin{aligned}
&\int_0^R dr_1 r_1^{n_1+l} \exp(-(\zeta_a+\zeta_b)r_1) \int_{r_1}^R dr_2 r_2^{n_3-(l+1)} \exp(\pm(\zeta_c+\zeta_d)r_2) \\
&\quad \times \int_{-1}^{1} dx_2 x_2^{n_4} \exp(-(\zeta_c+\zeta_d)r_2 x_2) \\
&\quad = \left\langle \int^R dr_1 r_1^{n_1+l} \exp(-(\zeta_a+\zeta_b)r_1) \right\rangle \\
&\quad \left\langle \int^R dr_2 r_2^{n_3-(l+1)} \exp(\pm(\zeta_c+\zeta_d)r_2) \int_{-1}^{1} dx_2 x_2^{n_4} \exp(-(\zeta_c+\zeta_d)r_2 x_2) \right\rangle \\
&\quad - \left\langle \int^R dr_1 r_1^{n_1+l} \exp(-(\zeta_a+\zeta_b)r_1) \int^{r_1} dr_2 r_2^{n_3-(l+1)} \exp(\pm(\zeta_c+\zeta_d)r_2) \right. \\
&\quad \left. \times \int_{-1}^{1} dx_2 x_2^{n_4} \exp(-(\zeta_c+\zeta_d)r_2 x_2) \right\rangle \\
&\quad - \left\langle \int^0 dr_1 r_1^{n_1+l} \exp(-(\zeta_a+\zeta_b)r_1) \right\rangle \\
&\quad \left\langle \int^R dr_2 r_2^{n_3-(l+1)} \exp(\pm(\zeta_c+\zeta_d)r_2) \int_{-1}^{1} dx_2 x_2^{n_4} \exp(-(\zeta_c+\zeta_d)r_2 x_2) \right\rangle \\
&\quad - \left\langle \int^0 dr_1 r_1^{n_1+l} \exp(-(\zeta_a+\zeta_b)r_1) \int^{r_1} dr_2 r_2^{n_3-(l+1)} \exp(\pm(\zeta_c+\zeta_d)r_2) \right. \\
&\quad \left. \times \int_{-1}^{1} dx_2 x_2^{n_4} \exp(-(\zeta_c+\zeta_d)r_2 x_2) \right\rangle
\end{aligned} \tag{3.128}$$

Coulomb-type integral 2–2 $(r_1 \le R \le r_2)$

$$\begin{aligned}
&\int_0^R dr_1 r_1^{n_1+l}\exp(-(\zeta_a+\zeta_b)r_1)\int_R^\infty dr_2 r_2^{n_3-(l+1)}\exp(-(\zeta_c+\zeta_d)r_2)\\
&\quad\times\int_{-1}^1 dx_2 x_2^{n_4}\exp(-(\zeta_c+\zeta_d)Rx_2)\\
&\quad= -\left\langle\int^R dr_1 r_1^{n_1+l}\exp(-(\zeta_a+\zeta_b)r_1)\right\rangle\\
&\qquad\left\langle\int^R dr_2 r_2^{n_3-(l+1)}\exp(-(\zeta_c+\zeta_d)r_2)\right\rangle\left\langle\int_{-1}^1 dx_2 x_2^{n_4}\exp(-(\zeta_c+\zeta_d)Rx_2)\right\rangle\\
&\qquad+\left\langle\int^0 dr_1 r_1^{n_1+l}\exp(-(\zeta_a+\zeta_b)r_1)\right\rangle\\
&\qquad\left\langle\int^R dr_2 r_2^{n_3-(l+1)}\exp(-(\zeta_c+\zeta_d)r_2)\right\rangle\left\langle\int_{-1}^1 dx_2 x_2^{n_4}\exp(-(\zeta_c+\zeta_d)Rx_2)\right\rangle
\end{aligned}\tag{3.129}$$

Coulomb-type integral 2–3 $(R \le r_1 \le r_2)$

$$\begin{aligned}
&\int_R^\infty dr_{A1} r_{A1}^{n_1+l}\exp(-(\zeta_a+\zeta_b)r_{A1})\int_{r_{A1}}^\infty dr_{A2} r_{A2}^{n_3-(l+1)}\exp(-(\zeta_c+\zeta_d)r_{A2})\\
&\quad\times\int_{-1}^1 dx_2 x_2^{n_4}\exp(-(\zeta_c+\zeta_d)Rx_2)\\
&\quad=\left\langle\int^R dr_1 r_1^{n_1+l}\exp(-(\zeta_a+\zeta_b)r_1)\right.\\
&\qquad\left.\times\int^{r_1} dr_2 r_2^{n_3-(l+1)}\exp(-(\zeta_c+\zeta_d)r_2)\right\rangle\left\langle\int_{-1}^1 dx_2 x_2^{n_4}\exp(-(\zeta_c+\zeta_d)Rx_2)\right\rangle\\
&\quad= Hh[R,n_1+l,\zeta_a+\zeta_b,n_3-(l+1),\zeta_c+\zeta_d](H[1,n_4,(\zeta_c+\zeta_d)R]-H[-1,n_4(\zeta_c+\zeta_d)R])
\end{aligned}\tag{3.130}$$

Hybrid-type integral 1–1 $(r_2 \le r_1 \le R)$

$$\begin{aligned}
&\int_0^R dr_1 r_1^{n_1-(l+1)}\exp(-(\zeta_a+\zeta_b)r_1)\int_0^{r_1} dr_2 r_2^{n_3+l}\exp(-(\zeta_c\pm\zeta_d)r_2)\\
&\quad\times\int_{-1}^1 dx_2 x_2^{n_4}\exp(-\zeta_d r_2 x_2)\\
&\quad=\left\langle\int^R dr_1 r_1^{n_1-(l+1)}\exp(-(\zeta_a+\zeta_b)r_1)\right.\\
&\qquad\left.\times\int^{r_1} dr_2 r_2^{n_3+l}\exp(-(\zeta_c\pm\zeta_d)r_2)\int_{-1}^1 dx_2 x_2^{n_4}\exp(-\zeta_d r_2 x_2)\right\rangle\\
&\qquad-\left\langle\int^R dr_1 r_1^{n_1-(l+1)}\exp(-(\zeta_a+\zeta_b)r_1)\right\rangle\left\langle\int^0 dr_2 r_2^{n_3+l}\exp(-(\zeta_c\pm\zeta_d)r_2)\right.\\
&\qquad\left.\times\int_{-1}^1 dx_2 x_2^{n_4}\exp(-\zeta_d r_2 x_2)\right\rangle\\
&\qquad-\left\langle\int^0 dr_1 r_1^{n_1-(l+1)}\exp(-(\zeta_a+\zeta_b)r_1)\right.\\
&\qquad\left.\times\int^{r_1} dr_2 r_2^{n_3+l}\exp(-(\zeta_c\pm\zeta_d)r_2)\int_{-1}^1 dx_2 x_2^{n_4}\exp(-\zeta_d r_2 x_2)\right\rangle\\
&\qquad+\left\langle\int^0 dr_1 r_1^{n_1-(l+1)}\exp(-(\zeta_a+\zeta_b)r_1)\right\rangle\left\langle\int^0 dr_2 r_2^{n_3+l}\exp(-(\zeta_c\pm\zeta_d)r_2)\right.\\
&\qquad\left.\times\int_{-1}^1 dx_2 x_2^{n_4}\exp(-\zeta_d r_2 x_2)\right\rangle
\end{aligned}\tag{3.131}$$

Hybrid-type integral 1–2 ($r_2 \le R \le r_1$)

$$\begin{aligned}
&\int_R^\infty dr_1 r_1^{n_1-(l+1)} \exp\left(-(\zeta_a+\zeta_b)r_1\right) \int_0^R dr_2 r_2^{n_3+l} \exp\left(-(\zeta_c \pm \zeta_d)r_2\right) \\
&\quad \times \int_{-1}^1 dx_2 x_2^{n_4} \exp\left(-\zeta_d r_2 x_2\right) \\
&\quad = -\left\langle \int^R dr_1 r_1^{n_1-(l+1)} \exp\left(-(\zeta_a+\zeta_b)r_1\right) \right\rangle \\
&\quad \left\langle \int^R dr_2 r_2^{n_3+l} \exp\left(-(\zeta_c \pm \zeta_d)r_2\right) \int_{-1}^1 dx_2 x_2^{n_4} \exp\left(-\zeta_d r_2 x_2\right) \right\rangle \\
&\quad + \left\langle \int^R dr_1 r_1^{n_1-(l+1)} \exp\left(-(\zeta_a+\zeta_b)r_1\right) \right\rangle \\
&\quad \left\langle \int^0 dr_2 r_2^{n_3+l} \exp\left(-(\zeta_c \pm \zeta_d)r_2\right) \int_{-1}^1 dx_2 x_2^{n_4} \exp\left(-\zeta_d r_2 x_2\right) \right\rangle
\end{aligned} \tag{3.132}$$

Hybrid-type integral 1–3 ($R \le r_2 \le r_1$)

$$\begin{aligned}
&\int_R^\infty dr_1 r_1^{n_1-(l+1)} \exp(-(\zeta_a+\zeta_b)r_1) \int_R^{r_1} dr_2 r_2^{n_3+l} \exp(-(\zeta_c + \zeta_d)r_2) \\
&\quad \times \int_{-1}^1 dx_2 x_2^{n_4} \exp(-\zeta_d R x_2) \\
&\quad = -\left\langle \int^R dr_1 r_1^{n_1-(l+1)} \exp(-(\zeta_a+\zeta_b)r_1) \right. \\
&\quad \left. \times \int^{r_1} dr_2 r_2^{n_3+l} \exp(-(\zeta_c + \zeta_d)r_2) \int_{-1}^1 dx_2 x_2^{n_4} \exp(-\zeta_d R x_2) \right\rangle \\
&\quad + \left\langle \int^R dr_1 r_1^{n_1-(l+1)} \exp(-(\zeta_a+\zeta_b)r_1) \right\rangle \\
&\quad \left\langle \int^R dr_2 r_2^{n_3+l} \exp(-(\zeta_c + \zeta_d)r_2) \int_{-1}^1 dx_2 x_2^{n_4} \exp(-\zeta_d R x_2) \right\rangle
\end{aligned} \tag{3.133}$$

Hybrid-type integral 2–1 ($r_1 \le r_2 \le R$)

$$\begin{aligned}
&\int_0^R dr_1 r_1^{n_1+1} \exp\left(-(\zeta_a+\zeta_b)r_1\right) \int_{r_1}^R dr_2 r_2^{n_3-(l+1)} \exp\left(-(\zeta_c \pm \zeta_d)r_2\right) \\
&\quad \times \int_{-1}^1 dx_2 x_2^{n_4} \exp\left(-\zeta_d r_2 x_2\right) \\
&\quad = \left\langle \int^R dr_1 r_1^{n_1+1} \exp\left(-(\zeta_a+\zeta_b)r_1\right) \right\rangle \\
&\quad \left\langle \int^R dr_2 r_2^{n_3-(l+1)} \exp\left(-(\zeta_c \pm \zeta_d)r_2\right) \int_{-1}^1 dx_2 x_2^{n_4} \exp\left(-\zeta_d r_2 x_2\right) \right\rangle \\
&\quad - \left\langle \int^R dr_1 r_1^{n_1+1} \exp\left(-(\zeta_a+\zeta_b)r_1\right) \int^{r_1} dr_2 r_2^{n_3-(l+1)} \exp\left(-(\zeta_c \pm \zeta_d)r_2\right) \right. \\
&\quad \left. \times \int_{-1}^1 dx_2 x_2^{n_4} \exp\left(-\zeta_d r_2 x_2\right) \right\rangle \\
&\quad - \left\langle \int^0 dr_1 r_1^{n_1+1} \exp\left(-(\zeta_a+\zeta_b)r_1\right) \right\rangle \\
&\quad \left\langle \int^R dr_2 r_2^{n_3-(l+1)} \exp\left(-(\zeta_c \pm \zeta_d)r_2\right) \int_{-1}^1 dx_2 x_2^{n_4} \exp\left(-\zeta_d r_2 x_2\right) \right\rangle \\
&\quad + \left\langle \int^0 dr_1 r_1^{n_1+1} \exp\left(-(\zeta_a+\zeta_b)r_1\right) \int^{r_1} dr_2 r_2^{n_3-(l+1)} \exp\left(-(\zeta_c \pm \zeta_d)r_2\right) \right. \\
&\quad \left. \times \int_{-1}^1 dx_2 x_2^{n_4} \exp\left(-\zeta_d r_2 x_2\right) \right\rangle
\end{aligned} \tag{3.134}$$

Hybrid-type integral 2–2 ($r_1 \le R \le r_2$)

$$\begin{aligned}
&\int_0^R dr_1 r_1^{n_1+1} \exp(-(\zeta_a+\zeta_b)r_1) \int_R^\infty dr_2 r_2^{n_3-(l+1)} \exp(-(\zeta_c+\zeta_d)r_2) \\
&\quad \times \int_{-1}^1 dx_2 x_2^{n_4} \exp(-\zeta_d R x_2) \\
&\quad = -\left\langle \int^R dr_1 r_1^{n_1+1} \exp(-(\zeta_a+\zeta_b)r_1) \right\rangle \\
&\quad \left\langle \int^R dr_2 r_2^{n_3-(l+1)} \exp(-(\zeta_c+\zeta_d)r_2) \right\rangle \left\langle \int_{-1}^1 dx_2 x_2^{n_4} \exp(-\zeta_d R x_2) \right\rangle \\
&\quad + \left\langle \int^0 dr_1 r_1^{n_1+1} \exp(-(\zeta_a+\zeta_b)r_1) \right\rangle \left\langle \int^R dr_2 r_2^{n_3-(l+1)} \exp(-(\zeta_c+\zeta_d)r_2) \right\rangle \\
&\quad \left\langle \int_{-1}^1 dx_2 x_2^{n_4} \exp(-\zeta_d R x_2) \right\rangle
\end{aligned} \tag{3.135}$$

Hybrid-type integral 2–3 ($R \le r_1 \le r_2$)

$$\begin{aligned}
&\int_R^\infty dr_1 r_1^{n_1+1} \exp(-(\zeta_a+\zeta_b)r_1) \int_{r_1}^\infty dr_2 r_2^{n_3-(l+1)} \exp(-(\zeta_c+\zeta_d)r_2) \\
&\quad \times \int_{-1}^1 dx_2 x_2^{n_4} \exp(-\zeta_d R x_2) \\
&\quad = \left\langle \int^R dr_1 r_1^{n_1+1} \exp(-(\zeta_a+\zeta_b)r_1) \right. \\
&\quad \left. \times \int^{r_1} dr_2 r_2^{n_3-(l+1)} \exp(-(\zeta_c+\zeta_d)r_2) \right\rangle \left\langle \int_{-1}^1 dx_2 x_2^{n_4} \exp(-\zeta_d R x_2) \right\rangle
\end{aligned} \tag{3.136}$$

Exchange-type integral 1–1 ($r_2 \le r_1 \le R$)

$$\begin{aligned}
&\int_0^R dr_1 r_1^{n_1-(l+1)} \exp(-(\zeta_a \pm \zeta_b)r_1) \int_{-1}^1 dx_1 x_1^{n_2} \exp(-\zeta_b r_1 x_1) \\
&\quad \times \int_0^{r_1} dr_2 r_2^{n_3+l} \exp(-(\zeta_c \pm \zeta_d)r_2) \int_{-1}^1 dx_2 x_2^{n_4} \exp(-\zeta_d r_2 x_2) \\
&\quad = \left\langle \int^R dr_1 r_1^{n_1-(l+1)} \exp(-(\zeta_a \pm \zeta_b)r_1) \int_{-1}^1 dx_1 x_1^{n_2} \exp(-\zeta_b r_1 x_1) \right. \\
&\quad \left. \times \int^{r_1} dr_2 r_2^{n_3+l} \exp(-(\zeta_c \pm \zeta_d)r_2) \int_{-1}^1 dx_2 x_2^{n_4} \exp(-\zeta_d r_2 x_2) \right\rangle \\
&\quad - \left\langle \int^R dr_1 r_1^{n_1-(l+1)} \exp(-(\zeta_a \pm \zeta_b)r_1) \int_{-1}^1 dx_1 x_1^{n_2} \exp(-\zeta_b r_1 x_1) \right\rangle \\
&\quad \times \left\langle \int^0 dr_2 r_2^{n_3+l} \exp(-(\zeta_c \pm \zeta_d)r_2) \int_{-1}^1 dx_2 x_2^{n_4} \exp(-\zeta_d r_2 x_2) \right\rangle \\
&\quad - \left\langle \int^0 dr_1 r_1^{n_1-(l+1)} \exp(-(\zeta_a \pm \zeta_b)r_1) \int_{-1}^1 dx_1 x_1^{n_2} \exp(-\zeta_b r_1 x_1) \right. \\
&\quad \left. \times \int^{r_1} dr_2 r_2^{n_3+l} \exp(-(\zeta_c \pm \zeta_d)r_2) \int_{-1}^1 dx_2 x_2^{n_4} \exp(-\zeta_d r_2 x_2) \right\rangle \\
&\quad + \left\langle \int_0^0 dr_1 r_1^{n_1-(l+1)} \exp(-(\zeta_a \pm \zeta_b)r_1) \int_{-1}^1 dx_1 x_1^{n_2} \exp(-\zeta_b r_1 x_1) \right\rangle \\
&\quad \times \left\langle \int^0 dr_2 r_2^{n_3+l} \exp(-(\zeta_c \pm \zeta_d)r_2) \int_{-1}^1 dx_2 x_2^{n_4} \exp(-\zeta_d r_2 x_2) \right\rangle
\end{aligned} \tag{3.137}$$

Exchange-type integral 1–2 ($r_2 \leq R \leq r_1$)

$$\begin{aligned}
&\int_R^\infty dr_1 r_1^{n_1-(l+1)} \exp(-(\zeta_a+\zeta_b)r_1) \int_{-1}^1 dx_1 x_1^{n_2} \exp(-\zeta_b r_1 x_1) \\
&\quad \times \int_0^R dr_2 r_2^{n_3+l} \exp(-(\zeta_c \pm \zeta_d)r_2) \int_{-1}^1 dx_2 x_2^{n_4} \exp(-\zeta_d r_2 x_2) \\
&\quad = -\left\langle \int^R dr_1 r_1^{n_1-(l+1)} \exp(-(\zeta_a+\zeta_b)r_1) \int_{-1}^1 dx_1 x_1^{n_2} \exp(-\zeta_b r_1 x_1) \right\rangle \\
&\quad \left\langle \int^R dr_2 r_2^{n_3+l} \exp(-(\zeta_c \pm \zeta_d)r_2) \int_{-1}^1 dx_2 x_2^{n_4} \exp(-\zeta_d r_2 x_2) \right\rangle \\
&\quad + \left\langle \int^R dr_1 r_1^{n_1-(l+1)} \exp(-(\zeta_a+\zeta_b)r_1) \int_{-1}^1 dx_1 x_1^{n_2} \exp(-\zeta_b r_1 x_1) \right\rangle \\
&\quad \left\langle \int^0 dr_2 r_2^{n_3+l} \exp(-(\zeta_c \pm \zeta_d)r_2) \int_{-1}^1 dx_2 x_2^{n_4} \exp(-\zeta_d r_2 x_2) \right\rangle
\end{aligned} \tag{3.138}$$

Exchange-type integral 1–3 ($R \leq r_2 \leq r_1$)

$$\begin{aligned}
&\int_R^\infty dr_1 r_1^{n_1-(l+1)} \exp(-(\zeta_a+\zeta_b)r_1) \int_{-1}^1 dx_1 x_1^{n_2} \exp(-\zeta_b R x_1) \\
&\quad \times \int_R^{r_1} dr_2 r_2^{n_3+l} \exp(-(\zeta_c + \zeta_d)r_2) \int_{-1}^1 dx_2 x_2^{n_4} \exp(-\zeta_d R x_2) \\
&\quad = -\left\langle \int^R dr_1 r_1^{n_1-(l+1)} \exp(-(\zeta_a+\zeta_b)r_1) \int^{r_1} dr_2 r_2^{n_3+l} \exp(-(\zeta_c+\zeta_d)r_2) \right\rangle \\
&\quad \left\langle \int_{-1}^1 dx_1 x_1^{n_2} \exp(-\zeta_b R x_1) \right\rangle \left\langle \int_{-1}^1 dx_2 x_2^{n_4} \exp(-\zeta_d R x_2) \right\rangle \\
&\quad + \left\langle \int^R dr_1 r_1^{n_1-(l+1)} \exp(-(\zeta_a+\zeta_b)r_1) \right\rangle \left\langle \int_{-1}^1 dx_1 x_1^{n_2} \exp(-\zeta_b R x_1) \right\rangle \\
&\quad \left\langle \int^R dr_2 r_2^{n_3+l} \exp(-(\zeta_c \pm \zeta_d)r_2) \right\rangle \left\langle \int_{-1}^1 dx_2 x_2^{n_4} \exp(-\zeta_d R x_2) \right\rangle
\end{aligned} \tag{3.139}$$

Exchange-type integral 2–1 $(r_1 \leq r_2 \leq R)$

$$\begin{aligned}
&\int_0^R dr_1 r_1^{n_1-(l+1)} \exp(-(\zeta_a \pm \zeta_b) r_1) \int_{-1}^1 dx_1 x_1^{n_2} \exp(-\zeta_b r_1 x_1) \\
&\quad \times \int_{r_1}^R dr_2 r_2^{n_3-(l+1)} \exp(-(\zeta_c \pm \zeta_d) r_2) \int_{-1}^1 dx_2 x_2^{n_4} \exp(-\zeta_d r_2 x_2) \\
&\quad = \left\langle \int^R dr_1 r_1^{n_1-(l+1)} \exp(-(\zeta_a \pm \zeta_b) r_1) \int_{-1}^1 dx_1 x_1^{n_2} \exp(-\zeta_b r_1 x_1) \right\rangle \\
&\quad \left\langle \int^R dr_2 r_2^{n_3-(l+1)} \exp(-(\zeta_c \pm \zeta_d) r_2) \int_{-1}^1 dx_2 x_2^{n_4} \exp(-\zeta_d r_2 x_2) \right\rangle \\
&\quad - \left\langle \int^R dr_1 r_1^{n_1-(l+1)} \exp(-(\zeta_a \pm \zeta_b) r_1) \int_{-1}^1 dx_1 x_1^{n_2} \exp(-\zeta_b r_1 x_1) \right. \\
&\quad \left. \times \int^{r_1} dr_2 r_2^{n_3-(l+1)} \exp(-(\zeta_c \pm \zeta_d) r_2) \int_{-1}^1 dx_2 x_2^{n_4} \exp(-\zeta_d r_2 x_2) \right\rangle \\
&\quad - \left\langle \int^0 dr_1 r_1^{n_1-(l+1)} \exp(-(\zeta_a \pm \zeta_b) r_1) \int_{-1}^1 dx_1 x_1^{n_2} \exp(-\zeta_b r_1 x_1) \right\rangle \\
&\quad \left\langle \int^R dr_2 r_2^{n_3-(l+1)} \exp(-(\zeta_c \pm \zeta_d) r_2) \int_{-1}^1 dx_2 x_2^{n_4} \exp(-\zeta_d r_2 x_2) \right\rangle \\
&\quad + \left\langle \int^0 dr_1 r_1^{n_1-(l+1)} \exp(-(\zeta_a \pm \zeta_b) r_1) \int_{-1}^1 dx_1 x_1^{n_2} \exp(-\zeta_b r_1 x_1) \right. \\
&\quad \left. \times \int^{r_1} dr_2 r_2^{n_3-(l+1)} \exp(-(\zeta_c \pm \zeta_d) r_2) \int_{-1}^1 dx_2 x_2^{n_4} \exp(-\zeta_d r_2 x_2) \right\rangle
\end{aligned} \tag{3.140}$$

Exchange-type integral 2–2 $(r_1 \leq R \leq r_2)$

$$\begin{aligned}
&\int_0^R dr_1 r_1^{n_1+l} \exp(-(\zeta_a \pm \zeta_b) r_1) \int_{-1}^1 dx_2 x_2^{n_4} \exp(-\zeta_b r_1 x_1) \\
&\quad \times \int_R^\infty dr_2 r_2^{n_3-(l+1)} \exp(-(\zeta_c + \zeta_d) r_2) \int_{-1}^1 dx_2 x_2^{n_4} \exp(-\zeta_d R x_2) \\
&\quad = - \left\langle \int^R dr_1 r_1^{n_1+l} \exp(-(\zeta_a \pm \zeta_b) r_1) \int_{-1}^1 dx_2 x_2^{n_4} \exp(-\zeta_b r_1 x_1) \right\rangle \\
&\quad \left\langle \int^R dr_2 r_2^{n_3-(l+1)} \exp(-(\zeta_c \pm \zeta_d) r_2) \right\rangle \left\langle \int_{-1}^1 dx_2 x_2^{n_4} \exp(-\zeta_d R x_2) \right\rangle \\
&\quad + \left\langle \int^0 dr_1 r_1^{n_1+l} \exp(-(\zeta_a \pm \zeta_b) r_1) \int_{-1}^1 dx_2 x_2^{n_4} \exp(-\zeta_b r_1 x_1) \right\rangle \\
&\quad \left\langle \int^R dr_2 r_2^{n_3-(l+1)} \exp(-(\zeta_c \pm \zeta_d) r_2) \right\rangle \left\langle \int_{-1}^1 dx_2 x_2^{n_4} \exp(-\zeta_d R x_2) \right\rangle
\end{aligned} \tag{3.141}$$

Exchange-type integral 2–3 ($R \le r_1 \le r_2$)

$$\begin{aligned}
&\int_R^\infty dr_1 r_1^{n_1+l}\exp(-(\zeta_a+\zeta_b)r_1)\int_{-1}^1 dx_2 x_2^{n_4}\exp(-\zeta_b R x_1)\\
&\quad\times\int_{r_1}^\infty dr_2 r_2^{n_3-(l+1)}\exp(-(\zeta_c+\zeta_d)r_2)\int_{-1}^1 dx_2 x_2^{n_4}\exp(-\zeta_d R x_2)\\
&\quad=\left\langle\int^R dr_1 r_1^{n_1+l}\exp(-(\zeta_a+\zeta_b)r_1)\int^{r_1} dr_2 r_2^{n_3-(l+1)}\exp(-(\zeta_c+\zeta_d)r_2)\right\rangle\\
&\quad\left\langle\int_{-1}^1 dx_2 x_2^{n_4}\exp(-\zeta_b R x_1)\right\rangle\left\langle\int_{-1}^1 dx_2 x_2^{n_4}\exp(-\zeta_d R x_2)\right\rangle
\end{aligned}\tag{3.142}$$

3.3 Summary of Formulation

We derived all types of one-center and two-center molecular integrals as a function of atomic distance and orbital exponents of STFs.

One-electron two-center molecular integrals, overlap integral S_{AB}^{ab} in Eq. (3.59), kinetic energy integral T_{AB}^{ab} in Eq. (3.62), and nuclear electron attraction integrals $V_{AB/A}^{ab}$ in Eq. (3.64), and $V_{AB/C}^{ab}$ in Eq. (3.72) are deduced using auxiliary functions, *Hb* in Eq. (3.D1) and *B* in Eq. (3.C1). This means that these molecular integrals belong to the same family of auxiliary functions. This is the same for dipole moment molecular integrals and other one electron integrals. We can calculate these auxiliary functions symbolically using formulas given in Appendices C and D. Detailed formulation was demonstrated for overlap integral and nuclear electron attraction integrals.

Two-electron two-center molecular integrals, Coulomb-type integral in Eq. (3.93), hybrid-type integral in Eq. (3.95) and exchange-type integral in Eq. (3.96) are deduced as integrations which are expressed by five kinds of half definite integral, *H* in Eq. (3.B2), *Hh* in Eq. (3.B14), *Hb* in Eq. (3.D1), *Hhb* in Eq. (3.D12), and *Hbhb* in Eq. (3.D13), defined in six regions, all of which can be calculated symbolically using formulas given in Appendices B and D. Detailed formulation was demonstrated for Coulomb-type repulsion integral.

As a result, all one-center and two-center molecular integrals derived in this article are expressed by the sum of factorized definite integrals. It will be easy to obtain polynomial expression of molecular integrals using the formulas in this article.

3.4 Discussion

Our aim to formulate analytically all types of molecular integrals over STFs for one-center and two-center integrals is attained. The results lead the polynomial molecular integrals by which we are to construct the algebraic MO equation and the seamless effective hamiltonian over a wide range of physical dimension. We consider the multivariable algebraic MO theory and the seamless effective hamiltonian will bring a new theoretical framework of quantum chemistry. This is our answer to Dirac's insight into chemistry as a fundamental problem after the discovery of quantum mechanics.

Advantage of analytical molecular integrals itself can be seen in many MO theoretical calculations such as in spectroscopy concerning electron transition probability which depends on the transition moment. By the similar analytical formulation demonstrated for overlap integrals, we can obtain an analytical molecular integral for transition moment as a function atomic distance. Using optimized STFs obtained by Mukoyama's method in which numerical basis function calculated by the DV-Xα method are fitted by STFs (Mukoyama and Yasui 1992), we can calculate two-center transition moment analytically as a function of atomic distance. In slow atomic collisions, for example, we can calculate two-center overlap integral between any STFs as a function of atomic distance. It is interesting where wave functions start effective contact between separated atoms, i.e., where effective overlap comes to bring energy separation between MOs. Long tail behavior of STF works effectively not only in the MO calculation for large molecular system but also in the study of atomic collision. For a nonadiabatic electron transition in slow atomic collisions, it is interesting where a nonadiabatic electron transition occurs. We can use analytical molecular integrals for radial coupling and rotational coupling as a function of atomic distance in the study of nonadiabatic chemical reaction. Analytical expression of molecular integrals over STFs provides the relation between molecular property and atomic configuration.

In Region 1–1 ($r_2 \le r_1 \le R$) and Region 2–1 ($r_1 \le r_2 \le R$), electrons move inside the sphere of radius R. In Region 1–2 ($r_2 \le R \le r_1$) and Region 2–2 ($r_1 \le R \le r_2$), one electron moves inside the sphere and the other electron moves outside the sphere. In Region 1–3 ($R \le r_2 \le r_1$) and Region 2–3 ($R \le r_1 \le r_2$), electrons move outside the sphere. This is the order of electron density from high to low.

It is most fundamental in molecular science to study the relation between molecular property and atomic configuration. From this point of view, we shall discuss the advantage of analytical expression of one-center and two-center molecular integrals over STFs derived in this article. Analytical expression of molecular integral is useful not only for numerical calculations of the Fock matrix elements but also for construction of algebraic MO equation through transformation from analytical expression to polynomial expression of molecular integrals by using the Taylor expansion method,

$$f(x) = \sum_{p=0}^{\infty} \frac{1}{p!} \sum_{k=0}^{p} \binom{p}{k} (-x_0)^{p-k} \left[\left(\frac{d}{dx} \right)^p f(x) \right]_{x=x_0} x^k \cong \sum_{i=0}^{N} A_i(x_0) x^i$$

where x_0 means the center of the Taylor expansion distributed so as to keep the polynomials in the same order in any region of variables. The last term is the approximate expression of single variable functions. This method can be easily extended to multivariable functions. Anyway, to derive analytical expression of molecular integrals is the first step for construction of multivariable algebraic MO equation as a simultaneous polynomial equation with respect to atomic distance and orbital exponent. We discussed about the advantage of polynomial molecular integrals in algebraic MO theory from several points of view in the previous chapter.

Acknowledgments I would like to express my heartfelt thanks to the following people. The late Professor A. Saika gave me the theme of molecular integrals over STFs to study the electron correlation effect in NMR chemical shifts by MBPT in my graduate school. He perceived the important nature of STF. For a long time from my graduate school, I have being inspired by Professor Shigeru Huzinaga. His book sparked my interest in molecular orbital theory. I got many ideas in reading his book. Professor Takeshi Mukoyama showed us the versatile advantage of STF in his study of electron excitation process and atomic collision. He has been supported me for a long time. Professor Tsutomu Watanabe has discussed this work enthusiastically with me for a long time. I have derived much from his discussion about slow atomic collision in which electron charge transfer and nonadiabatic transition occur. Professor Hirohiko Adachi introduced me the DV-Xα calculation method. I noticed the importance of nonlinear optimization of basis function from the DV-Xα calculation. Professor Tai-Ichi Shibuya introduced me Shibuya-Wolfman integral defined in momentum space. He has been encouraging me for a long time.

Appendices

Appendix A: Real Spherical Harmonic Function

This appendix contains many relations with respect to normalized real spherical harmonics used in this article. Some duplicates in the body are shown to help following formulations.

$$Y_l^m(\theta_A, \phi_A) = p_l^m(\cos\theta_A) f_m(\phi) \tag{3.A1}$$

Normalized Real Function with Respect to ϕ

$$f_m(\phi) = N_\phi(m)\left\{\begin{array}{c}\cos m\phi\\ \sin|m|\phi\end{array}\right\}\begin{array}{l}for\ m \geq 0\\ for\ m < 0\end{array} \tag{3.A2}$$

$$N_\phi(m) = (-1)^m[\pi(1+\delta_{m,0})]^{-\frac{1}{2}} \tag{3.A3}$$

$$\int_0^{2\pi} d\phi f_{m_a}(\phi) f_{m_b}(\phi) = \delta_{m_a,m_b} \tag{3.A4}$$

$$\int_0^{2\pi} d\phi f_m(\phi) = \sqrt{2\pi}\delta_{m,0} \tag{3.A5}$$

$$\Phi(m_a, m_b, m_c) \equiv \int_0^{2\pi} d\phi f_{m_a}(\phi) f_{m_b}(\phi) f_{m_c}(\phi) \tag{3.A6}$$

$$I(m_a, m_b, m_c) = \int_0^{2\pi} d\phi \begin{Bmatrix} \cos m_a\phi \\ \sin |m_a|\phi \end{Bmatrix} \begin{Bmatrix} \cos m_b\phi \\ \sin |m_b|\phi \end{Bmatrix} \begin{Bmatrix} \cos m_c\phi \\ \sin |m_c|\phi \end{Bmatrix} \tag{3.A7}$$

$$\Phi(m_a, m_b, m_c) = N_\phi(m_a) N_\phi(m_b) N_\phi(m_c) I(m_a, m_b, m_c) \tag{3.A8}$$

Normalized Associated Legendre Function

$$p_l^m(x) = N_x(l, m) P_l^m(x) \tag{3.A9}$$

$$N_x(l, m) = \left[\frac{2l+1}{2} \frac{(l-|m|)!}{(l+|m|)!} \right]^{1/2} \tag{3.A10}$$

$$P_l^m(x) = \left(1 - x^2\right)^{\frac{|m|}{2}} \sum_{\nu=0}^{\left[\frac{l-|m|}{2}\right]} \omega_\nu^{l,m} x^{l-|m|-2\nu} \tag{3.A11}$$

$$\omega_\nu^{l,m} = \frac{(-1)^\nu (2l - 2\nu - 1)!}{\nu! 2^\nu (l - |m| - 2\nu)!} \tag{3.A12}$$

$$\int_{-1}^{1} dx p_{l_a}^{m_a}(x) p_{l_b}^{m_b}(x) = \delta_{m_a, m_b} \delta_{l_a, l_b} \tag{3.A13}$$

$$\int_{-1}^{1} dx p_l^0(x) = \delta_{l, even} 2N_x(l, 0) \sum_{\nu=0}^{\left[\frac{l}{2}\right]} \frac{\omega_\nu^{l,0}}{l - 2\nu + 1} \tag{3.A14}$$

$$\Theta\begin{pmatrix} m_a & m_b & m_c \\ l_a & l_b & l_c \end{pmatrix} \equiv \int_{-1}^{1} p_{l_a}^{m_a}(x) p_{l_b}^{m_b}(x) p_{l_c}^{m_c}(x) dx \tag{3.A15}$$

$$|m_a| + |m_b| + |m_c| = 2m \tag{3.A16}$$

$$\int_{-1}^{1} x^{l_a + l_b + l_c - 2m - 2(\nu_a + \nu_b + \nu_c) + 2i} dx = \frac{2}{l_a + l_b + l_c - 2m - 2(\nu_a + \nu_b + \nu_c) + 2i + 1} \quad for \, \mathrm{mod}(l_a + l_b + l_c, 2) = 0 \tag{3.A17}$$

$$\int_{-1}^{1} x^{l_a + l_b + l_c - 2m - 2(\nu_a + \nu_b + \nu_c) + 2i} dx = 0 \quad for \, \mathrm{mod}(l_a + l_b + l_c, 2) \neq 0 \tag{3.A18}$$

$$\Theta\begin{pmatrix} m_a & m_b & m_c \\ l_a & l_b & l_c \end{pmatrix} = N_x(l_a, m_a) N_x(l_b, m_b) N_x(l_c, m_c) \times \sum_{\nu_a=0}^{\left[\frac{l_a-|m_a|}{2}\right]} \omega_{\nu_a}^{l_a, m_a} \sum_{\nu_b=0}^{\left[\frac{l_b-|m_b|}{2}\right]} \omega_{\nu_b}^{l_b, m_b} \sum_{\nu_c=0}^{\left[\frac{l_c-|m_c|}{2}\right]} \omega_{\nu_c}^{l_c, m_c} \sum_i^m \begin{pmatrix} m \\ i \end{pmatrix} (-1)^i \times \int_{-1}^{1} x^{l_a + l_b + l_c - 2m - 2(\nu_a + \nu_b + \nu_c) + 2i} dx \tag{3.A19}$$

Normalized Real Spherical Harmonic Function

$$\int d\Omega Y_l^m(\theta,\phi) = \sqrt{2\pi}\delta_{m,0}\int_{-1}^{1} dx p_l^0(x) \tag{3.A20}$$

$$\int d\Omega Y_{l_a}^{m_a}(\theta,\phi) Y_{l_a}^{m_a}(\theta,\phi) = \delta_{l_a,l_b}\delta_{m_a,m_b} \tag{3.A21}$$

$$Y_{l_a}^{m_a}(\theta,\phi)Y_{l_b}^{m_b}(\theta,\phi) = \sum_{m_{ab}}\Phi(m_a,m_b,m_{ab})\sum_{l_{ab}}\Theta\begin{pmatrix} m_a & m_b & m_{ab} \\ l_a & l_b & l_{ab}\end{pmatrix} Y_{l_{ab}}^{m_{ab}}(\theta,\phi) \tag{3.A22}$$

where

$$\max(|m|,|l_a - l_b|) + n \le l_{ab} \le l_a + l_b \tag{3.A23}$$

$$n = \begin{cases} 0, & (m \equiv |l_a - l_b| (\mathrm{mod} 2)) \\ 1, & (otherwise)\end{cases} \tag{3.A24}$$

$$\begin{aligned} & G\begin{pmatrix} m_a & m_b & m_{ab} \\ l_a & l_b & l_{ab}\end{pmatrix} \\ & \equiv \int d\Omega Y_{l_a}^{m_a}(\theta,\phi)Y_{l_b}^{m_b}(\theta,\phi)Y_{l_c}^{m_c}(\theta,\phi) \\ & = \sum_{m_{ab}}\Phi(m_a,m_b,m_{ab})\sum_{l_{ab}}\Theta\begin{pmatrix} m_a & m_b & m_{ab} \\ l_a & l_b & l_{ab}\end{pmatrix}\int d\Omega Y_{l_{ab}}^{m_{ab}}(\theta,\phi)Y_{l_c}^{m_c}(\theta,\phi) \\ & = \sum_{m_{ab}}\Phi(m_a,m_b,m_{ab})\sum_{l_{ab}}\Theta\begin{pmatrix} m_a & m_b & m_{ab} \\ l_a & l_b & l_{ab}\end{pmatrix}\delta_{m_{ab},m_c}\delta_{l_{ab},l_c} \end{aligned} \tag{3.A25}$$

Triple Product of Normalized Real Spherical Harmonic Function (1)

$$\begin{aligned} & Y_{l_a}^{m_a}(\theta,\phi)Y_{l_b}^{m_b}(\theta,\phi)Y_{l_c}^{m_c}(\theta,\phi) \\ & \quad = \sum_{m_{ab}}\Phi(m_a,m_b,m_{ab})\sum_{l_{ab}}\Theta\begin{pmatrix} m_a & m_b & m_{ab} \\ l_a & l_b & l_{ab}\end{pmatrix} \\ & \quad \sum_{m_{abc}}\Phi(m_a,m_b,m_{abc})\sum_{l_{ab}}\Theta\begin{pmatrix} m_a & m_b & m_{abc} \\ l_a & l_b & l_{abc}\end{pmatrix} Y_{l_{abc}}^{m_{abc}}(\theta,\phi) \end{aligned} \tag{3.A26}$$

$$\begin{aligned} & \int d\Omega Y_{l_a}^{m_a}(\theta,\phi)Y_{l_b}^{m_b}(\theta,\phi)Y_{l_c}^{m_c}(\theta,\phi) \\ & \quad = \sum_{m_{ab}}\Phi(m_a,m_b,m_{ab})\sum_{l_{ab}}\Theta\begin{pmatrix} m_a & m_b & m_{ab} \\ l_a & l_b & l_{ab}\end{pmatrix} \\ & \quad \sum_{m_{abc}}\Phi(m_a,m_b,m_{abc})\sum_{l_{ab}}\Theta\begin{pmatrix} m_a & m_b & m_{abc} \\ l_a & l_b & l_{abc}\end{pmatrix} \\ & \quad \int d\Omega Y_{l_{abc}}^{m_{abc}}(\theta,\phi) \end{aligned} \tag{3.A27}$$

$$= \sum_{m_{ab}} \Phi(m_a, m_b, m_{ab}) \sum_{l_{ab}} \Theta \begin{pmatrix} m_a & m_b & m_{ab} \\ l_a & l_b & l_{ab} \end{pmatrix}$$
$$\times \sum_{m_{abc}} \Phi(m_a, m_b, m_{abc}) \sum_{l_{ab}} \Theta \begin{pmatrix} m_a & m_b & m_{abc} \\ l_a & l_b & l_{abc} \end{pmatrix} \tag{3.A28}$$
$$\times \delta_{l_{abc},even} 2N_x(l_{abc}, 0) \sum_{\nu=0}^{\left[\frac{l_{abc}}{2}\right]} \frac{\omega_\nu^{l_{abc},0}}{l_{abc} - 2\nu + 1}$$

Transfer of Origin of Spherical Harmonics from *B* to *A*

$$Y_l^m(\theta_B, \phi) = \sum_{k=|m|}^{l} T_k^{lm} R^{l-k} r_A^k r_B^{-l} Y_k^m(\theta_A, \phi) \tag{3.A29}$$

$$T_k^{lm} = (-1)^{k+m} \frac{N_x(l,m)}{N_x(k,m)} \frac{(l+|m|)!}{(l-k)!(k+|m|)!} \tag{3.A30}$$

Triple Product of Normalized Real Spherical Harmonic Function (2)

$$Y_{l_a}^{m_a}(\theta_A, \phi) Y_{l_b}^{m_b}(\theta_B, \phi)$$
$$= \sum_{k_b=|m_b|}^{l_b} T_{k_b}^{l_b m_b} \sum_{m_{ab}} \Phi(m_a, m_b, m_{ab}) \sum_{l_{ab}} \Theta \begin{pmatrix} m_a & m_b & m_{ab} \\ l_a & k_b & l_{ab} \end{pmatrix} \tag{3.A31}$$
$$\times R^{l_b - k_b} r_A^{k_b} r_B^{-l_b} Y_{l_{ab}}^{m_{ab}}(\theta_A, \phi)$$

$$Y_{l_a}^{m_a}(\theta_A, \phi) Y_{l_b}^{m_b}(\theta_B, \phi) Y_{l_c}^{m_c}(\theta_A, \phi)$$
$$= \sum_{k_b=|m_b|}^{l_b} T_{k_b}^{l_b m_b} R^{l_b - k_b} r_A^{k_b} r_B^{-l_b} Y_{l_a}^{m_a}(\theta_A, \phi) Y_{k_b}^{m_b}(\theta_A, \phi) Y_{l_c}^{m_c}(\theta_A, \phi)$$
$$= \sum_{k_b=|m_b|}^{l_b} T_{k_b}^{l_b m_b} R^{l_b - k_b} r_A^{k_b} r_B^{-l_b} \sum_{m_{ab}} \Phi(m_a, m_b, m_{ab}) \sum_{l_{ab}} \Theta \begin{pmatrix} m_a & m_b & m_{ab} \\ l_a & k_b & l_{ab} \end{pmatrix}$$
$$\times \sum_{m_{abc}} \Phi(m_{ab} \quad m_c \quad m_{abc}) \sum_{l_{abc}} \Theta \begin{pmatrix} m_{ab} & m_c & m_{abc} \\ l_{ab} & l_c & l_{abc} \end{pmatrix} Y_{l_{abc}}^{m_{abc}}(\theta_A, \phi) \tag{3.A32}$$

$$\begin{aligned}
&\int_0^{2\pi} d\phi Y_{l_a}^{m_a}(\theta_A,\phi)Y_{l_b}^{m_b}(\theta_B,\phi)Y_{l_c}^{m_c}(\theta_A,\phi)\\
&= \sum_{k_b=|m_b|}^{l_b} T_{k_b}^{l_b m_b} R^{l_b-k_b} r_A^{k_b} r_B^{-l_b} \int_0^{2\pi} d\phi Y_{l_a}^{m_a}(\theta_A,\phi)Y_{k_b}^{m_b}(\theta_A,\phi)Y_{l_c}^{m_c}(\theta_A,\phi)\\
&= 2\pi N_\phi(0) \sum_{k_b=|m_b|}^{l_b} T_{k_b}^{l_b m_b} R^{l_b-k_b} r_A^{k_b} r_B^{-l_b}\\
&\times \sum_{m_{ab}} \Phi(m_a, m_b, m_{ab}) \sum_{l_{ab}} \Theta\begin{pmatrix} m_a & m_b & m_{ab} \\ l_a & k_b & l_{ab} \end{pmatrix}\\
&\times \sum_{m_{abc}} \Phi(\, m_{ab} \quad m_c \quad m_{abc}\,) \sum_{l_{abc}} \Theta\begin{pmatrix} m_{ab} & m_c & m_{abc} \\ l_{ab} & l_c & l_{abc} \end{pmatrix}\\
&\times p_{l_{abc}}^0(\,\cos\theta_A)
\end{aligned} \tag{3.A33}$$

Triple Product of Normalized Real Spherical Harmonic Function (3)

$$\begin{aligned}
&Y_{l_a}^{m_a}(\theta_B,\phi)Y_{l_b}^{m_b}(\theta_B,\phi)\\
&= \sum_{m_{ad}} \Phi(m_a, m_b, m_{ab}) \sum_{k_{cd}} \Theta\begin{pmatrix} m_a & m_b & m_{ab} \\ l_a & l_b & l_{ab} \end{pmatrix} Y_{l_{ab}}^{m_{ab}}(\theta_B,\phi)\\
&= \sum_{m_{ad}} \Phi(m_a, m_b, m_{ab}) \sum_{k_{cd}} \Theta\begin{pmatrix} m_a & m_b & m_{ab} \\ l_a & l_b & l_{ab} \end{pmatrix}\\
&\times \sum_{k_{ab}=|m_{ab}|}^{l_{ab}} T_{k_c}^{l_{ab} m_{ab}} R^{l_{ab}-k_{ab}} r_A^{k_{ab}} r_B^{-l_{ab}} Y_{k_{ab}}^{m_{ab}}(\theta_A,\phi)
\end{aligned} \tag{3.A34}$$

$$\begin{aligned}
&Y_{l_a}^{m_a}(\theta_B,\phi)Y_{l_b}^{m_b}(\theta_B,\phi)Y_{l_c}^{m_c}(\theta_A,\phi)\\
&= \sum_{m_{ab}} \Phi(m_a, m_b, m_{ab}) \sum_{l_{ab}} \Theta\begin{pmatrix} m_a & m_b & m_{ab} \\ l_a & l_b & l_{ab} \end{pmatrix} Y_{l_{ab}}^{m_{ab}}(\theta_B,\phi)Y_{l_c}^{m_c}(\theta_A,\phi)\\
&= \sum_{m_{ab}} \Phi(m_a, m_b, m_{ab}) \sum_{l_{ab}} \Theta\begin{pmatrix} m_a & m_b & m_{ab} \\ l_a & l_b & l_{ab} \end{pmatrix}\\
&\times \sum_{k_{ab}=|m_{ab}|}^{l_{ab}} T_{k_c}^{l_{ab} m_{ab}} R^{l_{ab}-k_{ab}} r_A^{k_{ab}} r_B^{-l_{ab}} Y_{k_{ab}}^{m_{ab}}(\theta_A,\phi)Y_{l_c}^{m_c}(\theta_A,\phi)
\end{aligned} \tag{3.A35}$$

$$
\begin{aligned}
&= \sum_{m_{ab}} \Phi(m_a, m_b, m_{ab}) \sum_{l_{ab}} \Theta\begin{pmatrix} m_a & m_b & m_{ab} \\ l_a & l_b & l_{ab} \end{pmatrix} \\
&\times \sum_{k_{ab}=|m_{ab}|}^{l_{ab}} T_{k_c}^{l_{ab} m_{ab}} R^{l_{ab}-k_{ab}} r_A^{k_{ab}} r_B^{-l_{ab}} \\
&\times \sum_{m_{abc}} \Phi(m_{ab}, m_c, m_{abc}) \sum_{l_{abc}} \Theta\begin{pmatrix} m_{ab} & m_c & m_{abc} \\ k_{ab} & l_c & l_{abc} \end{pmatrix} \\
&\times Y_{l_{abc}}^{m_{abc}}(\theta_A, \phi)
\end{aligned}
\tag{3.A36}
$$

$$
\begin{aligned}
&\int d\Omega Y_{l_a}^{m_a}(\theta_B, \phi) Y_{l_b}^{m_b}(\theta_B, \phi) Y_{l_c}^{m_c}(\theta_A, \phi) \\
&= 2\pi N_\phi(0) \sum_{m_{ad}} \Phi(m_a, m_b, m_{ab}) \sum_{k_{cd}} \Theta\begin{pmatrix} m_a & m_b & m_{ab} \\ l_a & l_b & l_{ab} \end{pmatrix} \\
&\times \sum_{k_{ab}=|m_{ab}|}^{l_{ab}} T_{k_c}^{l_{ab} m_{ab}} R^{l_{ab}-k_{ab}} r_A^{k_{ab}} r_B^{-l_{ab}} \\
&\times \sum_{m_{abc}} \Phi(m_{ab}, m_c, m_{abc}) \sum_{l_{abc}} \Theta\begin{pmatrix} m_{ab} & m_c & m_{abc} \\ k_{ab} & l_c & l_{abc} \end{pmatrix} \\
&\times \int_0^\pi \sin\theta_A d\theta_A \, p_{l_{abc}}^0(\cos\theta_A)
\end{aligned}
\tag{3.A37}
$$

$$
\begin{aligned}
&= 2\pi N_\phi(0) \sum_{m_{ad}} \Phi(m_a, m_b, m_{ab}) \sum_{k_{cd}} \Theta\begin{pmatrix} m_a & m_b & m_{ab} \\ l_a & l_b & l_{ab} \end{pmatrix} \\
&\times \sum_{k_{ab}=|m_{ab}|}^{l_{ab}} T_{k_c}^{l_{ab} m_{ab}} R^{l_{ab}-k_{ab}} r_A^{k_{ab}} r_B^{-l_{ab}} \\
&\times \sum_{m_{abc}} \Phi(m_{ab}, m_c, m_{abc}) \sum_{l_{abc}} \Theta\begin{pmatrix} m_{ab} & m_c & m_{abc} \\ k_{ab} & l_c & l_{abc} \end{pmatrix} \\
&\times \int_0^\pi \sin\theta_A d\theta_A \\
&\times \sum_{\nu=0}^{\left[\frac{l_{abc}}{2}\right]} \sum_{i=0}^{l_{abc}-2\nu} \sum_{j=0}^{l_{abc}-2\nu-i} \mathrm{P}(l_{abc}, \nu, i, j) R^{-l_{abc}+2(\nu+i)} r_A^{l_{abc}-2(\nu+i+j)} r_B^{2i}
\end{aligned}
\tag{3.A38}
$$

$$
\begin{aligned}
&\int d\Omega Y_{l_a}^{m_a}(\theta_B, \phi) Y_{l_b}^{m_b}(\theta_B, \phi) Y_{l_c}^{m_c}(\theta_A, \phi) \\
&= 2\pi N_\phi(0) \sum_{m_{ad}} \Phi(m_a, m_b, m_{ab}) \sum_{k_{cd}} \Theta\begin{pmatrix} m_a & m_b & m_{ab} \\ l_a & l_b & l_{ab} \end{pmatrix} \\
&\times \sum_{k_{ab}=|m_{ab}|}^{l_{ab}} T_{k_c}^{l_{ab} m_{ab}} \sum_{m_{abc}} \Phi(m_{ab}, m_c, m_{abc}) \sum_{l_{abc}} \Theta\begin{pmatrix} m_{ab} & m_c & m_{abc} \\ k_{ab} & l_c & l_{abc} \end{pmatrix} \\
&\times \int_0^\pi \sin\theta_A d\theta_A \\
&\times \sum_{\nu=0}^{\left[\frac{l_{abc}}{2}\right]} \sum_{i=0}^{l_{abc}-2\nu} \sum_{j=0}^{l_{abc}-2\nu-i} \mathrm{P}(l_{abc}, \nu, i, j) R^{l_{ab}-k_{ab}-l_{abc}+2(\nu+i)} r_A^{k_{ab}+l_{abc}-2(\nu+i+j)} r_B^{2i-l_{ab}}
\end{aligned}
\tag{3.A39}
$$

Appendix B: Half Definite Integral, H and Hh

We define a definite integral, H and Hh, in this appendix.

$$H[X,n,\alpha] \equiv \int^X x^n \exp(-\alpha x)dx \tag{3.B1}$$

We begin formulation for H in case a$\neq$0.

$$H\left[X,n,\alpha \middle| n \geq 0, \alpha \neq 0\right] = -\sum_{k=0}^{n} A(n,k)\alpha^{-(k+1)}X^{n-k}\exp(-\alpha X) \tag{3.B2}$$

$$A(n,k) = \frac{n!}{(n-k)!} \tag{3.B3}$$

$$\left\langle H\left[0,n,\alpha \middle| n \geq 0, \alpha \neq 0\right]\right\rangle = -n!\alpha^{-(n+1)} \tag{3.B4}$$

$$\left\langle H\left[\infty,n,\alpha \middle| n \geq 0, \alpha \neq 0\right]\right\rangle = 0 \tag{3.B5}$$

$$\left\langle H\left[0,n,\alpha \middle| n \geq 0, \alpha \neq 0\right]\right\rangle = -n!\alpha^{-(n+1)} \tag{3.B6}$$

$$\left\langle H\left[1,n,\alpha \middle| n \geq 0, \alpha \neq 0\right]\right\rangle = -\sum_{k=0}^{n} A(n,k)\alpha^{-(k+1)}\exp(-\alpha) \tag{3.B7}$$

$$\left\langle H\left[R,n,\alpha \middle| n \geq 0, \alpha \neq 0\right]\right\rangle = -\sum_{k=0}^{n} A(n,k)\alpha^{-(k+1)}R^{n-k}\exp(-\alpha R) \tag{3.B8}$$

For $n = -1$

$$H\left[X,n,\alpha \middle| n = -1, \alpha \neq 0\right] = Ei(-\alpha X) \tag{3.B9}$$

$$Ei(-\alpha X) = \int^X \frac{1}{x}\exp(-\alpha x)dx \tag{3.B10}$$

For $n \leq -2$,

$$\begin{aligned} &H\left[X,n,\alpha \middle| n \leq -2, \alpha \neq 0\right] \\ &= \frac{(-\alpha)^{-(n+1)}}{(-(n+1))!} Ei(-\alpha X) \\ &\quad - \sum_{k=0}^{-(n+1)} (-1)^{k-1} \frac{(-(n+1+k))}{(-(n+1))!} \alpha^{k-1} X^{n+k} \exp(-\alpha X) \end{aligned} \tag{3.B11}$$

In case $\alpha = 0$ and $n \neq -1$,

$$H\left[X,n,\alpha \middle| n \neq -1, \alpha = 0\right] = \int^X x^n dx = \frac{1}{n+1}X^{n+1} \tag{3.B12}$$

In case $\alpha = 0$ and $n = -1$,

$$H\left[X, n, \alpha \middle| n = -1, \alpha = 0\right] = \int^{X} x^{-1} dx = \log X \tag{3.B13}$$

The half definite integral Hh is defined using H as follows,

$$\begin{aligned} &Hh[X, n_1, \alpha_1, n_2, \alpha_2] \\ &= \int^{X} dr_1 r_1^{n_1} \exp(-\alpha_1 r_1) \quad \int^{r_1} dr_2 r_2^{n_2} \exp(-\alpha_2 r_2) \\ &= \int^{X} dr_1 r_1^{n_1} \exp(-\alpha_1 r_1) \quad H[r_1, n_2, \alpha_2] \end{aligned} \tag{3.B14}$$

$$\begin{aligned} &Hh\left[X, n_1, \alpha_1, n_2, \alpha_2 \middle| n_2 \geq 0, \alpha_2 \neq 0\right] \\ &= \int^{X} dr_1 r_1^{n_1} \exp(-\alpha_1 r_1) H\left[r_1, n_2, \alpha_2 \middle| n_2 \geq 0, \alpha_2 \neq 0\right] \\ &= -\sum_{k_2=0}^{n_2} A(n_2, k_2) \alpha_2{}^{-(k_2+1)} \int^{X} dr_1 r_1^{n_1+n_2-k_2} \exp(-(\alpha_1 + \alpha_2) r_1) \\ &= -\sum_{k_2=0}^{n_2} A(n_2, k_2) \alpha_2{}^{-(k_2+1)} H[X, n_1 + n_2 - k_2, \alpha_1 + \alpha_2] \end{aligned} \tag{3.B15}$$

$$\begin{aligned} &Hh\left[X, n_1, \alpha_1, n_2, \alpha_2 \middle| n_1 + n_2 \geq 0, n_2 \geq 0, \alpha_2 \neq 0\right] \\ &= -\sum_{k_2=0}^{n_2} A(n_2, k_2) \alpha_2{}^{-(k_2+1)} H[X, n_1 + n_2 - k_2, \alpha_1 + \alpha_2] \\ &= \sum_{k_2=0}^{n_2} A(n_2, k_2) \alpha_2{}^{-(k_2+1)} \sum_{k_1=0}^{n_1+n_2-k_2} A(n_1 + n_2 - k_2, k_1)(\alpha_1 + \alpha_2)^{-(k_1+1)} \\ &\times X^{n_1+n_2-(k_1+k_2)} \exp(-(\alpha_1 + \alpha_2) X) \end{aligned} \tag{3.B16}$$

$$\begin{aligned} &Hh\left[X, n_1, \alpha_1, n_2, \alpha_2 \middle| n_2 = -1, \alpha_2 \neq 0\right] \\ &= \int^{X} dr_1 r_1^{n_1} \exp(-\alpha_1 r_1) H\left[r_1, -1, \alpha_2 \middle| \alpha_2 \neq 0\right] \\ &= \int^{X} dr_1 r_1^{n_1} \exp(-\alpha_1 r_1) Ei(-\alpha_2 r_1) \end{aligned} \tag{3.B17}$$

$$\begin{aligned} &Hh\left[X, n_1, \alpha_1, n_2, \alpha_2 \middle| n_2 \leq 2, \alpha_2 \neq 0\right] \\ &= \int^{X} dr_1 r_1^{n_1} \exp(-\alpha_1 r_1) H\left[r_1, n_2, \alpha_2 \middle| n_2 \leq 2, \alpha_2 \neq 0\right] \end{aligned} \tag{3.B18}$$

$$\begin{aligned} &Hh\left[X, n_1, \alpha_1, n_2, \alpha_2 \middle| n_2 \neq -1, \alpha_2 = 0\right] \\ &= \int^{X} dr_1 r_1^{n_1} \exp(-\alpha_1 r_1) H\left[r_1, n_2, \alpha_2 \middle| n_2 \neq -1, \alpha_2 = 0\right] \\ &= \frac{1}{n_2 + 1} \int^{X} dr_1 r_1^{n_1+n_2+1} \exp(-\alpha_1 r_1) \\ &= \frac{1}{n_2 + 1} H[X, n_1 + n_2 + 1, \alpha_1] \end{aligned} \tag{3.B19}$$

$$\begin{aligned} &Hh\left[X, n_1, \alpha_1, n_2, \alpha_2 \middle| n_2 = -1, \alpha_2 = 0\right] \\ &= \int^X dr_1 r_1^{n_1} \exp(-\alpha_1 r_1) H\left[r_1, n_2, \alpha_2 \middle| n_2 = -1, \alpha_2 = 0\right] \\ &= \int^X dr_1 r_1^{n_1} \exp(-\alpha_1 r_1) \log r_1 \end{aligned} \tag{3.B20}$$

Appendix C: B Function

We define B function in this appendix. B function is an integral in which STF centered on B center is looked from A center.

$$\begin{aligned} B(n, \zeta, R, r_A) &\equiv \int_0^\pi \sin\theta_A d\theta_A r_B^n \exp(-\zeta r_B) \\ &= \int_{-1}^1 J(r_A, R) dx r_B^n \exp(-\zeta r_B) \end{aligned} \tag{3.C1}$$

If $r_A \le R$, $r_B = R + r_A x$. We obtain B function in Eq. (3.C1) as follows

$$\begin{aligned} r_B^n \exp(-\zeta r_B) &= r_B^n \exp(-\zeta r_B) \\ &= \exp(-\zeta R) \sum_{s=0}^n \binom{n}{s} R^s r_A^{n-s} \{x^{n-s} \exp(-\zeta r_A x)\} \end{aligned} \tag{3.C2}$$

$$\begin{aligned} &J\left(r_A, R \middle| r_A \le R\right) r_B^n \exp(-\zeta r_B) \\ &= R^{-1} r_B r_B^n \exp(-\zeta r_B) \\ &= \exp(-\zeta R) \sum_{s=0}^{n+1} \binom{n+1}{s} R^{s-1} r_A^{n+1-s} \{x^{n+1-s} \exp(-\zeta r_A x)\} \end{aligned} \tag{3.C3}$$

$$\begin{aligned} &B\left(n, \zeta, R, r_A \middle| r_A \le R\right) \\ &= \exp(-\zeta R) \sum_{s=0}^{n+1} \binom{n+1}{s} R^{s-1} r_A^{n+1-s} \left\{ \int_{-1}^1 dx x^{n+1-s} \exp(-\zeta r_A x) \right\} \\ &= \sum_{s=0}^{n+1} \binom{n+1}{s} R^{s-1} \exp(-\zeta R) r_A^{n+1-s} \\ &\quad \{H[1, n+1-s, \zeta r_A] - H[-1, n+1-s, \zeta r_A]\} \end{aligned} \tag{3.C4}$$

$$H[1, n+1-s, \zeta r_A] = -\sum_{k=0}^{n+1-s} A(n+1-s, k) \zeta^{-(k+1)} r_A^{-(k+1)} \exp(-\zeta r_A) \tag{3.C5}$$

$$\begin{aligned} H[-1, n+1-s, \zeta r_A] = -&\sum_{k=0}^{n+1-s} (-1)^{n+1-(s+k)} A(n+1-s, k) \zeta^{-(k+1)} \\ &\times r_A{}^{-(k+1)} \exp(\zeta r_A) \end{aligned} \tag{3.C6}$$

If $r_A \geq R$, we obtain B function in Eq. (3.C1) as follows

$$r_B^n \exp(-\zeta r_B) = \exp(-\zeta r_A) \sum_{s=0}^{n} \binom{n}{s} R^{n-s} r_A^s \{x^{n-s} \exp(-\zeta R x)\} \tag{3.C7}$$

$$\begin{aligned} &J(r_A, R | r_A \geq R) r_B^n \exp(-\zeta r_B) \\ &\quad = r_A^{-1} r_B r_B^n \exp(-\zeta r_B) \\ &\quad = \exp(-\zeta r_A) \sum_{s=0}^{n+1} \binom{n+1}{s} R^{n+1-s} r_A^{s-1} \{x^{n+1-s} \exp(-\zeta R x)\} \end{aligned} \tag{3.C8}$$

$$\begin{aligned} &B(n, \zeta, R, r_A | r_A \geq R) \\ &\quad = \exp(-\zeta r_A) \sum_{s=0}^{n+1} \binom{n+1}{s} R^{n+1-s} r_A^{s-1} \left\{ \int_{-1}^{1} dx x^{n+1-s} \exp(-\zeta R x) \right\} \\ &\quad = \sum_{s=0}^{n+1} \binom{n+1}{s} R^{n+1-s} r_A^s \exp(-\zeta r_A) \\ &\qquad \times \{\langle H[1, n+1-s, \zeta R] \rangle - \langle H[-1, n+1-s, \zeta R] \rangle\} \end{aligned} \tag{3.C9}$$

$$\langle H[1, n+1-s, \zeta R] \rangle = -\sum_{k=0}^{n+1-s} A(n+1-s, k) \zeta^{-(k+1)} R^{-(k+1)} \exp(-\zeta R) \tag{3.C10}$$

$$\begin{aligned} &\langle H[-1, n+1-s, \zeta R] \rangle \\ &\quad = -\sum_{k=0}^{n+1-s} (-1)^{n+1-(s+k)} A(n+1-s, k) \zeta^{-(k+1)} R^{-(k+1)} \exp(\zeta R) \end{aligned} \tag{3.C11}$$

where $<\ldots\ldots>$ means a definite integral. We obtain B function in Eq. (3.C1) as

$$\begin{aligned} &B(n, \zeta, R, r_A | r_A \leq R) \\ &= -\sum_{s=0}^{n+1} \binom{n+1}{s} R^{s-1} \exp(-\zeta R) \sum_{k=0}^{n+1-s} A(n+1-s, k) \zeta^{-(k+1)} r_A^{n-(s+k)} \exp(-\zeta r_A) \\ &\quad + \sum_{s=0}^{n+1} \binom{n+1}{s} R^{s-1} \exp(-\zeta R) \sum_{k=0}^{n+1-s} (-1)^{n+1-(s+k)} A(n+1-s, k) \zeta^{-(k+1)} \\ &\quad \times r_A^{n-(s+k)} \exp(\zeta r_A) \end{aligned} \tag{3.C12}$$

$$\begin{aligned} &B(n, \zeta, R, r_A | r_A \geq R) \\ &\quad = -\sum_{s=0}^{n+1} \binom{n+1}{s} \sum_{k=0}^{n+1-s} A(n+1-s, k) \zeta^{-(k+1)} R^{n-(s+k)} \exp(-\zeta R) \\ &\quad \times r_A^s \exp(-\zeta r_A) \\ &\quad + \sum_{s=0}^{n+1} \binom{n+1}{s} \sum_{k=0}^{n+1-s} (-1)^{n+1-(s+k)} A(n+1-s, k) \zeta^{-(k+1)} R^{n-(s+k)} \exp(\zeta R) \end{aligned} \tag{3.C13}$$

Appendix D: Half Definite Integral, Hb, Hbhb and Hhb

We define three half definite integrals, *Hb*, *Hbhb* and *Hhb* in this appendix.

$$\begin{aligned} &Hb[X, n_1, \zeta_a, n_2, \zeta_b] \\ &\equiv \int^X dr_A r_A^{n_1} \exp(-\zeta_a r_A) \int_0^\pi \sin\theta_A d\theta_A r_B^{n_2} \exp(-\zeta_b r_B) \\ &= \int^X dr_A r_A^{n_1} \exp(-\zeta_a r_A) \int_{-1}^1 J(r_A, R) dx r_B^{n_2} \exp(-\zeta_b r_B) \\ &= \int^X dr_A r_A^{n_1} \exp(-\zeta_a r_A) B(n_2, \zeta_b, R, r_A) \end{aligned} \tag{3.D1}$$

If $X \le R$,

$$\begin{aligned} &Hb\left[X, n_1, \zeta_a, n_2, \zeta_b \middle| \zeta_a \neq \zeta_b, X \le R\right] \\ &= \sum_{s=0}^{n_2+1} \binom{n_2+1}{s} R^{s-1} \exp(-\zeta_b R) \left[\sum_{k=0}^{n_2+1-s} A(n_2+1-s, k) \zeta_b^{-(k+1)} \right. \\ &\quad \times \sum_{t=0}^{n_1+n_2-(s+k)} A(n_1+n_2-(s+k), t)(\zeta_a+\zeta_b)^{-(t+1)} \\ &\quad \left. \times X^{n_1+n_2-(s+k)-t} \exp(-(\zeta_a+\zeta_b)X) \right] \\ &\quad - \sum_{s=0}^{n_2+1} \binom{n_2+1}{s} R^{s-1} \exp(-\zeta_b R) \sum_{k=0}^{n_2+1-s} (-1)^{n_2+1-(s+k)} A(n_2+1-s, k) \\ &\quad \times \sum_{t=0}^{n_1+n_2-(s+k)} A(n_1+n_2-(s+k), t)(\zeta_a+\zeta_b)^{-(t+1)} \\ &\quad \times X^{n_1+n_2-(s+k)-t} \exp(-(\zeta_a+\zeta_b)X) \end{aligned} \tag{3.D2}$$

$$\begin{aligned} &Hb\left[R, n_1, \zeta_a, n_2, \zeta_b \middle| \zeta_a \neq \zeta_b, X \le R\right] \\ &= \exp(-(\zeta_a+2\zeta_b)R) \left[\sum_{s=0}^{n_2+1} \binom{n_2+1}{s} \sum_{k=0}^{n_2+1-s} A(n_2+1-s, k) \zeta_b^{-(k+1)} \right. \\ &\quad \left. \times \sum_{t=0}^{n_1+n_2-(s+k)} A(n_1+n_2-(s+k), t)(\zeta_a+\zeta_b)^{-(t+1)} R^{n_1+n_2-(k+t+1)} \right] \\ &\quad - \exp(-(\zeta_a+2\zeta_b)R) \left[\sum_{s=0}^{n_2+1} \binom{n_2+1}{s} \sum_{k=0}^{n_2+1-s} (-1)^{n_2+1-(s+k)} A(n_2+1-s, k) \right. \\ &\quad \left. \times \sum_{t=0}^{n_1+n_2-(s+k)} A(n_1+n_2-(s+k), t)(\zeta_a+\zeta_b)^{-(t+1)} R^{n_1+n_2-(k+t+1)} \right] \end{aligned} \tag{3.D3}$$

$$\begin{aligned}
&Hb\left[0,n_1,\zeta_a,n_2,\zeta_b\middle|\zeta_a\neq\zeta_b,X\leq R\right]\\
&\quad=\exp(-\zeta_bR)\left[\sum_{s=0}^{n_2+1}\binom{n_2+1}{s}R^{s-1}\sum_{k=0}^{n_2+1-s}A(n_2+1-s,k)\zeta_b^{-(k+1)}\right.\\
&\qquad\left.\times(n_1+n_2-(s+k))!(\zeta_a+\zeta_b)^{n_1+n_2+1-(s+k)}\right]\\
&\qquad-\exp(-\zeta_bR)\left[\sum_{s=0}^{n_2+1}\binom{n_2+1}{s}R^{s-1}\sum_{k=0}^{n_2+1-s}(-1)^{n_2+1-(s+k)}A(n_2+1-s,k)\zeta_b^{-(k+1)}\right.\\
&\qquad\left.\times(n_1+n_2-(s+k))!(\zeta_a-\zeta_b)^{n_1+n_2+1-(s+k)}\right]
\end{aligned}\tag{3.D4}$$

$$\begin{aligned}
&Hb\left[X,n_1,\zeta_a,n_2,\zeta_b\middle|\zeta_a=\zeta_b,X\leq R\right]\\
&\quad=\sum_{s=0}^{n_2+1}\binom{n_2+1}{s}R^{s-1}\exp(-\zeta_bR)\left[\sum_{k=0}^{n_2+1-s}A(n_2+1-s,k)\zeta_b^{-(k+1)}\right.\\
&\qquad\times\sum_{t=0}^{n_1+n_2+2-(s+k)}A(n_1+n_2-(s+k),t)(\zeta_a+\zeta_b)^{-(t+1)}X^{n_1+n_2-(s+k+t)}\\
&\qquad\left.\times\exp(-(\zeta_a+\zeta_b)X)\right]+\sum_{s=0}^{n_2+1}\binom{n_2+1}{s}R^{s-1}\exp(-\zeta_bR)\\
&\qquad\times\sum_{k=0}^{n_2+1-s}(-1)^{n_2+1-(s+k)}A(n_2+1-s,k)\zeta_b^{-(k+1)}\frac{1}{n_1+n_2+1-(s+k)}X
\end{aligned}\tag{3.D5}$$

$$\begin{aligned}
&Hb\left[R,n_1,\zeta_a,n_2,\zeta_b\middle|\zeta_a=\zeta_b,X\leq R\right]\\
&\quad=\exp(-(\zeta_a+2\zeta_b)R)\left[\sum_{s=0}^{n_2+1}\binom{n_2+1}{s}\sum_{k=0}^{n_2+1-s}A(n_2+1-s,k)\zeta_b^{-(k+1)}\right.\\
&\qquad\times\left[\sum_{s=0}^{n_2+1}\binom{n_2+1}{s}\sum_{k=0}^{n_2+1-s}A(n_2+1-s,k)\zeta_b^{-(k+1)}\right.\\
&\qquad\left.\times\sum_{t=0}^{n_1+n_2+2-(s+k)}A(n_1+n_2-(s+k),t)(\zeta_a+\zeta_b)^{-(t+1)}R^{n_1+n_2-(k+t+1)}\right]\\
&\qquad+\exp(-\zeta_bR)\sum_{s=0}^{n_2+1}\binom{n_2+1}{s}\\
&\qquad\times\sum_{k=0}^{n_2+1-s}(-1)^{n_2+1-(s+k)}A(n_2+1-s,k)\zeta_b^{-(k+1)}\frac{1}{n_1+n_2+1-(s+k)}R^{n_1+n_2+1-(k+1)}
\end{aligned}\tag{3.D6}$$

$$Hb\left[0,n_1,\zeta_a,n_2,\zeta_b\middle|\zeta_a=\zeta_b,X\leq R\right]=0\tag{3.D7}$$

$$\begin{aligned}
&Hb\left[X, n_1, \zeta_a, n_2, \zeta_b \middle| X \geq R\right] \\
&\equiv \int^X dr_A r_A^{n_1} \exp(-\zeta_a r_A) \int_0^\pi \sin\theta_A d\theta_A r_B^n \exp(-\zeta_b r_B) \\
&= \int^X dr_A r_A^{n_1} \exp(-\zeta_a r_A) \int_{-1}^1 J\left(r_A, R \middle| r_A \geq R\right) dx r_B^n \exp(-\zeta_b r_B) \\
&= \int^X dr_A r_A^{n_1} \exp(-\zeta_a r_A) B\left(n_2, \zeta_b, R, r_A \middle| r_A \geq R\right)
\end{aligned} \tag{3.D8}$$

$$\begin{aligned}
&Hb\left[X, n_1, \zeta_a, n_2, \zeta_b \middle| X \geq R\right] \\
&= \sum_{s=0}^{n_2+1} \binom{n_2+1}{s} \left\{ \sum_{k=0}^{n_2+1-s} A(n_2+1-s, k) \zeta_b^{-(k+1)} R^{n_2-(s+k)} \exp(-\zeta_b R) \right\} \\
&\quad \times \sum_{t=0}^{n_1+s} A(n_1+s, t)(\zeta_a+\zeta_b)^{-(t+1)} X^{n_1+s-t} \exp(-(\zeta_a+\zeta_b)X) \\
&\quad - \sum_{s=0}^{n_2+1} \binom{n_2+1}{s} \left\{ \sum_{k=0}^{n_2+1-s} (-1)^{n_2+1-(s+k)} A(n_2+1-s, k) \zeta_b^{-(k+1)} R^{n_2-(s+k)} \exp(\zeta_b R) \right\} \\
&\quad \times \sum_{t=0}^{n_1+s} A(n_1+s, t)(\zeta_a+\zeta_b)^{-(t+1)} X^{n_1+s-t} \exp(-(\zeta_a+\zeta_b)X)
\end{aligned} \tag{3.D9}$$

$$\begin{aligned}
&Hb\left[R, n_1, \zeta_a, n_2, \zeta_b \middle| X \geq R\right] \\
&= \exp(-(\zeta_a+2\zeta_b)R) \sum_{s=0}^{n_2+1} \binom{n_2+1}{s} \left\{ \sum_{k=0}^{n_2+1-s} A(n_2+1-s, k) \zeta_b^{-(k+1)} R^{n_2-(s+k)} \right\} \\
&\quad \left\{ \sum_{t=0}^{n_1+s} A(n_1+s, t)(\zeta_a+\zeta_b)^{-(t+1)} R^{n_1+s-t} \right\} \\
&\quad - \exp(-\zeta_a R) \sum_{s=0}^{n_2+1} \binom{n_2+1}{s} \left\{ \sum_{k=0}^{n_2+1-s} (-1)^{n_2+1-(s+k)} A(n_2+1-s, k) \zeta_b^{-(k+1)} R^{n_2-(s+k)} \right\} \\
&\quad \left\{ \sum_{t=0}^{n_1+s} A(n_1+s, t)(\zeta_a+\zeta_b)^{-(t+1)} R^{n_1+s-t} \right\}
\end{aligned} \tag{3.D10}$$

$$\begin{aligned}
&Hb\left[X, n_1, \zeta_a, n_2, \zeta_b \middle| X \geq R\right] \\
&\equiv \int^X dr_A r_A^{n_1} \exp(-\zeta_a r_A) \int_0^\pi \sin\theta_A d\theta_A r_B^n \exp(-\zeta_b r_B) \\
&= \int^X dr_A r_A^{n_1} \exp(-\zeta_a r_A) \int_{-1}^1 J\left(r_A, R \middle| r_A \geq R\right) dx r_B^n \exp(-\zeta_b r_B) \\
&= \int^X dr_A r_A^{n_1} \exp(-\zeta_a r_A) B\left(n_2, \zeta_b, R, r_A \middle| r_A \geq R\right)
\end{aligned} \tag{3.D11}$$

$$\begin{aligned}
&Hhb[X, n_1, \zeta_a, n_2, \zeta_b, n_3, \zeta_c] \\
&\quad \equiv \int^X dr_{A_1} r_{A_1}^{n_1} \exp(-\zeta_a r_{A_1}) \int^{r_{A_1}} dr_{A_2} r_{A_2}^{n_2} \exp(-\zeta_a r_{A_2}) \\
&\quad \times \int_0^\pi \sin\theta_{A_2} d\theta_{A_2} r_{B_2}^{n_3} \exp(-\zeta_c r_{B_2}) \\
&\quad = \int^X dr_A r_A^n \exp(-\zeta_a r_A) Hb[r_A, n_2, \zeta_b, n_3, \zeta_c]
\end{aligned} \tag{3.D12}$$

$$\begin{aligned}
&Hbhb[X, n_1, \zeta_a, n_2, \zeta_b, n_3, \zeta_c, n_4, \zeta_d] \\
&\quad \equiv \int^X dr_{A_1} r_{A_1}^{n_1} \exp(-\zeta_a r_{A_1}) B(n_2, \zeta_b, R, r_{A1_1}) Hb[r_{A_1}, n_3, \zeta_c, n_4, \zeta_d]
\end{aligned} \tag{3.D13}$$

In some cases of multiple integral, two types of integral appear.

$$\int^X x^n \exp(-ax) \log x \tag{3.D14}$$

and

$$\int^X x^n \exp(-ax) Ei(-bx) \tag{3.D15}$$

These integrals are expressed by the exponential integral function, the Euler gamma function, the generalized hypergeometric function, and log x.

References

Avery J (2000) Hyperspherical harmonics and generalized Sturmians. Kluwer Academic Publishers, Dordrecht/Boston

Barnett MP, Coulson CA (1951) The evaluation of integrals occurring in the theory of molecular structure. Parts I & II. Phil Trans Roy Soc London A243:221–249

Barnett MP, Capitani JF, von zur Gathen J, Gerhard J (2004) Symbolic calculation in chemistry, selected examples. Int J Quant Chem 100:80–104

Buehler RJ, Hirschfelder JO (1951) Bipolar expansion of Coulombic potentials. Phys Rev 83:628–633

Christoffersen RE, Ruedenberg K (1968) Hybrid integrals over Slater-type atomic orbitals. J Chem Phys 49:4285–4292

Dirac PAM (1929) Quantum mechanics of many-electron systems. Proc Roy Soc Lond A 123:714–733

Fernández Rico J, López R, Ramírez G, Tablero C (1994) Molecular integrals with Slater basis. V. Recurrence algorithm for the exchange integrals. J Chem Phys 101:9807–9816

Guseinov II (2007) Expansion formulae for two-center integer and noninteger n STO charge densities and their use in evaluation of multi-center integrals. J Math Chem 42:415–422

Hierse W, Oppeneer PM (1994) Fast and stable algorithm for the analytical computation of 2-center coulomb and overlap integrals over Slater-type orbitals. Int I Quant Chem 52:1249–1265

Hobson EW (1931) The theory of spherical and ellipsoidal harmonics. Cambridge U. P, Cambridge, Sec. 89
Huzinaga S (1980, 2000) Molecular orbital method [Japanese]. Iwanami Book Publisher, Tokyo. Cambridge University Press
Jones HW (1980) Computer-generated formulas for overlap integrals of Slater-type orbitals. Int J Quant Chem 18:709–713
Jones HW (1988) Analytical evaluation of multicenter molecular integrals over Slater-type orbitals using expanded Lowdin alpha functions. Phys Rev A 38:1065–1068
Jones HW (1994) Development in multicenter molecular integrals over STOs using expansions in spherical-harmonics. Int I Quant Chem 51:417–423
Kato T (1957) On the eigenfunctions of many-particle systems in quantum mechanics. Commun Pure Appl Math 10:151–177
Kohn W, Sham LL (1964) Self-consistent equations including exchange and correlation effects. Phys Rev A 140:1133–1138
Kotani M, Amemiya A, Ishiguro AE, Kimura T (1955) Table of molecular integrals. Maruzen, Tokyo
Löwdin PO (1956) Quantum theory of cohesive properties of solids. Adv Phys 5:1–172
Milleur MB, Twerdochlib M, Hirschfelder JO (1966) Bipolar angle averages and two-center, two-particle integrals involving r_{12}. J Chem Phys 45:13–20
Mukoyama T, Hock G (1994) L2 approximation of atomic continuum wave functions. Bull Inst Chem Res, Kyoto Univ 72:195–201
Mukoyama T, Kagawa T (1983) On the momentum representation of the relativistic Hartree-Fock-Roothaan wavefunctions. J Phys B: Atom Mol Phys 16:1875–1880
Mukoyama T, Yasui J (1992) Analytical expression of the Hartree-Fock wave functions. Bull Inst Chem Res, Kyoto Univ 70:385–391
Reinhardt P, Hoggan PE (2009) Cusps and derivatives for wave-functions expanded in Slater orbitals: a density study. Int J Quantum Chem 109:3191–3198
Roothaan CCJ (1951) New developments in molecular orbital theory. Rev Mod Phys 23:69–89
Sharma RR (1976) Expansion of a function about a displaced center for multicenter integrals: a general and closed expression of a Slater orbital and for overlap integrals. Phys Rev A13:517–527
Shibuya T, Wulfman CE (1965) Molecular orbitals in momentum space. Proc Roy Soc A 286:376–389
Silver DM (1971) Unified treatment of diatomic electron interaction integrals over Slater-type atomic orbitals. J Math Phys 12:1937–1943
Slater JC (1951) A simplification of the Hartree-Fock method. Phys Rev 81:385–390
Slater JC (1974) Quantum theory of molecules and solids. McGraw-Hill, New York
Slater JC (1979) The calculation of molecular orbitals. Wiley, New York
Wojnecki R, Modrak P (1993) Calculation of two-center integrals between Slater-type orbitals. Comput Chem 17:287–290
Yasui J (2010) Polynomial expressions of molecular integral functionals over Slater-type-orbitals and its application to the extension of Hartree-Fock-Roothaan equation. Bull Soc DV-Xα 23:54–59
Yasui J (2011) Algebraic molecular orbital equation. Bull Soc DV-Xα 24:47–54
Yasui J, Saika A (1982) Unified analytical evaluation of two-center, two-electron integrals over Slater-type orbitals. J Chem Phys 76:468–472
Yasui J, Mukoyama T, Shibuya T (1996) Analytical expressions of atomic wave functions and molecular integrals for the x-ray transition probabilities of molecules. Adv X-Ray Chem Anal Jpn 27:329–338

Chapter 4
Atom-Atom Interaction Potential from the Gaussian Quadrature Method and Classical Molecular Dynamics

Kimichika Fukushima

4.1 Introduction

Investigations into atomic and molecular dynamics (MD) (Haile 1997), which sometimes includes chemical reactions, represent important themes in science and technology. A quantum wave equation of nuclei and electrons describing their motion is separated into two parts, namely nuclei and electrons bound to the nuclei, in the Born–Oppenheimer approximation due to the light electron mass compared to the nucleus mass (Born and Oppenheimer 1927). The nuclei motion is described by quantum wave equation or classical Newtonian equations. In the nuclei motion equation the total electronic energy, including nuclear repulsions, enters into the interaction potential between constituent atoms/molecules. Hereafter, we consider atoms as composite particles for simplicity.

An early piece of research into molecular dynamics using the Newtonian equation is seen in literature (Alder and Wainwright 1957) featuring the use of the solid sphere model. Rahman and Stillinger performed molecular dynamics for water using the Lenard–Jones potential (Rahman and Stillinger 1973; Stillinger and Rahman 1972, 1974). The Newton equation describes the motion of atomic particles by specifying their positions and velocities as functions of time under the influence of forces acting on them. One practical way of approximating solutions to the differential Newtonian equation is via a numerical iterative method. At the beginning of time iterations, the initial positions and momentums of particles are specified, and sequential time values are determined by the previous time values using difference equations of differential Newtonian equations. This classical molecular dynamics reduces the calculation time compared to quantum molecular dynamics.

K. Fukushima (✉)
Advanced Reactor System Engineering Department, Toshiba Nuclear Engineering Service Corporation, 8, Shinsugita-cho, Isogo-ku, Yokohama 235-8523, Japan
e-mail: kimichika1a.fukushima@glb.toshiba.co.jp

T. Ishii et al. (eds.), *The DV-Xα Molecular-Orbital Calculation Method*,
DOI 10.1007/978-3-319-11185-8_4

Molecular dynamics is applicable to physics (atom, molecules and condensed matter), material science (inorganic and organic materials) as well as pharmaceutical chemistry (drug design) and so on. Investigated phenomena include, for example, collisions, atomic configurations, phase transitions, mass and energy transfers as well as chemical reactions.

Classical molecular dynamics is performed using an electronic potential energy hyper-surface acting on atoms derived by electron wave functions. One of the approximations used to determine electron wave functions is the Hartree–Fock method, which is sometimes insufficient to predict the potential energy surface and binding energy among atoms. The research was then oriented toward the introduction of electronic correlations, for which considerable calculation time is required. It was revealed, however, that the one-electron approximation via the local density functional method (Slater 1951, 1974, 1979; Gáspár 1954; Nozières and Pines 1958; Kohn and Sham 1965; Wigner 1934; von Barth and Hedin 1972; Janak et al. 1975; Vosko et al. 1980; Perdew and Zunger 1981; Ellis and Painter 1970; Adachi et al. 1978; Averill and Ellis 1973; Rosén et al. 1976; Satoko and Ohnishi 1994) predicts the potential energy surface to thus present reasonable equilibrium interatomic distances and binding energies. This is one of the reasons why the local density functional method has been widely accepted. The total energy was also applied to evaluate the potential energy surface in chemical reactions.

It was denoted in the previous paper that the total energy calculated highly accurately by the Gaussian quadrature (GQ) method is useful for classical molecular dynamics (Fukushima et al. 1982; Fukushima 2009). The matrices of the Hamiltonian, overlap integral and total energy are numerically integrated via the Gauss–Legendre and Gauss–Laguerre quadratures for finite and semi-infinite sections, respectively (Stroud and Secrest 1966). The author has then combined the total energy calculation method using Gaussian quadratures and classical molecular dynamics. The merit of this method is that it evaluates the interaction among many atoms, even in an approximation using atom-atom pair potentials, unlike the analysis of high energy atomic collisions on the order of kev–Mev in nuclear materials, which concerns a sequence of independent binary collisions between an incident atom and one target atom (Robinson and Torrens 1974). In this article, calculations are performed for systems composed of many atoms with a mutual pair interaction potential and an incident atom with a low energy below several hundred eV.

We finally note that the number density of atoms in materials is on the order of the Avogadro number. The calculation power is presently limited, and the total atoms treated are up to about several ten-thousands or so. Furthermore, the total time scale of calculations is approximately the number of total time steps multiplied by the time increment on an atomic unit scale, which restricts the analysis concerning time-dependent phenomena.

This chapter is organized as follows. Section 4.2 describes the Born–Oppenheimer approximation and formalism of classical molecular dynamics. Section 4.3 presents the derivation of the atom-atom interaction potential using the Gaussian quadrature molecular orbital method. Section 4.4 is devoted to the numerical calculation method of the Newtonian equation, while Sect. 4.5 reports on classical molecular dynamics calculations for carbon-carbon collisions, finally presenting conclusions.

4.2 Born–Oppenheimer Approximation and Formalism of Classical Molecular Dynamics

The Born–Oppenheimer approximation breaks the wavefunction $\Psi_{\mathrm{T}}(\boldsymbol{r}_1, \boldsymbol{r}_2 \cdots, \boldsymbol{R}_1, \boldsymbol{R}_2, \cdots)$ of a system, which is composed of electrons and nuclei located at $\boldsymbol{r}_k$ and $\boldsymbol{R}_I$, respectively, into

$$\Psi_{\mathrm{T}}(\boldsymbol{r}_1, \boldsymbol{r}_2 \cdots, \boldsymbol{R}_1, \boldsymbol{R}_2, \cdots) = \Psi_{\mathrm{e}}(\boldsymbol{r}_1, \boldsymbol{r}_2 \cdots)\Psi_{\mathrm{N}}(\boldsymbol{R}_1, \boldsymbol{R}_2, \cdots), \tag{4.1}$$

where Ψ_{e} $(\boldsymbol{r}_1, \boldsymbol{r}_2 \cdots)$ and Ψ_{N} $(\boldsymbol{R}_1, \boldsymbol{R}_2, \cdots)$ are electronic and nuclear wavefunctions, including other coordinates as constants. The energy eigenvalue of an electronic system is the total electronic energy E_{e}, which amounts in the density functional scheme in atomic units to

$$\begin{aligned} E_{\mathrm{e}} = & \sum_i \int \mathrm{d}\boldsymbol{r} \psi_i^*(\boldsymbol{r}) \left[-\frac{1}{2}\nabla_r^2 + V_{\mathrm{N}}(\boldsymbol{r}) + \frac{1}{2}V_{\mathrm{C}}(\boldsymbol{r}) \right] \psi_i(\boldsymbol{r}) + E_{\mathrm{xc}}[\rho(\boldsymbol{r})] \\ & + \sum_{I<J} \frac{Z_I Z_J}{|\boldsymbol{R}_I - \boldsymbol{R}_J|}, \end{aligned} \tag{4.2}$$

where V_{N} $(\boldsymbol{r})$ and V_{C} $(\boldsymbol{r})$ are nuclear and electronic Coulomb potentials, respectively, and Z_I is the nuclear charge. The quantity E_{xc} $[\rho(\boldsymbol{r})]$ is the exchange-correlation energy, which is a functional of electronic charge density $\rho(\boldsymbol{r})$. The one-electron wave function ψ_i $(\boldsymbol{r})$ is derived from the wave equation

$$H(\boldsymbol{r})\psi_i(\boldsymbol{r}) = \varepsilon_i \psi_i(\boldsymbol{r}). \tag{4.3}$$

Here,

$$H(\boldsymbol{r}) = -\frac{1}{2}\nabla_r^2 + V_{\mathrm{N}}(\boldsymbol{r}) + V_{\mathrm{C}}(\boldsymbol{r}) + V_{\mathrm{xc}}(\boldsymbol{r}), \tag{4.4}$$

where ε_i is the electronic energy eigenvalue and V_{xc} $(\boldsymbol{r})$ is the exchange-correlation potential. The nuclear wave equation is given by

$$\left[-i\frac{\partial}{\partial t} - \frac{1}{2}\nabla_R^2 + V_{\mathrm{e}}(\boldsymbol{R}_1, \boldsymbol{R}_2, \cdots) \right] \psi_I(\boldsymbol{R}) = 0, \tag{4.5}$$

where

$$V_{\mathrm{e}}(\boldsymbol{R}_1, \boldsymbol{R}_2, \cdots) = E_{\mathrm{e}}. \tag{4.6}$$

The equation above implies that the total electronic energy becomes the interaction potential among constituent atomic particles. Electronic equations are solved separately from nuclear equations.

One of the quantum molecular dynamics is the Car–Parrinello method (Car and Parrinello 1985), which expands an electron wavefunction in terms of plane waves using pseudopotentials in a periodic system. The full potential method, which deals with all electrons, includes significant contributions from core electrons to total energy. The plane wave method, meanwhile, using the pseudopotential, eliminates contributions from core electrons by substituting the potential for an effective potential to suppress the variation of valence electron wavefunctions. The motion of particles seems to be virtual if artificial masses are used, and the number of plane waves increases for heavy atoms.

The above nuclear wave equation, which depends on time t approaches to the Newtonian equation in the classical limit, is denoted as

$$\frac{\mathrm{d}}{\mathrm{d}t}\boldsymbol{p}_I = \boldsymbol{f}_I. \tag{4.7}$$

Here,

$$\boldsymbol{f}_I = -\nabla_{\boldsymbol{R}_I} V_{\mathrm{e}}(\boldsymbol{R}_1, \boldsymbol{R}_2, \cdots), \tag{4.8}$$

$$\boldsymbol{p}_I = M_I \boldsymbol{v}_I, \tag{4.9}$$

where $\boldsymbol{p}_I$, M_I, $\boldsymbol{v}_I$ are (in order) the momentum, mass and velocity of a nucleus and

$$\frac{\mathrm{d}\boldsymbol{R}_I}{\mathrm{d}t} = \frac{\boldsymbol{p}_I}{M_I}. \tag{4.10}$$

The decomposition of the Newtonian equation into two equations with the first order derivative is the Hamiltonian formalism, which is explicitly written as

$$\frac{\mathrm{d}}{\mathrm{d}t}\boldsymbol{p}_I = -\frac{\partial H_N}{\partial \boldsymbol{R}_I} = -\nabla_{\boldsymbol{R}_I} V_{\mathrm{e}}(\boldsymbol{R}_1, \boldsymbol{R}_2, \cdots), \tag{4.11}$$

$$\frac{\mathrm{d}}{\mathrm{d}t}\boldsymbol{R}_I = \frac{\partial H_N}{\partial \boldsymbol{p}_I}, \tag{4.12}$$

where

$$H_N = \sum_I \frac{\boldsymbol{p}_I^2}{2M_I} + V_{\mathrm{e}}(\boldsymbol{R}_1, \boldsymbol{R}_2, \cdots). \tag{4.13}$$

4.3 Derivation of Atom-Atom Interaction Potential Using the Gaussian Quadrature Quantum Molecular Orbital Method

The Newton equation in classical molecular dynamics contains forces among particles, in which the potential plays an important role and is decomposed into potentials of two-body, three-body, ... and so on. The classical molecular dynamics usually used employs two-body interatomic interaction (pair) potentials expressed by

$$V_{\mathrm{e}}(\boldsymbol{R}_1, \boldsymbol{R}_2, \cdots) \approx \sum_{I<J} V_{\mathrm{e}}^{(2)}(\boldsymbol{R}_I, \boldsymbol{R}_J). \tag{4.14}$$

The pair potential is further reduced in spherical coordinates to

$$V_{\mathrm{e}}^{(2)}(\boldsymbol{R}_I, \boldsymbol{R}_J) = V(r), \tag{4.15}$$

where

$$r = |\boldsymbol{R}_I - \boldsymbol{R}_J|. \tag{4.16}$$

The above potential between atoms is calculated non-empirically using the quantum molecular orbital method with the use of the Gaussian quadrature method.

In the molecular orbital method, quantum equations of electrons are solved by the linear combination of atomic orbitals (LCAO) by expanding the wave function in terms of atomic orbitals as

$$\psi_i(\boldsymbol{r}) = \sum_j C_{ij}\chi_j(\boldsymbol{r}). \tag{4.17}$$

Atomic basis functions in the Gaussian quadrature method, which originated from the DV (discrete variational) method, are derived via solution of the atomic wave equation in spherical coordinates. The molecular wave equation is then expressed by the secular equation in matrix form

$$\boldsymbol{HC} = \boldsymbol{\varepsilon SC}, \tag{4.18}$$

where $\boldsymbol{\varepsilon}$ are the matrix form of eigenvalues. Elements of the Hamiltonian and the overlap integral are represented by

$$H_{ij} = \int \mathrm{d}\boldsymbol{r}\chi_i^*(\boldsymbol{r})H(\boldsymbol{r})\chi_j(\boldsymbol{r}), \tag{4.19}$$

$$S_{ij} = \int \mathrm{d}\boldsymbol{r}\chi_i^*(\boldsymbol{r})\chi_j(\boldsymbol{r}), \tag{4.20}$$

respectively.

The Gaussian Quadrature method achieves highly accurate integrals for pair atoms in elliptic coordinates. These (ξ, η, ϕ) are expressed in terms of orthogonal coordinates as

$$\xi = \frac{|\boldsymbol{r} - \boldsymbol{R}_I| + |\boldsymbol{r} - \boldsymbol{R}_J|}{|\boldsymbol{R}_I - \boldsymbol{R}_J|}, \tag{4.21}$$

$$\eta = \frac{|\boldsymbol{r} - \boldsymbol{R}_I| - |\boldsymbol{r} - \boldsymbol{R}_J|}{|\boldsymbol{R}_I - \boldsymbol{R}_J|}, \tag{4.22}$$

where

$$1 \leq \xi < \infty, \tag{4.23}$$

$$-1 \leq \eta \leq 1, \tag{4.24}$$

and ϕ is an angle around an axis connecting atoms I and J. The matrix element of the Hamiltonian becomes

$$H_{ij} = \int_{-\pi}^{\pi} \mathrm{d}\phi \int_{-1}^{1} \mathrm{d}\eta \int_{1}^{\infty} \mathrm{d}\xi \left[(J) \chi_i^*(\boldsymbol{r}) H(\boldsymbol{r}) \chi_j(\boldsymbol{r}) \right], \tag{4.25}$$

where $J = \dfrac{\partial(x, y, z)}{\partial(\xi, \eta, \phi)}$ is the Jacobian matrix.

The aforementioned integrals in a finite region are implemented by transformation of the variable $\phi' = (1/2\pi)\phi$ as well as the integral section, and using the Gauss–Legendre formula given by

$$\int_{-1}^{1} \mathrm{d}s f(s) \cong \sum_{i'}^{n-1} w_{\mathrm{e}}(s_{i'}) f(s_{i'}), \tag{4.26}$$

where f is the integrand as a function of $s = \phi'$ or $s = \eta$ with

$$w_{\mathrm{e}}(s_{i'}) = \int \mathrm{d}s \prod\nolimits_{j' \neq i'} \frac{s - s_{j'}}{s_{i'} - s_{j'}}, \tag{4.27}$$

and s_i satisfies for $-1 \leq s \leq 1$,

$$w_{\mathrm{e}}(s_{i'}) P_n(s_{i'}) = 0. \tag{4.28}$$

Here, $P_n(s)$ is the Legendre function defined by

$$nP_n(s) - (2n-1)sP_{n-1}(s) + (n-1)P_{n-2}(s) = 0, \tag{4.29}$$

with

$$P_0(s) = 1, \tag{4.30}$$

$$P_1(s) = s, \tag{4.31}$$

$$P_2(s) = \frac{3}{2}s^2 - \frac{1}{2}. \tag{4.32}$$

Similarly, integrals in an infinite region use transformation of the variable $\xi' = (\xi - 1)/a_{\xi'}$ with $a_{\xi'}$ being an arbitrary scaling parameter around 4, and the Gauss–Laguerre formula given by

$$\int_0^\infty \mathrm{d}u g(u) = \int_0^\infty \mathrm{d}u \exp(-u)[\exp(u)g(u)] \cong \sum_{i'}^{n-1} w_\mathrm{a}(u_{i'})[\exp(u_{i'})g(u_{i'})], \tag{4.33}$$

where g is the integrand as a function of $u = \xi'$, and

$$w_\mathrm{a}(u_{i'}) = \int_0^\infty \mathrm{d}u \exp(-u) \prod_{j' \neq i'} \frac{u - u_{j'}}{u_{i'} - u_{j'}}. \tag{4.34}$$

The matrix element of the Hamiltonian is transformed, with the help of the variable transformation of $\phi' = \phi/2\pi$ and $\xi' = (\xi - 1)/a_{\xi'}$, as

$$\begin{aligned} H_{ij} &= 2\pi(a_{\xi'}) \int_{-1}^{1} \mathrm{d}\phi' \int_{-1}^{1} \mathrm{d}\eta \int_0^\infty \mathrm{d}\xi' \left[(J)\chi_i^*(\mathbf{r})H(\mathbf{r})\chi_j(\mathbf{r})\right] \\ &= 2\pi(a_{\xi'}) \int_{-1}^{1} \mathrm{d}\phi' \int_{-1}^{1} \mathrm{d}\eta \int_0^\infty \mathrm{d}\xi' \exp(-\xi') \left[\exp(\xi')(J)\chi_i^*(\mathbf{r})H(\mathbf{r})\chi_j(\mathbf{r})\right]. \end{aligned} \tag{4.35}$$

Using the Gaussian quadratures for the integral above, we get

$$\begin{aligned} H_{ij} &= 2\pi(a_{\xi'}) \\ &\times \Big[\sum_{i'j'k'} w_\mathrm{a}(\xi'_{i'}) w_\mathrm{e}(\eta_{j'}) w_\mathrm{e}(\phi'_{k'}) \left[\exp(\xi'_{i'}) \left(J|_{\xi'=\xi'_{i'}, \eta=\eta_{j'}, \phi'=\phi'_{k'}}\right) \chi_i^*(\mathbf{r}_{i'j'k'}) H(\mathbf{r}_{i'j'k'}) \chi_j(\mathbf{r}_{i'j'k'})\right], \end{aligned} \tag{4.36}$$

where the quantities such as $\mathbf{r}_{i'j'k'}$ are at the value of $(\xi'_{i'}, \eta_{j'}, \phi'_{k'})$.

For multi-atomic molecules, the integral is decomposed into elementary integrals between pair atoms. Integrals for these decomposed pair atoms are carried out directly. Another multicenter integral is performed by the decomposition using a projection operator $\hat{P}_I$ to the I-th constituent atom including cases of $I \geq 3$ and associated coordinates $(\Xi_I, \Theta_I, \Omega_I)$ as

$$H_{ij} = \int \mathrm{d}\boldsymbol{r}\chi_i^*(\boldsymbol{r})H(\boldsymbol{r})\chi_j(\boldsymbol{r}) = \sum_I \int \mathrm{d}\Xi_I \mathrm{d}\Theta_I \mathrm{d}\Omega_I \frac{\partial(x,y,z)}{\partial(\Xi_I,\Theta_I,\Omega_I)} \hat{P}_I(\Xi_I,\Theta_I,\Omega_I)$$
$$\times \left[\chi_i^*(\Xi_I,\Theta_I,\Omega_I)H(\Xi_I,\Theta_I,\Omega_I)\chi_j(\Xi_I,\Theta_I,\Omega_I)\right], \tag{4.37}$$

under the condition

$$\sum_I \hat{P}_I(\Xi_I,\Theta_I,\Omega_I) = 1. \tag{4.38}$$

The projection above is the Becke's decomposition procedure (Becke 1988; Delley 1990) but the fuzzy cell is not used. The numerical integrals utilize the Gaussian quadratures.

4.4 Numerical Calculation Method of the Newtonian Equation

The differential Newtonian equation is solved by numerical iterations. Momentum $\boldsymbol{p}_I(t_n+\Delta t)$ at the next time $t_n+\Delta t$ is expanded around the sequential time t_n yielding (Ashcer and Petzold 1998)

$$\boldsymbol{p}_I(t_n+\Delta t) = \boldsymbol{p}_I(t_n) + (\Delta t)\frac{\mathrm{d}}{\mathrm{d}t}\boldsymbol{p}_I(t)\big|_{t=t_n}. \tag{4.39}$$

We then have

$$\frac{\mathrm{d}}{\mathrm{d}t}\boldsymbol{p}_I(t)\big|_{t=t_n} = \frac{\boldsymbol{p}_I(t_n+\Delta t) - \boldsymbol{p}_I(t_n)}{\Delta t}. \tag{4.40}$$

By denoting the momentum and force as

$$\boldsymbol{p}_I^{n+1} = \boldsymbol{p}_I(t_n+\Delta t), \tag{4.41}$$
$$\boldsymbol{p}_I^{n} = \boldsymbol{p}_I(t_n), \tag{4.42}$$
$$\boldsymbol{f}_I^{n} = \boldsymbol{f}_I(t_n), \tag{4.43}$$

a finite difference form of the Newtonian equation is expressed as

$$\frac{\boldsymbol{p}_I^{n+1} - \boldsymbol{p}_I^{n}}{\Delta t} = \boldsymbol{f}_I^{n}, \tag{4.44}$$

giving a solution

$$\boldsymbol{p}_I^{n+1} = \boldsymbol{p}_I^{n} + \Delta t \boldsymbol{f}_I^{n}, \tag{4.45}$$

The solution at the time step t_{n+1} is a function of the quantities at t_n. The other part of the Newtonian equation for $\boldsymbol{R}_I(t)$ including $\boldsymbol{p}_I(t)$ is transformed to a finite difference equation arriving at a solution

$$\boldsymbol{R}_I^{n+1} = \boldsymbol{R}_I^n + \Delta t \frac{\boldsymbol{p}_I^n}{M_I}. \tag{4.46}$$

Many numerical methods of classical molecular dynamics utilize the above or modified procedure. The popular Verlet method (Verlet 1967) numerically derives the $(n+1)$th variable from the nth and $(n-1)$th variables as

$$\boldsymbol{R}_I^{n+1} = 2\boldsymbol{R}_I^n - 2\boldsymbol{R}_I^{n-1} + \Delta t^2 \frac{\boldsymbol{f}_I^n}{M_I}, \tag{4.47}$$

$$\boldsymbol{p}_I^n = \frac{M_I}{2\Delta t}\left(\boldsymbol{R}_I^{n+1} - \boldsymbol{R}_I^{n-1}\right). \tag{4.48}$$

Additionally, as an option, we here show a full implicit method, which is numerically stable. The difference equation of the Newtonian equation for $\boldsymbol{p}_I(t)$ containing $\boldsymbol{f}_I(t)$ is replaced by (Crank and Nicolson 1947)

$$\frac{\boldsymbol{p}_I^{n+1} - \boldsymbol{p}_I^n}{\Delta t} = \theta \boldsymbol{f}_I^n + (1-\theta)\boldsymbol{f}_I^{n+1}, \tag{4.49}$$

to give

$$\boldsymbol{p}_I^{n+1} - \Delta t(1-\theta)\boldsymbol{f}_I^{n+1} = \boldsymbol{p}_I^n + \Delta t \theta \boldsymbol{f}_I^n, \tag{4.50}$$

where θ is a suitable parameter (frequently used value is 0.5) and $\boldsymbol{f}_I$ at the nth time step is a function of a set of $\boldsymbol{R}_I$. The equation above is written as

$$\boldsymbol{p}_I^{n+1} - \Delta t(1-\theta)\boldsymbol{f}_I\left(\boldsymbol{R}_1^{n+1}, \boldsymbol{R}_2^{n+1}, ..\right) = \boldsymbol{p}_I^n + \Delta t \theta \boldsymbol{f}_I\left(\boldsymbol{R}_1^n, \boldsymbol{R}_2^n, ..\right). \tag{4.51}$$

The other Newtonian equation for $\boldsymbol{R}_I(t)$ with $\boldsymbol{p}_I(t)$ results in a similar difference equation

$$\boldsymbol{R}_I^{n+1} - \Delta t(1-\theta)\frac{\boldsymbol{p}_I^{n+1}}{M_I} = \boldsymbol{R}_I^n + \Delta t \theta \frac{\boldsymbol{p}_I^n}{M_I}. \tag{4.52}$$

Since the force $\mathbf{f}_I(\mathbf{R}_1^{n+1}, \mathbf{R}_2^{n+1},..)$ at the next time step is a nonlinear function of $\mathbf{R}_I^{n+1}$, the force is expanded linearly for sequential repetitions to approach real solutions (Tjalling 1995). During the main iteration from the nth time step to $(n+1)$ th time step, we insert a set of subiteration (repetition) steps, where each repetition is like from kth step to $(k+1)$th step toward the $(n+1)$th convergence. By denoting the position at the kth repetition step as $\mathbf{R}_I^{n+1,k}$, we derive the following relation

$$f_I\left(R_1^{n+1,k+1}, R_2^{n+1,k+1}, ..\right) = f_I\left(R_1^{n+1,k}, R_2^{n+1,k}, ..\right) + \sum_J \left(R_J^{n+1,k+1} - R_J^{n+1,k}\right) \bullet \frac{\partial}{\partial R_J^k} f_I\left(R_1^{n+1,k}, R_2^{n+1,k}, ..\right). \tag{4.53}$$

By substituting the approximate force $\mathbf{f}_I(\mathbf{R}_1^{n+1,k+1}, \mathbf{R}_2^{n+1,k+1}, ..)$ for the real force $\mathbf{f}_I(\mathbf{R}_1^{n+1}, \mathbf{R}_2^{n+1}, ..)$ and manipulating $\mathbf{p}_I^{n+1}$ in a similar but trivial way, we arrive at the final difference form of the Newtonian equation for $p_I(t)$ containing $f_I(t)$ expressed as

$$\begin{aligned} & p_I^{n+1,k+1} - \Delta t(1-\theta)\sum_J \left[R_J^{n+1,k+1} \bullet \frac{\partial}{\partial R_J^k} f_I\left(R_1^{n+1,k}, R_2^{n+1,k}, ..\right)\right] \\ & = p_I^n + \Delta t\theta f_I\left(R_1^n, R_2^n, ..\right) + \Delta t(1-\theta) f_I\left(R_1^{n+1,k}, R_2^{n+1,k}, ..\right) \\ & - \Delta t(1-\theta)\sum_J \left[R_J^{n+1,k} \bullet \frac{\partial}{\partial R_J^k} f_I\left(R_1^{n+1,k}, R_2^{n+1,k}, ..\right)\right]. \end{aligned} \tag{4.54}$$

By solving these multi-variable linear algebraic equations in a matrix form, all solutions of $(k+1)$th variables in terms of kth variables are derived sequentially.

4.5 Classical Molecular Dynamics Calculations Using the Gaussian Quadrature Method and Pair Potentials

Classical molecular dynamics calculations were performed for an energetic carbon (C) atom injected into a cluster composed of C atoms. The interatomic potential used was the pair potential derived in elliptic coordinate by Gaussian quadrature (GQ) method with the formulae of Gauss–Legendre and Gauss–Laguerre. It was reported in our previous papers that the total energy error of the GQ result with (20,16) integration points in (ξ, η) coordinates for the axis-symmetric CO was less than 10^{-3} Ry, while the corresponding error of the DV result was 3 Ry. (Fukushima et al. 1982; Fukushima 2009). The interatomic potential calculated was fitted to the Morse potential as

$$V(r) = V_0\left\{[1 - \exp(-a_r(r - r_0))]^2 - 1\right\}, \tag{4.55}$$

where V_0 (=5.50 eV) is the dissociation energy, r_0(=2.63 a.u.) is the bond distance and $2a_r^2V_0$ (=8.26 × 10^5 dyn/cm) is the force constant for the present case. Molecular dynamics calculations were performed using the Verlet method.

Figure 4.1 shows an initial incident atom with the kinetic energy of 400 eV along the x axis located at ($-2a_c$, 0) in two-dimensional (x,y) coordinates and

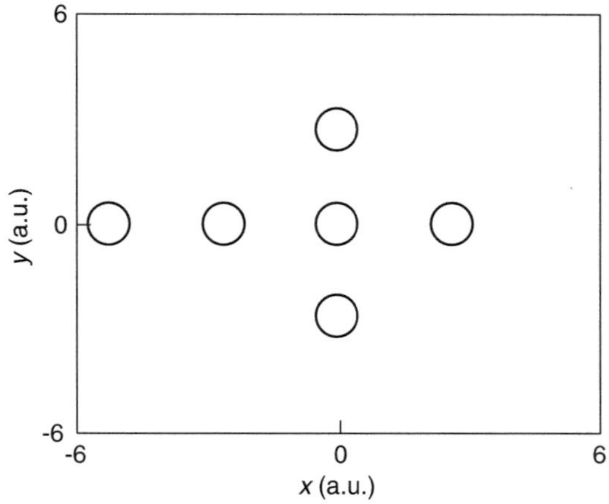

Fig. 4.1 Positions in a.u. of an incident atom with 400 eV on the left hand side and five rest target atoms at 0 s

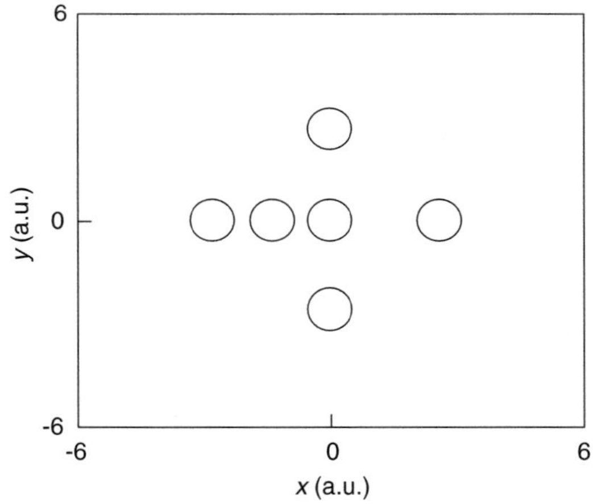

Fig. 4.2 Positions in a.u. of the incident atom with 400 eV on the left hand side and five target atoms at 1.76×10^{-15} s

five rest target atoms at $(-a_c,0)$, $(0,0)$, $(a_c,0)$, $(0,-a_c)$ and $(0,a_c)$. Here, a_c was set to a value of 2.63 a.u.

Figures 4.2, 4.3 and 4.4 show the positions of the incident atom with 400 eV and five target atoms at 1.76×10^{-15}, 3.16×10^{-15} and 4.10×10^{-15} s, respectively. It is clear that the energy is transferred sequentially through collisions. Figures 4.5 and 4.6 show the position and velocity in a.u. (including time units) of the incident atom as a function of time in units with second. Most of the energy of the incident atom is transferred during the initial collision. We note that the interatomic potential includes an attractive interaction between atoms, hence the positions and velocities of the atoms differ slightly from those derived by a solid sphere model.

Figures 4.7, 4.8 and 4.9 indicate the positions of atoms when the energy of an incident atom is 1 eV. Although some of this energy is transferred by collisions, most is not. The motion of the incident atom shows oscillations and is relatively complicated compared with the case of the 400 eV incident energy. Figures 4.10 and 4.11, which are the position and velocity of the incident atom with 1 eV as a

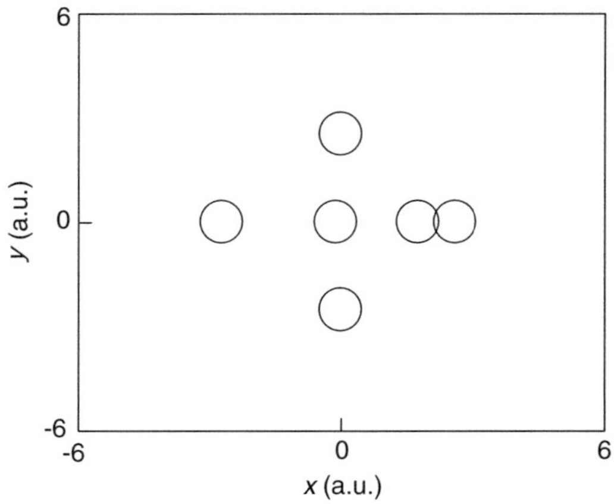

Fig. 4.3 Positions in a.u. of the incident atom with 400 eV on the left hand side and five target atoms at 3.16×10^{-15} s

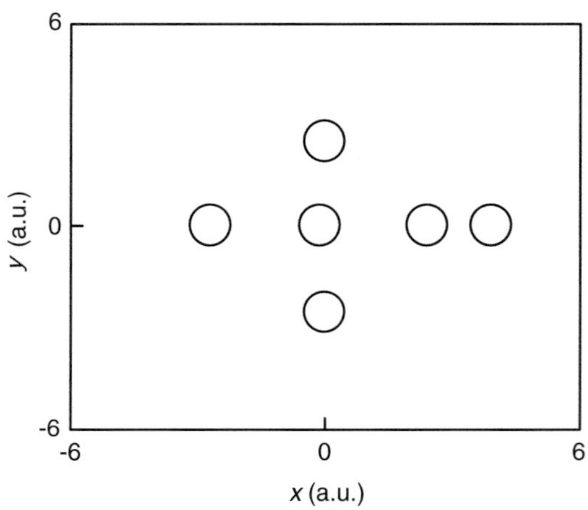

Fig. 4.4 Positions in a.u. of the incident atom with 400 eV on the left hand side and five target atoms at 4.10×10^{-15} s

function of time, confirm such complicated motion. These two types of collisions are seen in an experiment, where an ion injected with an accelerator into fullerene is trapped in a cage by the substitution for a constituent atom (Ohtsuki et al. 1995).

We subsequently analyzed the channeling in collisions, where an incident atom was injected along the x axis. Figures 4.12, 4.13, 4.14 and 4.15 show the positions of a 400 eV incident atom and the other target atoms as a function of time. Figures 4.16 and 4.17 are the time dependence of the position and the velocity of the incident atom, respectively. The incident atom passes through the interstitial regions around atoms. Figures 4.18, 4.19, 4.20 and 4.21 show the corresponding results for the 50 eV incident atom, while Figs. 4.22, 4.23, 4.24, 4.25 and 4.26 indicate the case of the 1 eV incident atom. The injected atom cannot pass through the interstitial region and the atom almost stops when the incident energy is 50 eV. Conversely, the incident atom oscillates when the energy of an incident atom is 1 eV.

The channeling analysis further added six atoms to the original six target atoms used in the previous calculation with the calculation conditions unchanged. Figures 4.27, 4.28, 4.29, 4.30, 4.31 and 4.32 show the results for the 400 eV incident

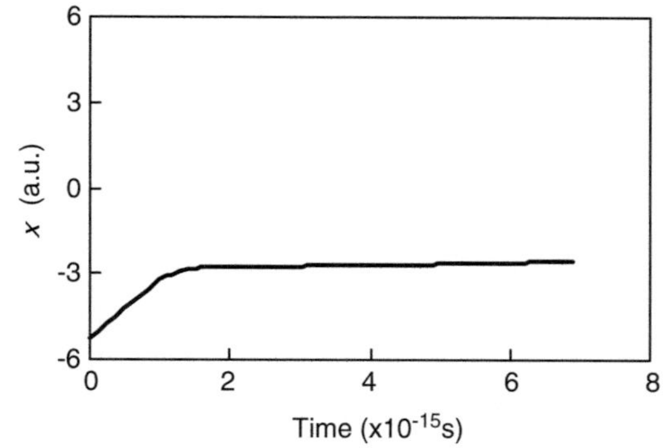

Fig. 4.5 Position in a.u. of the incident atom with 400 eV along the *x* axis as a function of time

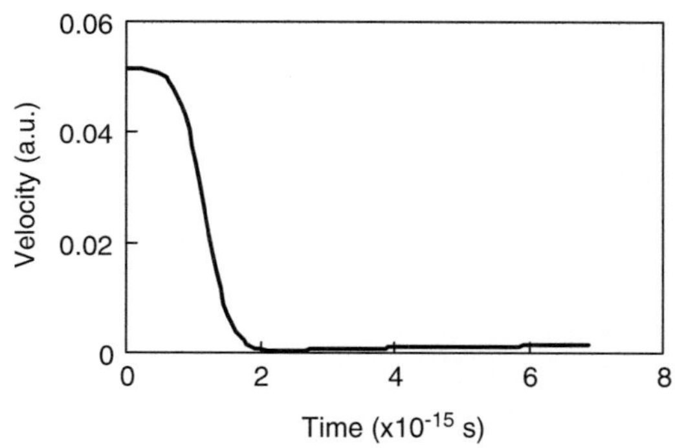

Fig. 4.6 Velocity in a.u. of the incident atom with 400 eV as a function of time

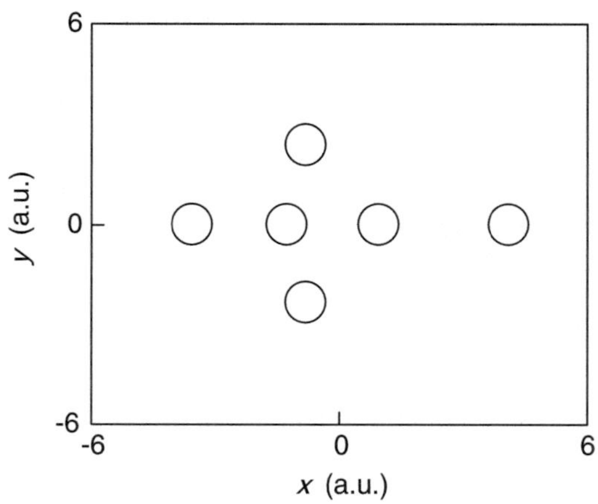

Fig. 4.7 Positions in a.u. of the incident atom with 1 eV on the left hand side and five target atoms at 3.75×10^{-14} s

energy, while Figs. 4.33, 4.34, 4.35 and 4.36 are for the 50 eV incident energy, and Figs. 4.37, 4.38, 4.39, 4.40 and 4.41 are for the 1 eV incident energy. The motions of twelve target atoms are suppressed compared to the previous case of six target atoms. Experimentally, such channeling is observed in ion bombardment on a material with accelerators (Feldman et al. 1982).

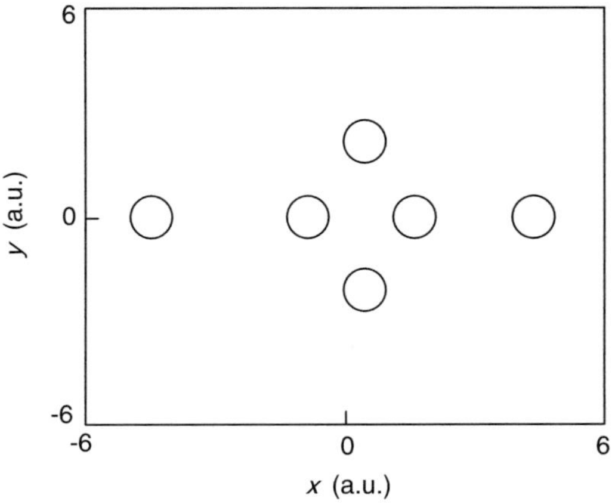

Fig. 4.8 Positions in a.u. of the incident atom with 1 eV on the left hand side and five target atoms at 6.56×10^{-14} s

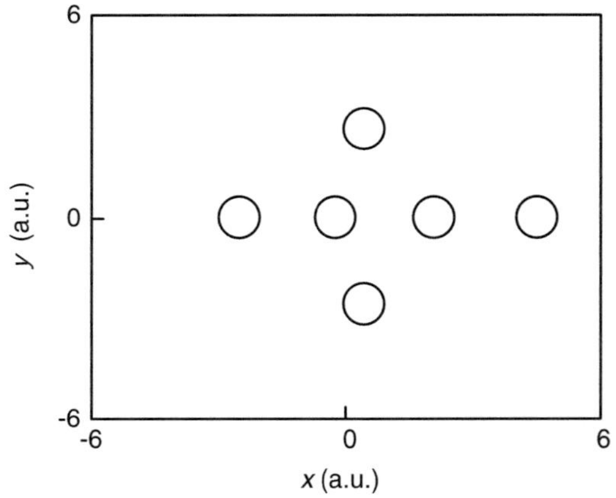

Fig. 4.9 Positions in a.u. of the incident atom with 1 eV on the left hand side and five target atoms at 9.36×10^{-14} s

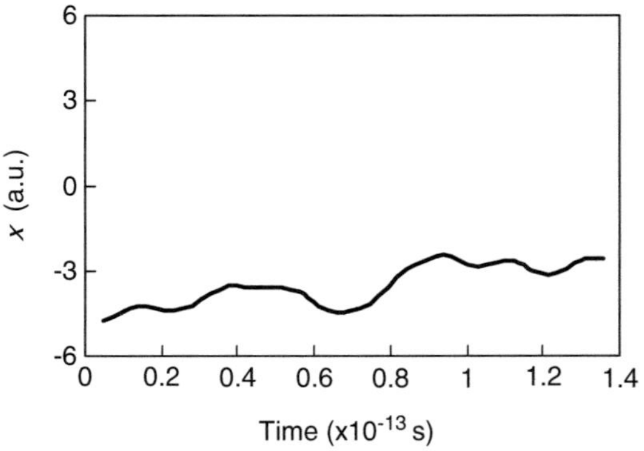

Fig. 4.10 Position in a.u. of the incident atom with 1 eV along the *x* axis as a function of time

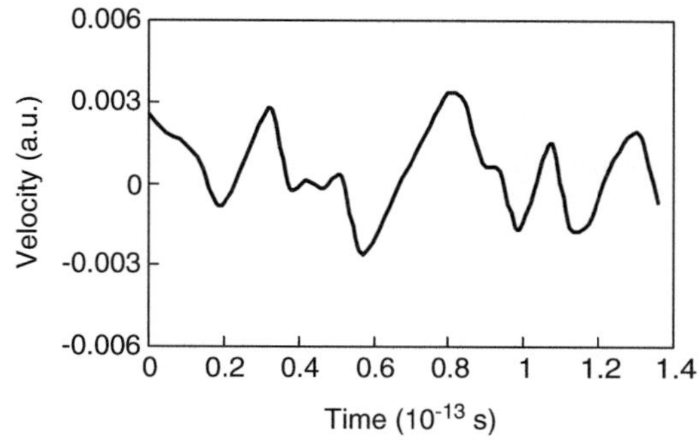

Fig. 4.11 Velocity in a.u. of the incident atom with 1 eV as a function of time

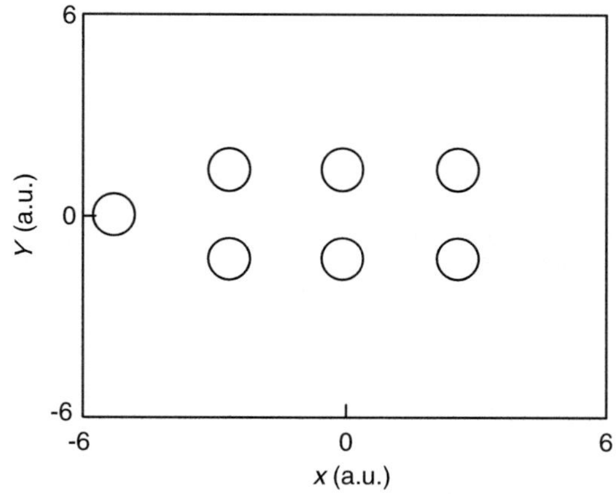

Fig. 4.12 Positions in a.u. of the incident atom with 400 eV on the left hand side and six target atoms at rest at 0 s in channeling

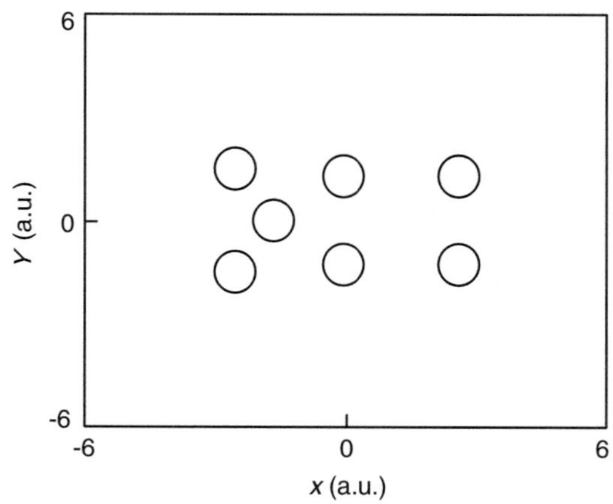

Fig. 4.13 Positions in a.u. of the incident atom with 400 eV and six target atoms at 1.76×10^{-15} s in channeling

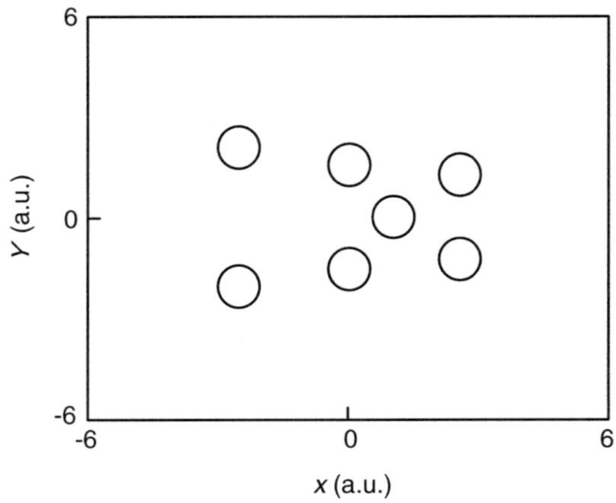

Fig. 4.14 Positions in a.u. of the incident atom with 400 eV and six target atoms at 3.16×10^{-15} s in channeling

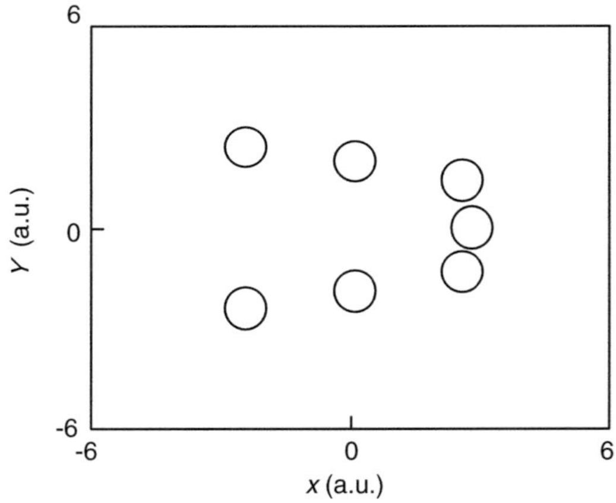

Fig. 4.15 Positions in a.u. of the incident atom with 400 eV and six target atoms at 4.10×10^{-15} s in channeling

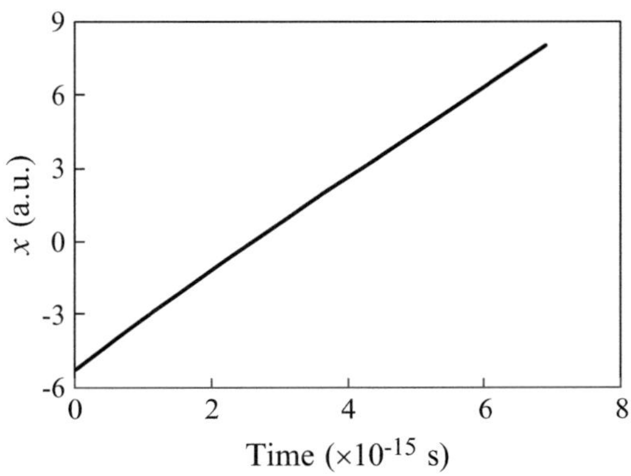

Fig. 4.16 Position in a.u. of the incident atom with 400 eV along the x axis in six target atoms as a function of time in channeling

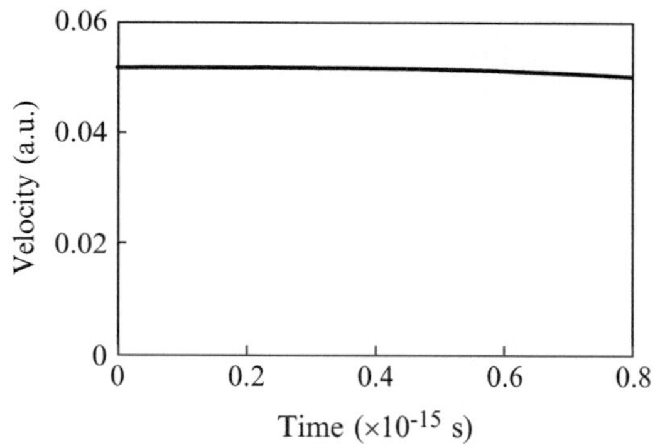

Fig. 4.17 Velocity in a.u. of the incident atom with 400 eV along the x axis in six target atoms as a function of time in channeling

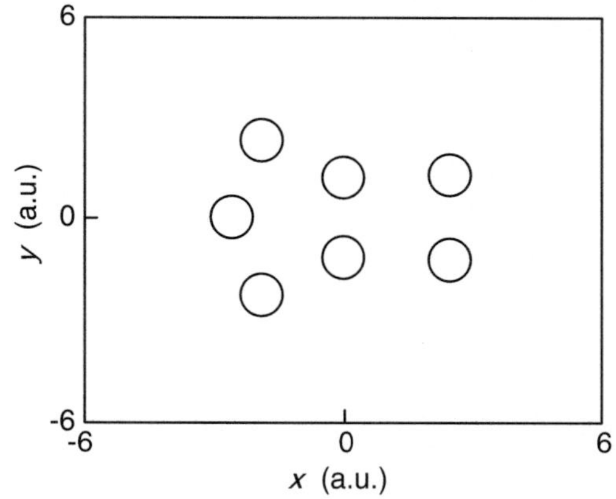

Fig. 4.18 Positions in a.u. of the incident atom with 50 eV and six target atoms at 5.30×10^{-15} s in channeling

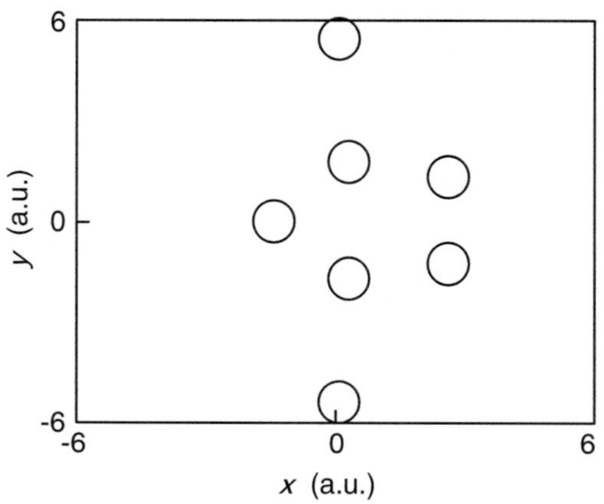

Fig. 4.19 Positions in a.u. of the incident atom with 50 eV and six target atoms at 1.32×10^{-14} s in channeling

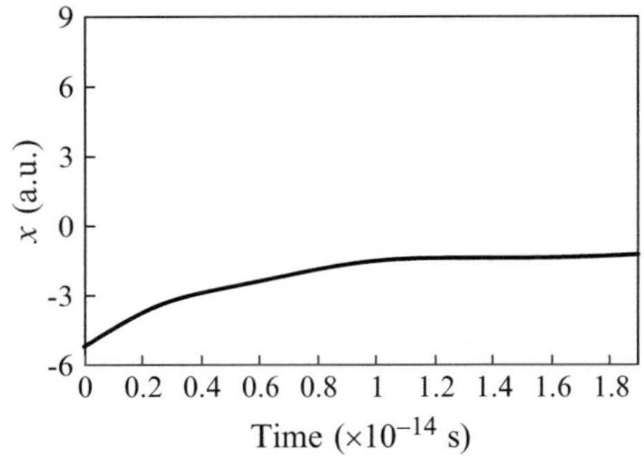

Fig. 4.20 Position in a.u. of the incident atom with 50 eV along the x axis in six target atoms as a function of time in channeling

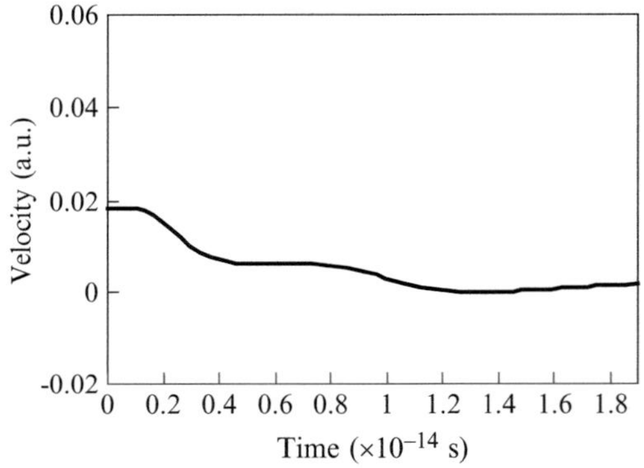

Fig. 4.21 Velocity in a.u. of the incident atom with 50 eV along the x axis in six target atoms as a function of time in channeling

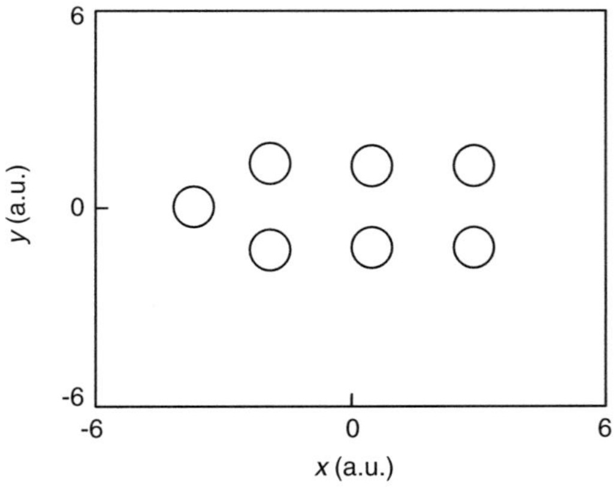

Fig. 4.22 Positions in a.u. of the incident atom with 1 eV and six target atoms at 4.45×10^{-14} s in channeling

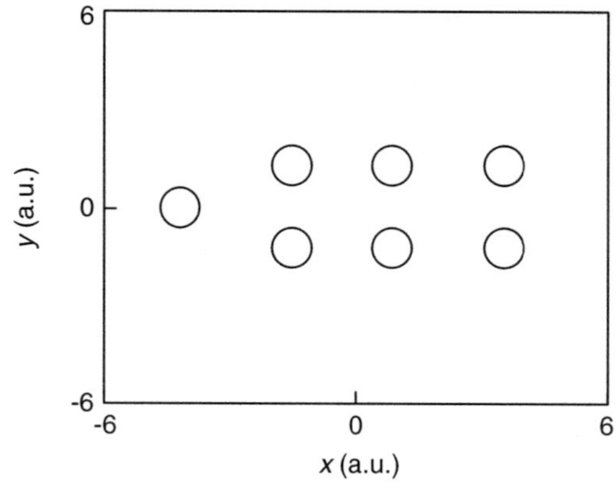

Fig. 4.23 Positions in a.u. of the incident atom with 1 eV and six target atoms at 6.56×10^{-14} s in channeling

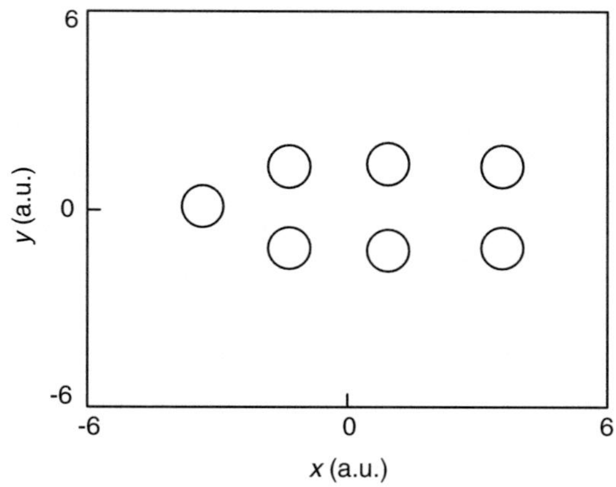

Fig. 4.24 Positions in a.u. of the incident atom with 1 eV and six target atoms at 7.96×10^{-14} s in channeling

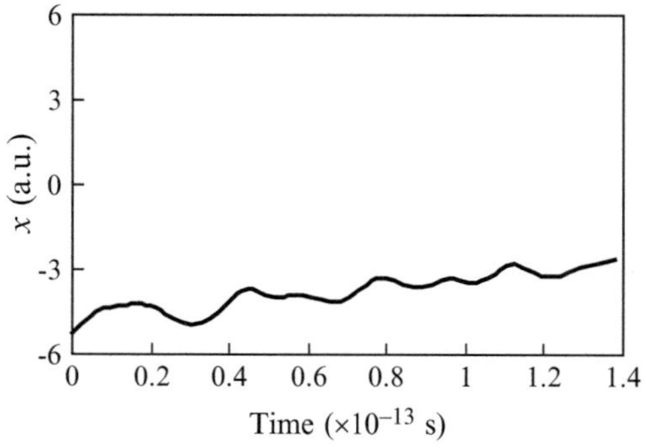

Fig. 4.25 Position in a.u. of the incident atom with 1 eV along the x axis in six target atoms as a function of time in channeling

Fig. 4.26 Velocity in a.u. of the incident atom with 1 eV along the *x* axis in six target atoms as a function of time in channeling

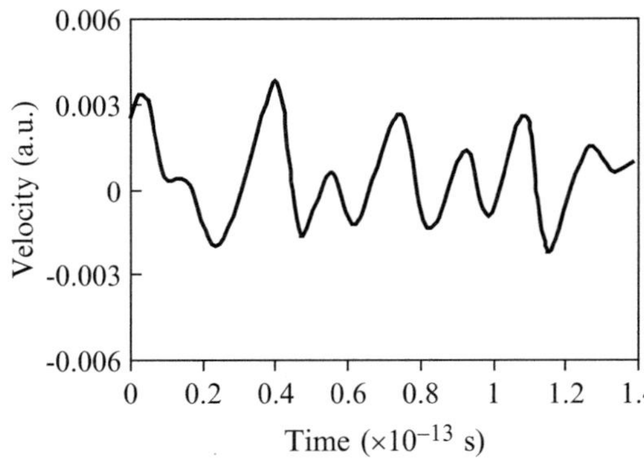

Fig. 4.27 Positions in a.u. of the incident atom with 400 eV on the left hand side and 12 target atoms at rest at 0 s in channeling

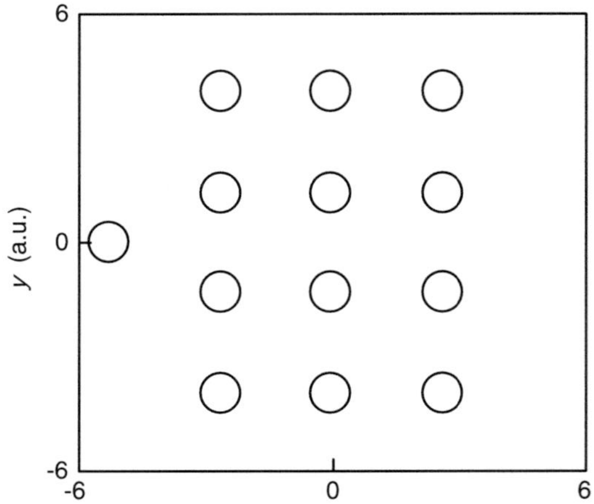

Fig. 4.28 Positions in a.u. of the incident atom with 400 eV and 12 target atoms at 1.76×10^{-15} s in channeling

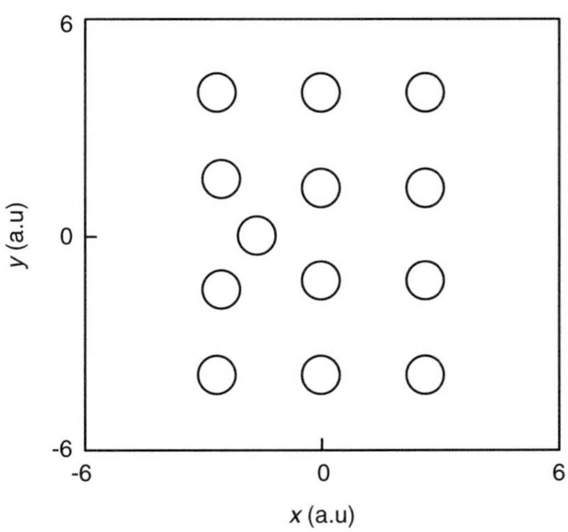

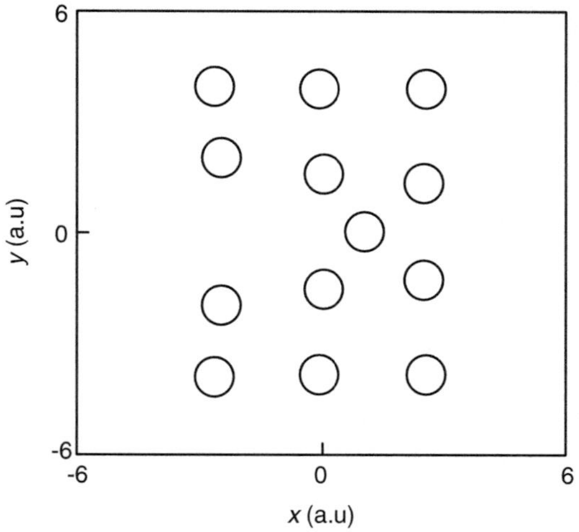

Fig. 4.29 Positions in a.u. of the incident atom with 400 eV and 12 target atoms at 3.16×10^{-15} s in channeling

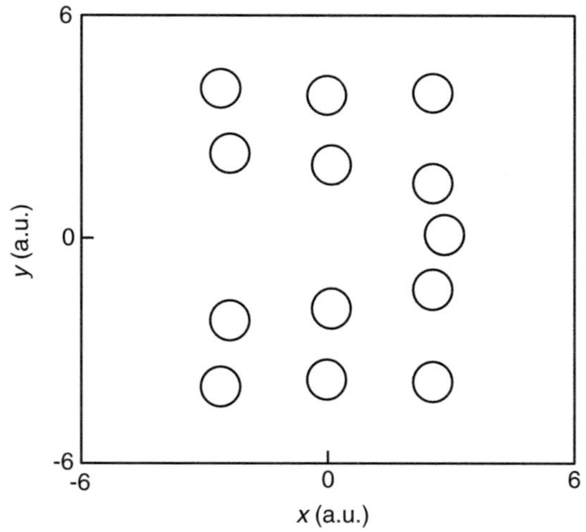

Fig. 4.30 Positions in a.u. of the incident atom with 400 eV and 12 target atoms at 4.10×10^{-15} s in channeling

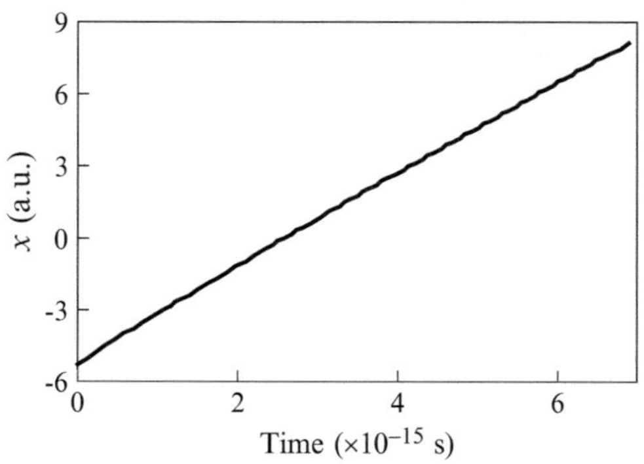

Fig. 4.31 Position in a.u. of the incident atom with 400 eV along the x axis in 12 target atoms as a function of time in channeling

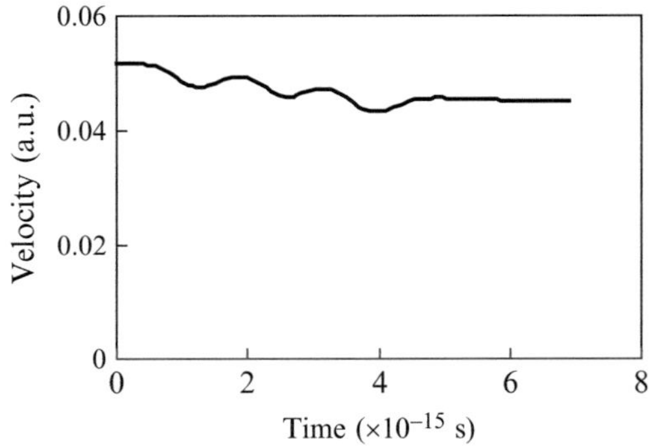

Fig. 4.32 Velocity in a.u. of the incident atom with 400 eV along the *x* axis in 12 target atoms as a function of time in channeling

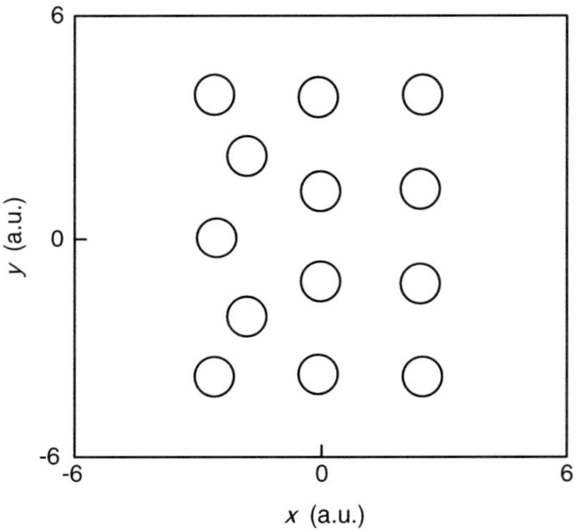

Fig. 4.33 Positions in a.u. of the incident atom with 50 eV and 12 target atoms at 5.30×10^{-15} s in channeling

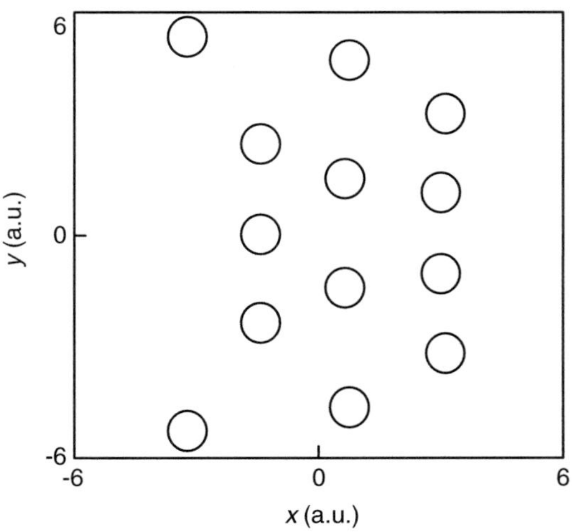

Fig. 4.34 Positions in a.u. of the incident atom with 50 eV and 12 target atoms at 1.32×10^{-14} s in channeling

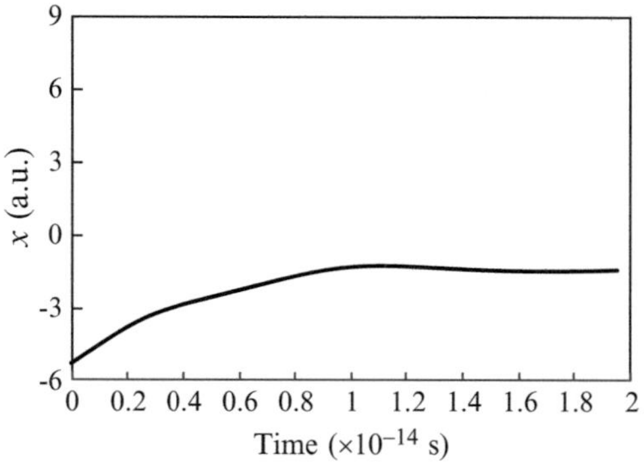

Fig. 4.35 Position in a.u. of the incident atom with 50 eV along the x axis in 12 target atoms as a function of time in channeling

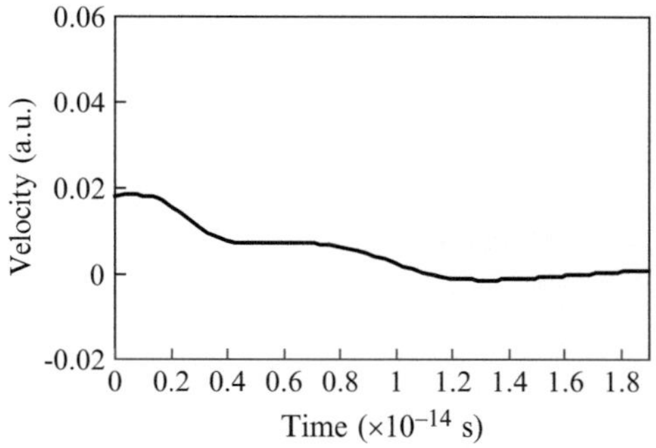

Fig. 4.36 Velocity in a.u. of the incident atom with 50 eV along the x axis in 12 target atoms as a function of time in channeling

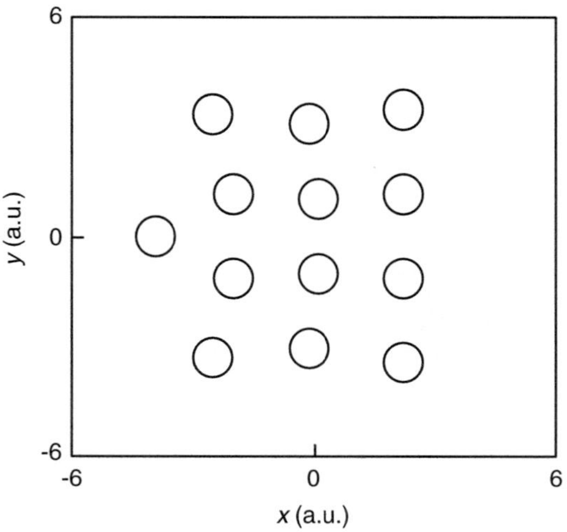

Fig. 4.37 Positions in a.u. of the incident atom with 1 eV and 12 target atoms at 1.40×10^{-14} s in channeling

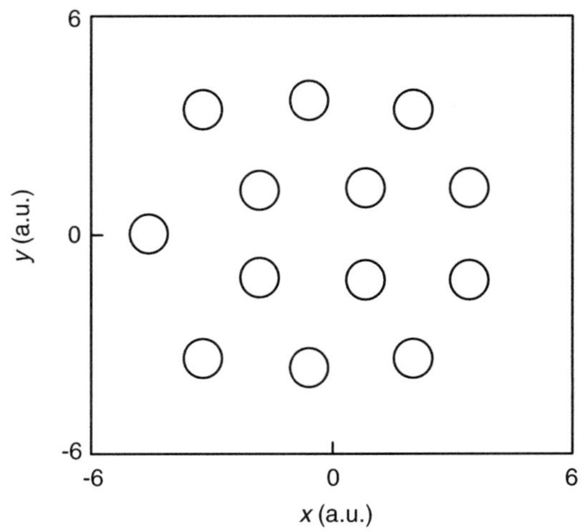

Fig. 4.38 Positions in a.u. of the incident atom with 1 eV and 12 target atoms at 2.34×10^{-14} s in channeling

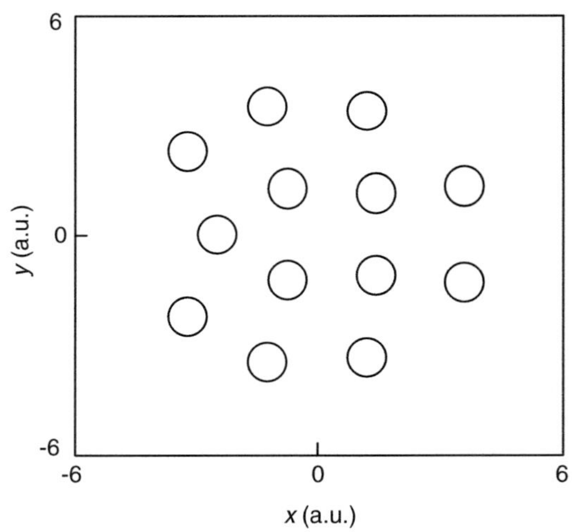

Fig. 4.39 Positions in a.u. of the incident atom with 1 eV and 12 target atoms at 4.45×10^{-14} s in channeling

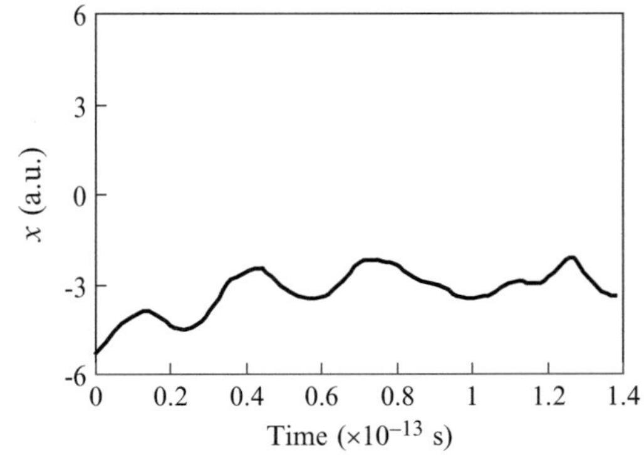

Fig. 4.40 Position in a.u. of the incident atom with 1 eV along the x axis in 12 target atoms as a function of time in channeling

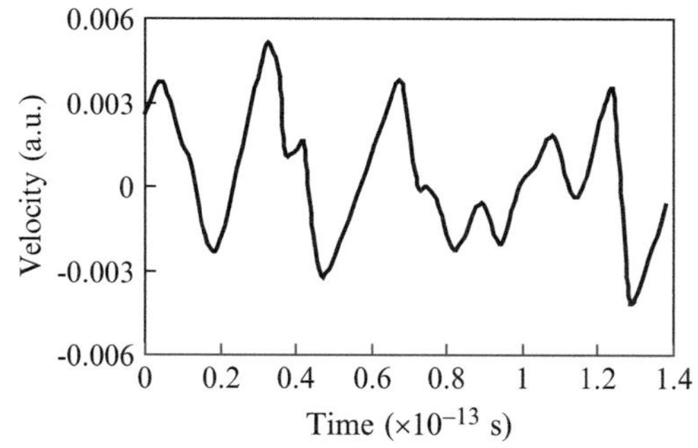

Fig. 4.41 Velocity in a.u. of the incident atom with 1 eV along the *x* axis in 12 target atoms as a function of time in channeling

Conclusions

The Born–Oppenheimer approximation enables an independent description of the motion of nuclei and electrons bound by the same particles. The total electronic energy, including repulsions among nuclei, amounts to the potential acting on nucleus. Classical molecular dynamics traces the motion of nuclei with electrons using the Newtonian equation. We developed a highly accurate method of calculating total electronic energy using Gaussian quadratures (GQ). This article has reported a method combining the Gaussian Quadrature method of total energy calculation with classical molecular dynamics. The author examined an energetic atom injected into a target object composed of atoms. Conventional high energy atomic calculations for nuclear materials consider a sequence of binary collisions between an incident atom and one of the target atoms. The present molecular dynamics, however, evaluated the interaction between an incident atom and surrounding target atoms. Even if the total interaction is derived by summing interactions between pair atoms, the derived potential differs from the binary potential between an incident atom and one target atom. When an injected atom is bound even in a short time by the potential produced by target atoms, it may be regarded as trapped or react with the surround atoms. When the incident atom finally goes away from the target atoms, we derive the scattering properties.

References

Adachi H, Tsukada M, Satoko C (1978) Discrete variational Xα cluster calculations. I. J Phys Soc Jpn 45:875–883

Alder BJ, Wainwright TE (1957) Phase transition for a hard sphere system. J Chem Phys 27:1208–1209

Ashcer UM, Petzold LR (1998) Computer methods for ordinary differential equations and differential-algebraic equations. SIAM, Philadelphia

Averill FW, Ellis DE (1973) An efficient numerical multicenter basis set for molecular orbital calculations: application to $FeCl_4$. J Chem Phys 59:6412–6418
Becke AD (1988) A multicenter numerical integration scheme for polyatomic molecules. J Chem Phys 88:2547–2553
Born M, Oppenheimer JR (1927) Zur Quantentheorie der Molekeln. Ann Phys (Leipzig) 84:457–484
Car R, Parrinello M (1985) Unified approach for molecular dynamics and density-functional theory. Phys Rev Lett 55:2471–2474
Crank J, Nicolson P (1947) A practical method for numerical evaluation of solution of partial differential equations of the heat conduction type. Proc Camb Philos Soc 43:50–67
Delley B (1990) An all-electron numerical method for solving the local density functional for polyatomic molecules. J Chem Phys 92:508–517
Ellis DE, Painter GS (1970) Discrete variational method for the energy-band problem with general crystal potentials. Phys Rev B 2:2887–2898
Feldman LC, Mayer JW, Picraux ST (1982) Material analysis by ion channelling. Academic Press, New York
Fukushima K, Mizuno J, Fujima K, Adachi H (1982) Gaussian quadrature Xα method for self-consistent Hartree-Fock-Slater equation. I. Application to the total energy calculation of diatomic molecule. J Phys Soc Jpn 51:4028–4035 doi: http://dx.doi.orgl.10.1143/JPSJ51.4028
Fukushima K (2009) A Gaussian quadrature method for total energy analysis in electronic state calculations. Int J Quantum Chem 109:2664–2671
Gáspár R (1954) Über eine approximation des Hartree-Fock schen potentials durch eine universelle potentialfunktion. Acta Phys Acad Sci Hung 3:263–286
Haile JM (1997) Molecular dynamics simulation. Wiley, New York
Janak JF, Moruzzi VL, Williams AR (1975) Ground-state thermomechanical properties of some cubic elements in the local-density formalism. Phys Rev B 12:1257–1261
Kohn W, Sham LJ (1965) Self-consistent equations including exchange and correlation effects. Phys Rev 140:A1133–A1138
Nozières P, Pines D (1958) Correlation energy of a free electron gas. Phys Rev 111:442–454
Ohtsuki T, Masumoto K, Sueki K, Kobayashi K, Kikuchi K (1995) Observation of radioactive fullerene families labeled with 11C. J Am Chem Soc 117:12869–12870
Perdew JP, Zunger A (1981) Self-interaction correction to density-functional approximations for many-electron systems. Phys Rev B 23:5048–5079
Rahman A, Stillinger FH (1973) Hydrogen-bond patterns in liquid water. J Am Chem Soc 95:7943–7948
Robinson MT, Torrens IM (1974) Computer simulation of atomic displacement cascades in solids in the binary-collision approximation. Phys Rev B 9:5008–5024
Rosén A, Ellis DE, Adachi H, Averill FW (1976) Calculations of molecular ionization energies using a self-consistent-charge Hartree-Fock-Slater method. J Chem Phys 65:3629–3634
Satoko C, Ohnishi S (1994) Density functional theory and its applications: electronic state of molecules and clusters (in Japanese). Kodansha, Tokyo
Slater JC (1951) A simplification of the Hartree-Fock method. Phys Rev 81:385–390
Slater JC (1974) Quantum theory of molecules and solids, vol 4, The self-consistent field for molecules and solids. McGraw-Hill, New York
Slater JC (1979) The calculation of molecular orbitals. Wiley, New York
Stillinger FH, Rahman A (1972) Molecular dynamics study of temperature effects on water structure and kinetics. J Chem Phys 57:1281–1292
Stillinger FH, Rahman A (1974) Improve simulation of liquid water by molecular simulation. J Chem Phys 60:1545–1557
Stroud AH, Secrest DH (1966) Gaussian quadrature formulas. Prentice-Hall, Englewood Cliffs
Tjalling JY (1995) Historical development of the Newton–Raphson method. SIAM Rev 37:531–551

Verlet L (1967) Computer "experiments" on classical fluids. I. Thermodynamical properties of Lennard-Jones molecules. Phys Rev 159:98–103
von Barth U, Hedin L (1972) A local exchange-correlation potential for the spin polarized case. i. J Phys C 5:1629–1642
Vosko SH, Wilk L, Nusair M (1980) Accurate spin-dependent electron liquid correlation energies for local spin density calculations: a critical analysis. Can J Phys 58:1200–1211
Wigner E (1934) On the interaction of electrons in metals. Phys Rev 46:1002–1011

Chapter 5
Comparison of Contributions to Interatomic Interactions Between Covalent and Ionic Bonds from Total Energy Calculations

Kimichika Fukushima

5.1 Introduction

The density functional Theory (DFT) (Slater 1951, 1974, 1979; Gáspár 1954; Nozières and Pines 1958; Kohn and Sham 1965; Wigner 1934; von Barth and Hedin 1972; Janak et al. 1975; Vosko et al. 1980; Perdew and Zunger 1981) and calculation methods (Ellis and Painter 1970; Adachi et al. 1978; Averill and Ellis 1973; Rosén et al. 1976; Satoko and Ohnishi 1994) has greatly succeeded for atoms, molecules, solids as well as practical materials. The potential energy surface predicted by using DFT leads to reasonable equilibrium interatomic distances and binding energies. The present author and coworkers developed the Gaussian quadrature (GQ) method (Fukushima et al. 1982; Fukushima 2009) for total energy calculations. The matrices of the Hamiltonian, overlap integral as well as total energy are numerically integrated using the Gauss–Legendre and Gauss–Laguerre quadratures (Stroud and Secrest 1966).

The atom-atom interaction and interaction potential are useful quantities for properties of molecules and solids as well as practical materials. The simple and practical quantities used to evaluate the bond strength between atoms are the Mulliken overlap charge (Mulliken 1955a, b, c, d) and the long range Coulomb interaction. The former implies the covalent bond strength and the latter ionic interaction. Meanwhile, the total energy in the Born–Oppenheimer approximation expresses the atom-atom interaction potential. Total energy evaluations require a highly accurate integration of the matrix elements of the Hamiltonian and the overlap integral as well as the total energy. The present author and coworkers developed the Gaussian quadrature method deriving the highly accurate total energy. The nitrogen-nitrogen interaction is apparently dominated by the covalent bond, while the main lithium-fluorine interaction is

K. Fukushima (✉)
Advanced Reactor System Engineering Department, Toshiba Nuclear Engineering Service Corporation, 8, Shinsugita-cho, Isogo-ku, Yokohama 235-8523, Japan
e-mail: kimichika1a.fukushima@glb.toshiba.co.jp

T. Ishii et al. (eds.), *The DV-Xα Molecular-Orbital Calculation Method*,
DOI 10.1007/978-3-319-11185-8_5

probably the ionic interaction. It is not obvious to the present author which is the main interaction between boron and nitrogen atoms. The preliminary calculation using the total energy clarified the dominant interaction.

This chapter is organized as follows. Section 5.2 describes the formalism of the total energy and Mulliken populations. Section 5.3 compares contributions of the covalent bond and the ionic bond to the total energy elucidating the main contribution, followed by conclusions.

5.2 Formalism

The total electronic energy of an electronic system E_e, which amounts in the density functional scheme in atomic units to

$$E_e = \sum_i \int d\boldsymbol{r}\psi_i^*(\boldsymbol{r})\left[-\frac{1}{2}\nabla_r^2 + V_N(\boldsymbol{r}) + \frac{1}{2}V_C(\boldsymbol{r})\right]\psi_i(\boldsymbol{r}) + E_{xc}[\rho(\boldsymbol{r})] + \sum_{I<J}\frac{Z_I Z_J}{|\boldsymbol{R}_I - \boldsymbol{R}_J|}, \tag{5.1}$$

where $V_N(\boldsymbol{r})$ and $V_C(\boldsymbol{r})$ are nuclear and electronic Coulomb potentials, respectively, and Z_I is the nuclear charge. The quantity $E_{xc}[\rho(\boldsymbol{r})]$ is the exchange-correlation energy, which is a functional of electronic charge density $\rho(\boldsymbol{r})$. An electronic wave function is expanded in terms of atomic basis functions expressed by (Here, atomic orbitals are real numbers.)

$$\psi_i(\boldsymbol{r}) = \sum_{jI} C_{ijI}\chi_j(\boldsymbol{r} - \boldsymbol{R}_I). \tag{5.2}$$

The overlap charge between atoms I and J is denoted as

$$Q_{IJ} = 2\sum_{ijk} n_i C_{ijI} C_{ikJ}\chi_j(\boldsymbol{r} - \boldsymbol{R}_I)\chi_k(\boldsymbol{r} - \boldsymbol{R}_J), \tag{5.3}$$

where n_i is the number of electrons occupying the ith molecular orbital. The long range ionic interaction between ions I and J are

$$E_{IJ} = \frac{\Delta Q_I \Delta Q_J}{|\boldsymbol{R}_J - \boldsymbol{R}_J|}, \tag{5.4}$$

where the ionic charge ΔQ_I is defined as

$$\Delta Q_I = -\sum_J \sum_{ijk} n_i C_{ijI} C_{ikJ}\chi_j(\boldsymbol{r} - \boldsymbol{R}_I)\chi_k(\boldsymbol{r} - \boldsymbol{R}_J) + Z_I. \tag{5.5}$$

5.3 Results and Discussion

The overlap charge and total energy were calculated for a diatomic N_2 molecule. Table 5.1 shows the dependence of fractions of the relative overlap charge and the relative total energy on interatomic distances, where E_e^b is the total electronic energy, which does not include the nucleus-nucleus repulsion. Both the overlap charge and the total energy are normalized by the difference between values at the interatomic distances of the inner bound $d_i = 2.3$ a.u. and the outer bound $d_o = 3.2$ a.u. Fractions of the relative overlap charge and the relative total energy are then denoted as

$$Q_{12f} = \frac{Q_{12}(|\boldsymbol{R}_1 - \boldsymbol{R}_2|) - Q_{12}(d_o)}{Q_{12}(d_i) - Q_{12}(d_o)}, \tag{5.6}$$

and

$$E_{ef}^b = \frac{E_e^b(|\boldsymbol{R}_1 - \boldsymbol{R}_2|) - E_e^b(d_o)}{E_e^b(d_i) - E_e^b(d_o)}, \tag{5.7}$$

respectively. The dependence of the covalent bond strength on interatomic distances corresponds well to that of the total energy. It is seen that the main contribution to the interatomic interaction in N_2 is the covalent bond.

Next, the ionic interaction and total energy were calculated for the LiF diatomic molecule. Table 5.2 shows the dependence of fractions of the relative ionic interaction, which is the long-range Coulomb attraction, and the relative total energy on interatomic distances. The dependence of the fractional relative ionic interaction on interatomic distance is evaluated using a definition given by

$$E_{ionf} = \frac{E_{ion}(|\boldsymbol{R}_1 - \boldsymbol{R}_2|) - E_{ion}(d_o)}{E_{ion}(d_i) - E_{ion}(d_o)}. \tag{5.8}$$

Table 5.1 Fractions of the relative overlap charge and relative total energy excluding nucleus-nucleus repulsion as a function of the interatomic distance for N_2

Interatomic distance (a.u.)	Fractional relative overlap charge $Q_{12f} = \frac{Q_{12}(\|\boldsymbol{R}_1-\boldsymbol{R}_2\|)-Q_{12}(d_o)}{Q_{12}(d_i)-Q_{12}(d_o)}$	Fractional relative total energy (excluding nucleus-nucleus repulsion) $E_{ef} = \frac{E_e^b(\|\boldsymbol{R}_1-\boldsymbol{R}_2\|)-E_e^b(d_o)}{E_e^b(d_i)-E_e^b(d_o)}$
2.3	1.00	1.00
2.5	0.79	0.74
2.6	0.57	0.52
2.8	0.36	0.33
3.0	0.17	0.15
3.2	0.000	0.000

Table 5.2 Fractions of the relative ionic interaction and the relative total energy as a function of the interatomic distance for LiF

Interatomic distance (a.u.)	Fractional relative ionic interaction $E_{\text{ionf}} = \frac{E_{\text{ion}}(\lvert \boldsymbol{R}_1 - \boldsymbol{R}_2 \rvert) - E_{\text{ion}}(d_o)}{E_{\text{ion}}(d_i) - E_{\text{ion}}(d_o)}$	Fractional relative total energy $E_{\text{ef}} = \frac{E_e(\lvert \boldsymbol{R}_1 - \boldsymbol{R}_2 \rvert) - E_e(d_o)}{E_e(d_i) - E_e(d_o)}$
3.21	1.000	1.000
3.40	0.660	0.716
3.59	0.353	0.479
3.78	0.160	0.271
3.97	0.000	0.000

Table 5.3 Fractions of the relative overlap charge, the relative ionic interaction and the relative total energy as a function of the interatomic distance for BN

Interatomic distance (a.u.)	Fractional relative overlap charge	Fractional relative ionic interaction	Fractional relative total energy (excluding nucleus-nucleus repulsion)
2.6	1.00	1.00	1.00
2.7	0.76	0.63	0.73
2.8	0.51	0.36	0.47
2.9	0.26	0.16	0.23
3.0	0.00	0.00	0.00

The dependence of the fractional relative ionic interaction strength on interatomic distances corresponds well to that of the fractional relative total energy. Ionic interaction can thus be considered the main contribution to the interatomic bond in LiF.

To evaluate the main contribution to the bond in a diatomic BN molecule, the covalent overlap population, ionic interaction and total energy were calculated. Table 5.3 shows the dependence of fractions of the relative covalent bond, the relative ionic interaction and the relative total energy on interatomic distances. The dependence of the relative overlap charge on interatomic distances corresponds well to that of the relative total energy, hence the main contribution to the interatomic bond in BN is the covalent bond.

Conclusions

Using the GQ (Gaussian Quadrature) method with a highly accuracy for total energy calculations, this article has derived that the interaction between nitrogen-nitrogen atoms is the covalent bond, whose bond strength expressed in terms of Mulliken overlap charge is consistent with the total energy. Conversely, lithium-fluorine (LiF) interaction is the ionic interaction, which is preliminarily supported by the long rage Coulomb interaction with the Mulliken charges, which indicates the consistency to the total energy.

References

Adachi H, Tsukada M, Satoko C (1978) Discrete variational Xα cluster calculations. I. J Phys Soc Jpn 45:875–883

Averill FW, Ellis DE (1973) An efficient numerical multicenter basis set for molecular orbital calculations: Application to $FeCl_4$. J Chem Phys 59:6412–6418

Ellis DE, Painter GS (1970) Discrete variational method for the energy-band problem with general crystal potentials. Phys Rev B 2:2887–2898

Fukushima K, Mizuno J, Fujima K, Adachi H (1982) Gaussian quadrature Xα method for self-consistent Hartree-Fock-Slater equation. I. Application to the total energy calculation of diatomic molecule. J Phys Soc Jpn 51:4028–4035

Fukushima K (2009) A Gaussian quadrature method for total energy analysis in electronic state calculations. Int J Quantum Chem 109:2664–2671

Gáspár R (1954) Über eine approximation des Hartree-Fock schen potentials durch eine universelle potentialfunckion. Acta Phys Acad Sci Hung 3:263–286

Janak JF, Moruzzi VL, Williams AR (1975) Ground-state thermomechanical properties of some cubic elements in the local-density formalism. Phys Rev B 12:1257–1261

Kohn W, Sham LJ (1965) Self-consistent equations including exchange and correlation effects. Phys Rev 140:A1133–A1138

Mulliken RS (1955a) Electronic population analysis on LCAO-MO in molecular wave functions I. J Chem Phys 23:1833–1840

Mulliken RS (1955b) Electronic population analysis on LCAO-MO in molecular wave functions II. Overlap populations, bond orders, and covalent bond energies. J Chem Phys 23:1841–1846

Mulliken RS (1955c) Electronic population analysis on LCAO-MO in molecular wave functions III. Effects of hybridization on overlap and gross AO populations. J Chem Phys 23:2338–2342

Mulliken RS (1955d) Electronic population analysis on LCAO-MO in molecular wave functions IV. Bonding and antibonding in LCAO and valence-bond theories. J Chem Phys 23:2343–2346

Nozières P, Pines D (1958) Correlation energy of a free electron gas. Phys Rev 111:442–454

Perdew JP, Zunger A (1981) Self-interaction correction to density-functional approximations for many-electron systems. Phys Rev B 23:5048–5079

Rosén A, Ellis DE, Adachi H, Averill FW (1976) Calculations of molecular ionization energies using a self-consistent-charge Hartree-Fock-Slater method. J Chem Phys 65:3629–3634

Satoko C, Ohnishi S (1994) Density functional theory and its applications: electronic state of molecules and clusters (in Japanese). Kodansha, Tokyo

Slater JC (1951) A simplification of the Hartree-Fock method. Phys Rev 81:385–390

Slater JC (1974) Quantum theory of molecules and solids. The self-consistent field for molecules and solids, vol 4. McGraw-Hill, New York

Slater JC (1979) The calculation of molecular orbitals. Wiley, New York

Stroud AH, Secrest DH (1966) Gaussian quadrature formulas. Prentice-Hall, Englewood Cliffs

von Barth U, Hedin L (1972) A local exchange-correlation potential for the spin polarized case. i. J Phys C 5:1629–1642

Vosko SH, Wilk L, Nusair M (1980) Accurate spin-dependent electron liquid correlation energies for local spin density calculations: a critical analysis. Can J Phys 58:1200–1211

Wigner E (1934) On the interaction of electrons in metals. Phys Rev 46:1002–1011

Chapter 6
Total Energy Calculation by DV-Xα Method and Its Accuracy

Katsumi Nakagawa

The basis functions of DV-Xα method are prepared numerically for the environment proper to the molecule or the cluster to be calculated. They make it easier to apply DV-Xα method for wide variety of models but make it difficult to calculate total energy of the model with chemical accuracy.

The author improved the total energy calculating method based on the spin-DV-Xα method. In this article, the method is briefly described and applied for various organic or inorganic molecules comprising various atoms from H to Xe. Calculated results are compared with experimental data and their counterparts by various methods in the Gaussian package to evaluate their accuracy.

It was found that they are more accurate than expected from the simple calculating method, provided that molecular orbitals were prepared appropriately. The obtained results demonstrate the usefulness of numerical basis functions and the feasibility of DV-Xα method as a convenient tool to analyze quantitatively wide variety of molecules. At the same time, some unsatisfactory results suggest how to prepare more appropriate molecular orbitals in DV-Xα method.

6.1 Introduction

The DV-Xα (Discrete Variational Xα) method is based on Hartree-Fock-Slater (HFS) equation and numerical treatment of molecular integrals. In HFS equation, complicated electron exchange interaction of Hartree-Fock equation is replaced by

K. Nakagawa (✉)
MO BASICS Research, 3-1-1-201, Hiyoshi-Honcho, Kohoku-ku,
Yokohama-shi, Kanagawa 223-0062, Japan
e-mail: nakagawa.katsumi@nifty.com

T. Ishii et al. (eds.), *The DV-Xα Molecular-Orbital Calculation Method*,
DOI 10.1007/978-3-319-11185-8_6

simple Xα potential, which makes the calculations easier. And the numerical treatment makes it possible to adopt accurate atomic orbitals (AOs) as basis functions. In Japan, the DV-Xα program is developed by Adachi and his collaborators (Adachi et al. 1978). The DV-Xα program package including source codes, is distributed by the society for Discrete Variational Xα and it has been widely used to assign spectral lines for variety of molecules or crystals and to interpret structures or reactions of various molecules by the way of Mulliken's population analysis. The main program in the package that calculates spin-polarizable molecular orbitals (MOs)s and their eigenvalues has the name "sspin". Hereafter let's call it SSPIN.

Unfortunately the numerical treatment makes it difficult to calculate total energy with accuracies required for chemical application. Ziegler demonstrated that molecular bond formation energy can be calculated accurately by the transition state method (Ziegler and Rauk 1977). But his method can calculate only one bond formation energy in a molecule at a time and has limited application. Ogasawara developed the total energy calculation program "Coulomb" (Kowada et al. 2009). In their method, AOs calculated by SSPIN are used as basis functions but MOs are recalculated by using more accurate potential than SSPIN. It requires additional time to get SCF convergence again. The author developed "Total energy calculation by spin-DV-Xα" (TESDA) program (Nakagawa 2003). It calculates total energy using MOs by SSPIN as they are and is equipped with error cancelling procedure, which realized total energy calculation with relatively small number of sample points. Further, results by TESDA reflect the characteristics of MOs and will give suggestions to obtain more accurate physical or chemical properties by SSPIN.

In this article, the author described DV-Xα method and his total energy calculating method briefly and applied it for various atoms, organic or inorganic molecules including elements from H to Xe. For some cases, he compared obtained results with their counterparts by several methods in the Gaussian package and evaluated accuracy of results by his method.

6.2 DV-Xα Method and Its Characteristics

6.2.1 Hartree-Fock-Slater Equation

In Hartree-Fock method, electronic exchange interaction is included in the total energy by using a Slater determinant describing the electronic state. But it requires fairy large calculation cost. Instead of it, Slater defined a Fermi hole, which is formed by electronic exchange interaction around a specified point in the electron cloud and its influence is represented by Xα term. In his method (Hartree-Fock-Slater method), total energy ε of the system is represented like

$$\begin{aligned}\varepsilon = &\sum_{\sigma}\sum_{k} f_k^{\sigma} \int \phi_k^{\sigma}(r)\left(-\frac{1}{2}\nabla^2\right)\phi_k^{\sigma}(r)dr - \int \sum_{\nu} \frac{Z_\nu}{r_\nu}\rho(r)dr \\ &+ \frac{1}{2}\iint \rho(r_1)\frac{1}{r_{12}}\rho(r_2)dr_1 dr_2 - \frac{9}{4}\alpha\left(\frac{3}{4\pi}\right)^{1/3}\int \left\{\rho^{\alpha}(r)^{4/3} + \rho^{\beta}(r)^{4/3}\right\}dr.\end{aligned} \tag{6.1}$$

Here, the first term of right side represents kinetic energy, where $\phi_k^{\sigma}(r)$ denotes k-th MO with $\sigma = \alpha$-spin or β-spin, f_k^{σ} denotes its occupation number. The second term represents nuclear-electron interaction, where Z_ν denotes charge of ν-th nucleus. The third term represents electron–electron interaction and the fourth term Xα energy, which means electron–electron interaction reduced by Fermi hole formation. Here, $\rho(r)$ denotes electron density, or

$$\rho(r) = \rho^{\alpha}(r) + \rho^{\beta}(r) = \sum_{k} f_k^{\alpha}\phi_k(r)^2 + \sum_{k} f_k^{\beta}\phi_k(r)^2. \tag{6.2}$$

and α is a constant, which is usually set to 0.7.

HFS equation is obtained by applying the variational principle for Eq. (6.1) and represented as follows,

$$\begin{aligned}&\left\{-\frac{1}{2}\nabla^2 - \sum_{\nu}\frac{Z_\nu}{r_{1\nu}} + \int \frac{1}{r_{12}}\rho(r_2)dr_2 - 3\alpha\left(\frac{3}{4\pi}\rho^{\sigma}(r_1)\right)^{1/3}\right\}\phi_k^{\sigma}(r_1) \\ &= \varepsilon_k^{\sigma}\phi_k^{\sigma}(r_1).\end{aligned} \tag{6.3}$$

Here ε_k^{σ} is the eigenvalue of ϕ_k^{σ}. In HFS equation, electronic state is represented by a set of one electron orbital ϕ_k^{σ} ($k = 1, 2, 3, \ldots, k_{\text{max\_}\sigma}$). DV-Xα method is based on HFS equation (6.3), which also can be considered as Kohn-Sham type density functional equation having the primitive local functional.

6.2.2 Atomic Orbitals as Basis Functions

Because MOs are very similar to AOs at least near each nucleus of the molecule, it is desirable that AOs themselves are utilized as basis functions. But GTOs, which are usually used in ab-initio method and DFT, are far from AO-like and complicated combination of various types of GTO is indispensable to get accurate results. In DV-Xα method, basis set comprises actual AOs, radial components of which are calculated numerically according to the electron distribution for each atom at each SCF step and multiplied by spherical harmonic functions. Figure 6.1 shows radial components of AOs of Zn atom, where every orbital has correct number of nodes and every orbital has exact form near the nuclei.

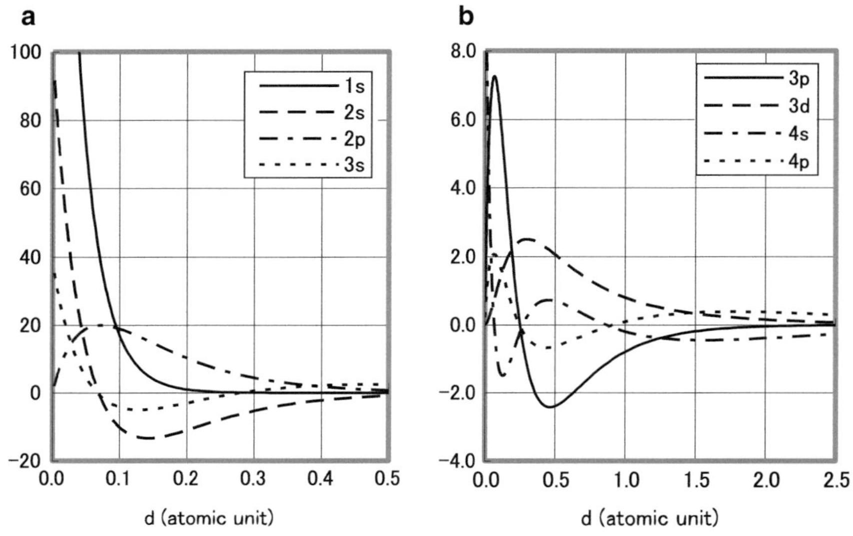

Fig. 6.1 (**a**) Radial components of 1s–3s orbitals of Zn atom. (**b**) Radial components of 3p–4p orbitals of Zn atom

6.2.3 Discrete Variational Method

For the numerical integration in 1D space, various simple and high performance methods can be used. For example, Simpson's method is used to calculate radial components of AOs of SSPIN. But for 3D space, Monte-Carlo method is almost the only practical method. In Monte-Carlo method, sample points are distributed uniformly in a unit cube by using quasi-random-number and mapped to around each nucleus. Figure 6.2 shows the sample point system generated by SSPIN for Zn atom. Because usually functions to be integrated have large absolute value near the nuclei, sample points are distributed densely near the nuclei and sparsely at peripheral area. In this case, almost all sample points are located inside the radius r = 6.0. Figure 6.3 shows the sample point system generated by SSPIN for $SiCl_4$ molecules. For molecule, sample points are shared randomly with all atoms constituting the molecule and each atom wears its own sample points. The sharing ratio of sample points for a heavier atom is adjusted to be relatively larger.

Each sample point r_i has the weight w_i inversely proportional to its appearance probability. The sum of weighted function value at each sample point corresponds to the integration of the function. To solve HFS equation by Loothaan's method, matrix elements of Fock operator F_{rs} and overlap integral S_{rs} have to be calculated numerically like

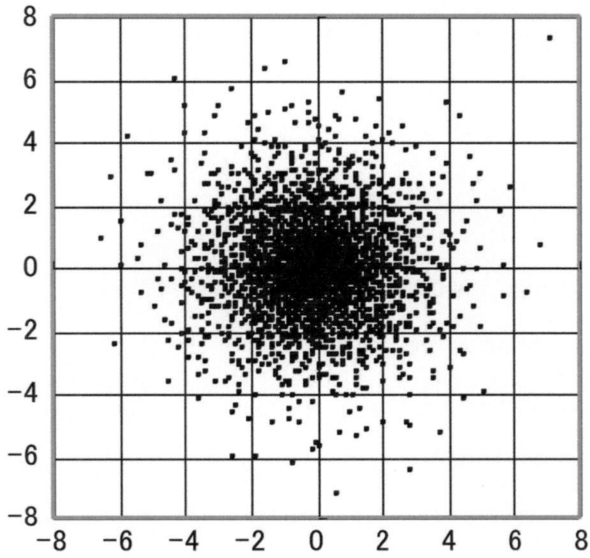

Fig. 6.2 Sample point system for a Zn atom

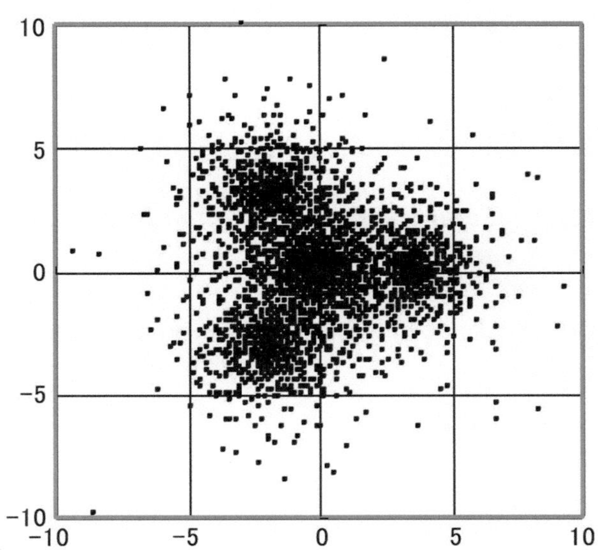

Fig. 6.3 Sample point system for a $SiCl_4$ molecule

$$F_{rs}^{\sigma} = \sum_i \chi_r^{\sigma}(r_i)\left\{-\frac{1}{2}\nabla^2 - \sum_{\nu}\frac{Z_{\nu}}{r_{i\nu}} + \sum_j \frac{\rho(r_j)w_j}{r_{ij}} - 3\alpha\left(\frac{3}{4\pi}\rho^{\sigma}(r_i)\right)^{1/3}\right\} \times \chi_s^{\sigma}(r_i)w_i \tag{6.4}$$

$$S_{rs}^{\sigma} = \sum_i \chi_r^{\sigma}(r_i)\chi_s^{\sigma}(r_i)w_i, \tag{6.5}$$

where χ_r^{σ} is r-th AO with spin σ. Using F_{rs}^{σ} and S_{rs}^{σ}, MO $\phi_k^{\sigma}(r) = \sum_r C_k^{\sigma}\chi_k^{\sigma}(r)$ and eigenvalue ε_k^{σ} are calculated.

Actually electron–electron Coulomb interaction is not calculated like the third term of Eq. (6.4) in SSPIN, but approximated as follows

$$\sum_j \frac{\rho(r_j)w_j}{r_{ij}} \approx \sum_{\nu}\left[\frac{4\pi}{r_{\nu i}}\int_0^{r_{\nu i}} \rho_{\nu}(r_{\nu}){r_{\nu}}^2 dr_{\nu} + 4\pi\int_{r_{\nu i}}^{\infty}\frac{\rho_{\nu}(r_{\nu}){r_{\nu}}^2}{r_{\nu}}dr_{\nu}\right] \tag{6.6}$$

Here $\rho(r)$ is approximated to the sum of the contribution of ρ_{ν}, which is spherically distributed around nucleus ν and $r_{\nu i}$ is the distance between the nuclei ν and the point r_i. The ρ_{ν} s were already obtained at the calculation of AOs and summation of quantities for all nucleus is far easier than that for all sample points. Equation (6.6) is called SCC (Self-Consistent-Charge) approximation. SCC approximation considerably reduces calculation cost, but is one of possible causes to degrade total energies.

6.3 The Calculating Method of Total Energy

6.3.1 Total Energy Calculation by Monte-Carlo Method and Its Errors

On the sample point system, total energy ε of (6.1) is represented as follows.

$$\begin{aligned}\varepsilon = &\sum_{\sigma}\sum_k f_k^{\sigma}\sum_i \phi_k^{\sigma}(r_i)\left(-\frac{1}{2}\nabla^2\right)\phi_k^{\sigma}(r_i)w_i - \sum_i\sum_{\nu}\frac{Z_{\nu}}{r_{\nu i}}\rho(r_i)w_i \\ &+\frac{1}{2}\sum_i \rho(r_i)\sum_j\frac{\rho(r_j)}{r_{ij}}w_jw_i - \frac{9}{4}\alpha\left(\frac{3}{4\pi}\right)^{1/3}\sum_{\sigma}\sum_i \rho^{\sigma}(r_i)^{4/3}w_i.\end{aligned} \tag{6.7}$$

TESDA calculates total energies according to Eq. (6.7) by using ϕ_k^{σ} s obtained by SSPIN. But electron–electron Coulomb interaction for $j = i$ in the third term, can't be calculated because $r_{ii} = 0$. Instead of it, TESDA uses mutual Coulomb interaction of uniformly distributed charge inside the sphere around the sample point r_i as described as

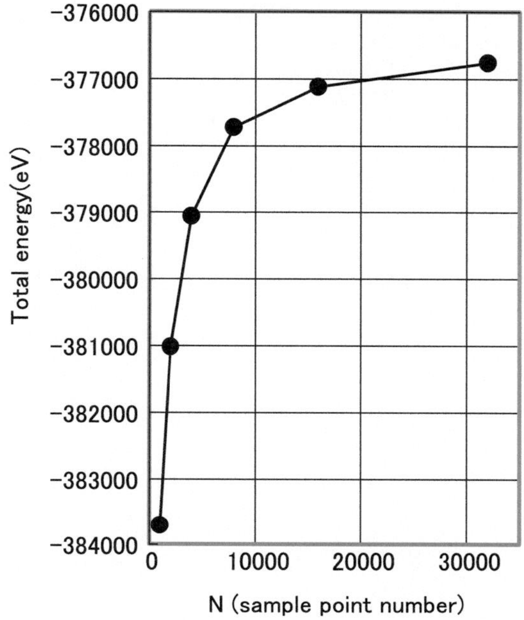

Fig. 6.4 Sample point number dependence of total energy (eV) of I_2 molecule before error cancelling. Number of sample points are 1,000, 2,000, 4,000, 8,000, 16,000 and 32,000 respectively

$$\frac{1}{2}\frac{\rho^{\sigma}(r_i)^2}{r_{ii}}w_i^{\,2} \rightarrow \frac{4\pi}{5}\left(\frac{3}{4\pi}\right)^{2/3}\rho^{\sigma}(r_i)^2 w_i^{\,2/3}. \tag{6.8}$$

where the sphere has the same volume as weight w_i.

In the case of non-relativistic DV-Xα method, it is required that molecules comprising elements to fifth row can be calculated with required accuracy. But as heavier atoms are accompanied with more electrons and has deeper potentials, it is difficult to calculate their total energies with required accuracy. As an example, sample point number N dependence of total energy of I_2 molecule is demonstrated in Fig. 6.4, where large systematic error is observed. And total energy vs. bond length (inter-nuclear distance) curve for 8,000 sample points (4,000 point per atom) is demonstrated in Fig. 6.5. The range of scattering is around 1 eV and the total energy minimum can't be determined.

Usually around 500 sample points per atom is used to obtain reasonable MOs and their eigenvalues. But for chemical applications, slight difference (typically 0.1 eV) between large total energies (typically $10^4 \sim 10^6$ eV) has importance or PPM-class-accuracy is required for total energy calculation. Result by Monte-Carlo method includes statistical error inversely proportional to N, provided that sample points are generated by quasi-random-numbers. Therefore the simplest way to improve this situation is to use more points. But it is limited by computation resource and unrealistic in large molecules.

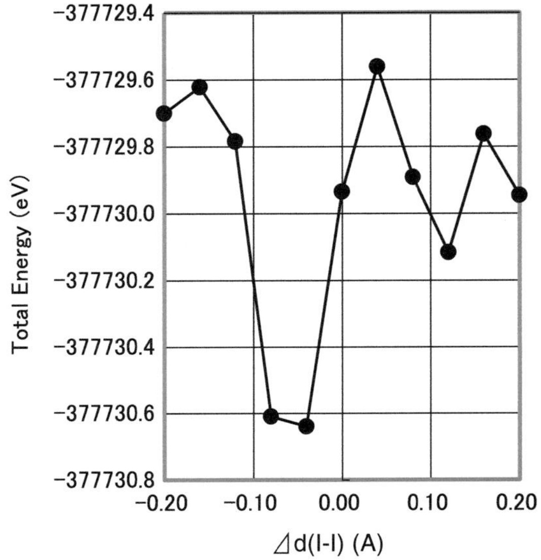

Fig. 6.5 Bond length (d) dependence of total energy (eV) of I_2 molecule before error cancelling. Here $\Delta d = 0$ means that d = the experimental equilibrium bond length. 8,000 points were used

6.3.2 Error Cancellation Procedure

Because the charge density is higher and the potential is deeper in the inner region of each atom of the molecule, major part of total energy error must be generated there. On the other hand, total energy of each atom in the molecule must include similar numerical error, provided that the same sampling point system is used. Therefore the difference between calculated total energy of the molecule E_{mol} and the sum of total energy of each single atom constructing the molecule $\sum_{\nu} E_{\nu}$ is expected not to include the major numerical error. Further if each acculate total energy of each single atom $E_{\nu}{}^{0}$ can be calculated, genuine total energy of the molecule $E_{mol}{}^{0}$ can be obtained, or

$$E_{mol}{}^{0} = E_{mol} - \sum_{\nu} \left(E_{\nu} - E_{\nu}{}^{0} \right). \tag{6.9}$$

It should be noted that $E_{\nu}{}^{0}$ is not the total energy of the atom standing alone, but that of the atom in the environment of the molecule. Fortunately it can be calculated easily by using the results of AO calculation in SSPIN and very accurate because it can be transformed into numerical integration in 1D space. The sample point number dependence of total energy of I_2 molecule after error cancelling is demonstrated in Fig. 6.6 and total energy of I_2 molecule vs. inter-nuclear distance curve for 8,000 sample points (4,000 point per atom) after error cancelling is demonstrated in Fig. 6.7. Comparing Fig. 6.6 vs Fig. 6.4 and Fig. 6.7 vs Fig. 6.5, the effect of error canceling procedure was found to be remarkable.

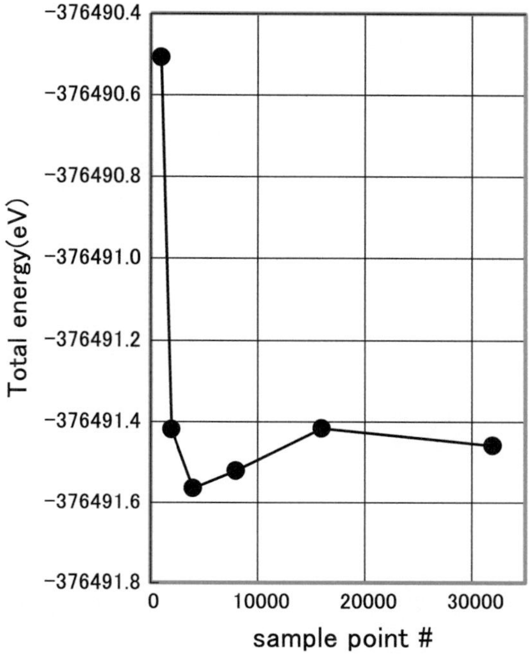

Fig. 6.6 Sample point number dependence of total energy (eV) of I_2 molecule after error cancelling. Number of sample points are 1,000, 2,000, 4,000, 8,000, 16,000 and 32,000 respectively

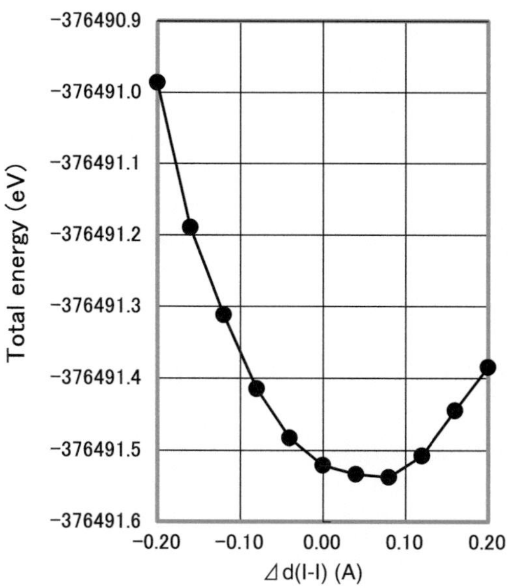

Fig. 6.7 Bond length (d) dependence of total energy (eV) of I_2 molecule after error cancelling. Here $\Delta d = 0$ means that d equals the experimental equilibrium bond length. 8,000 points were used

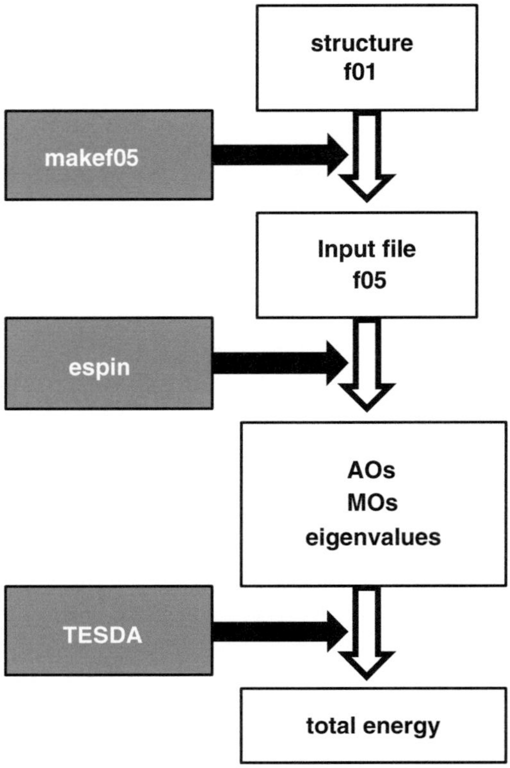

Fig. 6.8 Total energy calculation procedure

The total energy minimum in Fig. 6.7 is located at the position 0.08 A longer than experimental bond length. This error can not be eliminated even if more sample points were used and does not originate from numerical calculation error. Prescriptions for this problem will be discussed later.

6.3.3 Total Energy Calculation Procedure

Figure 6.8 shows total energy calculation procedure by DV-Xα method. At first one needs to prepare the simple text file "f01", which described structure of the molecule to be calculated. The pre-processor program "makef05" (hereafter MAKEF05) reads the f01 file and prepare the input file "f05". A f05 file includes various parameters for DV-Xα calculation. MAKEF05 sets these parameters to be default status. Some of them will be explained in detail later. The program "espin" (hereafter ESPIN) works in completely the same way as SSPIN until SCF convergence is obtained, but it generates the input files for TESDA. TESDA reads these files and generates output files including the total energy of the molecule.

ESPIN and TESDA are tentatively distributed for members of the society for Discrete Variational Xα and work in the environment of the program package distributed by the society.

6.3.4 Simple Examples of Structure Optimization

Structure optimization of molecule may be one of the most frequently used functions of molecular orbital method. And the similarity of the calculated and experimental equilibrium structures is the good measure for accuracy of the calculation. But TESDA can execute one point calculation of the total energy only, one has to calculate total energies for successively transformed structures and find the total energy minimum by hand.

Firstly let's consider diatomic molecules. In this case, Morse potential V_{M} (6,10) is well known as a function expressing binding energy in wide range of interatomic distance r.

$$V_M(r) = D_e\left(e^{-\alpha(r-r_0)} - 1\right)^2, \tag{6.10}$$

where D_e is a bond dissociation energy and r_0 is equilibrium bond length. Unless higher excitation is considered, Eq. (5.10) can be expanded around r_0 as follows,

$$V_M(r) = D_e\alpha^2(r - r_0)^2 - D_e\alpha^3(r - r_0)^3 + O\left[(r - r_0)^4\right]. \tag{6.11}$$

The first term of (6.11) is a harmonic term that determines molecular vibration wavenumber ω_{e}. The relation between its coefficient $D_e\alpha^2$ and ω_{e} will be explained in Sect. 6.7.2. The second term of (6.11) is the lowest anharmonic term indispensable to fit the curve to calculated total energies appropriately. Based on the expansion (6.11), V_{M} can be expressed in the general form like

$$V_M(r) = ar^3 + br^2 + cr + d. \tag{6.12}$$

Coefficients a, b, c, d in (6.12) are determined by fitting (6.12) to more than four calculated total energies. Then we can determine r_{o} by solving the analytical differentiation of (6.12).

For a multi-atomic molecule having more than two structural parameters, the author assumed that the dependence of its total energy on arbitrary parameter x has the expression similar to (6.12)

$$V(x) = ax^3 + bx^2 + cx + d, \tag{6.13}$$

and determined equilibrium parameter x_{o} by solving the analytical differentiation of (6.13). To check accuracy of the curve fitting, five or more results were used.

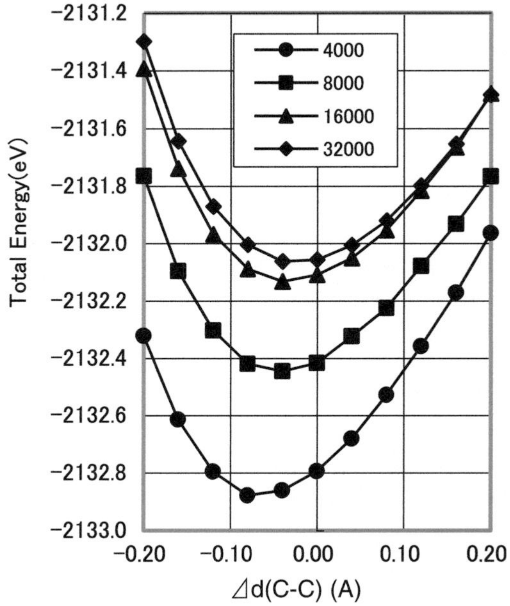

Fig. 6.9 C-C bond length dependence of total energy of C_2H_6 molecule, with four levels of sample point number 4,000, 8,000, 16,000 and 32,000 respectively

If the molecule has three structural parameters *x*, *y*, *z*, its total energies with successively changed parameter *x* were calculated and parameters *y* and *z* fixed to experimental ones at first. Then parameters *y* were optimized with fixed *x* and *z* to experimental ones, and so on. Practically, only small or symmetric molecules could be calculated and symmetries of molecules were assumed.

6.3.4.1 C_2H_6, C_2H_4 and C_2H_2 Molecules

Let's investigate structures of C_2H_2, C_2H_4 and C_2H_6 molecules and see the sample point number dependence. Here default f05 files were used without parameter adjusting. If $\mathbf{C_{3v}}$ structure is assumed, C_2H_6 molecule has three independent structural parameters d(C-C), d(C-H) and ∠(H-C-C) remain to be optimized. Figure 6.9 shows the bond length d(C-C) dependence of total energy of C_2H_6 molecule with other parameters fixed to experimental ones. Δd(C-C) in the horizontal axis designates the deviation of d(C-C) from the experimental bond length. Four curves correspond to ones calculated with 4,000, 8,000, 16,000 and 32,000 sample points respectively. Total sample points were shared by C atoms (14 %/atom) and by H atoms (12 %/atom) automatically by MAKEF05. The minimum by 32,000 sample points is located at d = 1.5110 A, which agrees well with experimental bond length d = 1.5351 A.

Figure 6.10 shows the bond length d(C-H) dependence of total energy of C_2H_6 molecule with other parameters fixed to experimental ones. Here all six C-H bond lengths were altered simultaneously and the molecule holds its symmetry. The minimum by 4,000 sample points is located at d = 1.1142 A, which agrees well with experimental bond length. d = 1.0940 A.

Fig. 6.10 C-H bond length dependence of total energy of C_2H_6 molecule, with four levels of sample point number

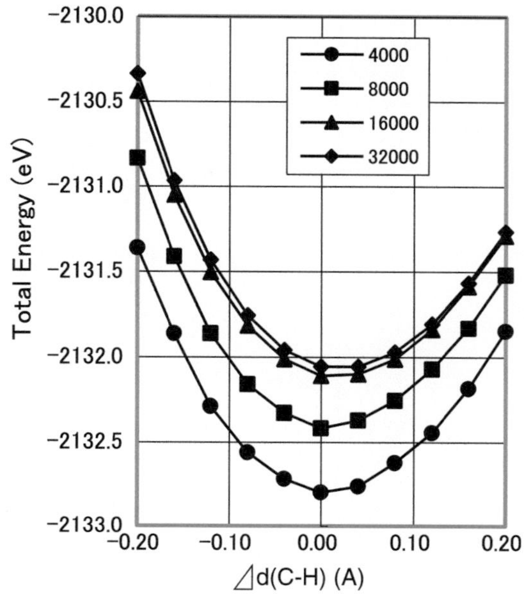

Fig. 6.11 H-C-C bend angle dependence of total energy of C_2H_6 molecule, with four levels of sample point number

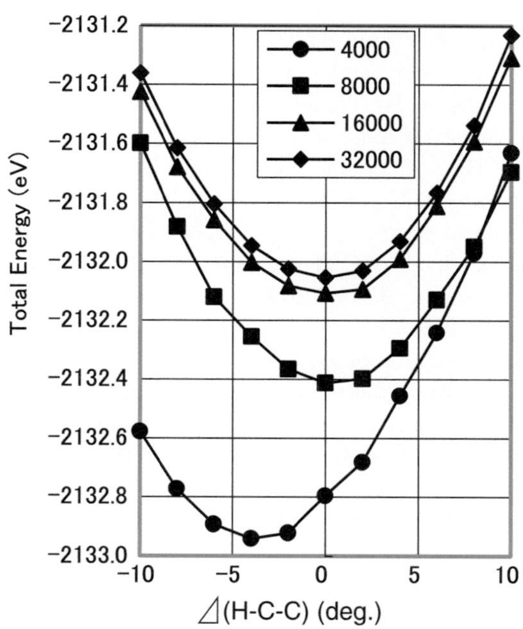

Figure 6.11 shows the bending angle∠(H-C-C) dependence. The minimum by 32,000 sample points is located at $\theta = 111.0°$, which agrees well with experimental bending angle $\theta = 111.7°$.

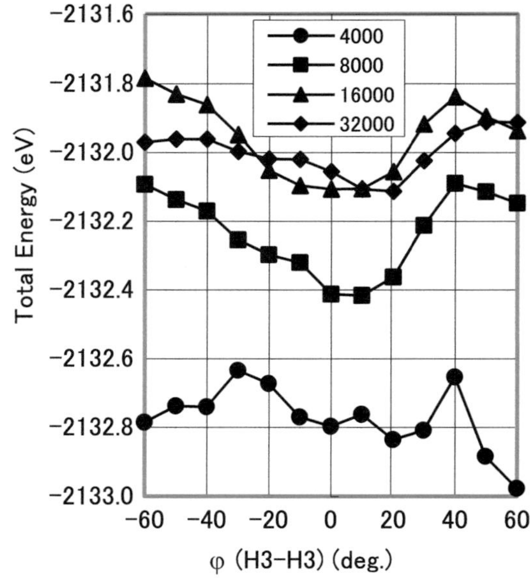

Fig. 6.12 Twist angle θ (H_3-H_3) dependence of total energy of C_2H_6 molecule, with four levels of sample point number

Although $\mathbf{C_{3v}}$ structure was assumed so far, the effect of conformation was investigated. Figure 6.12 shows twist angle ϕ (H_3-H_3) dependence of total energy, where $\phi = 0$ means the staggered conformation. The experimental height of conformational barrier is as small as 0.13 eV. The fluctuation of curves is rather dominant, but curves by more than 8,000 points seem to suggest that staggered structure is more stable.

The C-C double bond length dependences of C_2H_4 is shown in Fig. 6.13. The minimum by 24,000 sample points is located at d = 1.0864 A which agrees well with experimental bond length d = 1.087 A. Further the C-C triple bond length dependence of C_2H_2 are shown in Fig. 6.14. The minimum by 16,000 sample points is located at d = 1.0583 A which agrees well with experimental bond length d = 1.060 A.

Inspection of Figs. 6.9, 6.10, 6.11, 6.13, and 6.14 of C_2H_n may suggest that the condition of 2,000 sample points per atoms is enough for light atoms. But even with 4,000 points per atom, analysis of small total energy difference especially accompanied with conformation change was found to be difficult.

6.4 Adjustment of Preparation Parameters of MOs

6.4.1 Parameters in the Input File "f05"

In the case of calculations of C_2H_6, C_2H_4 and C_2H_2, default parameters in a f05 file prepared by MKEF05 were adopted without adjustment and excellent agreements

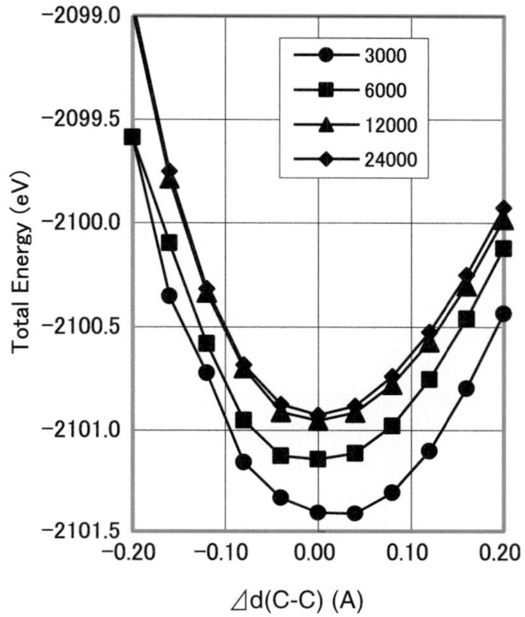

Fig. 6.13 C-C bond length dependence of total energy of C_2H_4 molecule, with four levels of sample point number

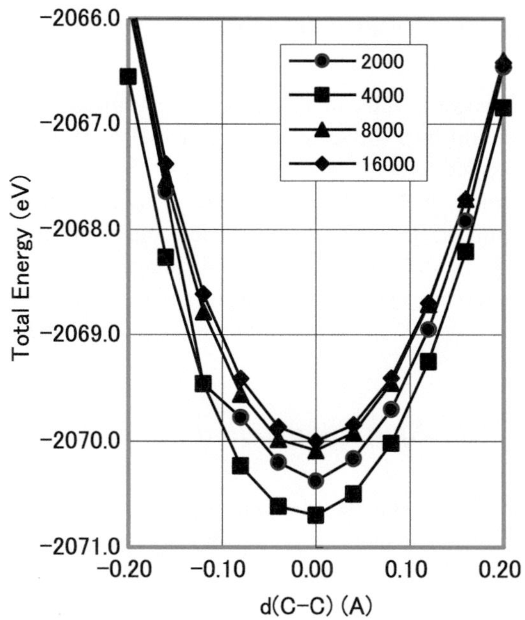

Fig. 6.14 C-C bond length dependence of total energy of C_2H_2 molecule, with four levels of sample point number

with experimental data were obtained except for twist angle ϕ (H_3-H_3) dependence of C_2H_6 molecule. But excellent agreement is not necessarily obtained in every case. And there is little room for adjustment in TESDA to improve accuracy. On the other hand, adjustment of parameters in f05 files often improves accuracy of results.

Table 6.1 Parameters influencing the accuracy of total energy

	Described in f01	Described only in f05
Reference atoms	O	
Spin polarization in AO		O
Spin polarization in MO		O
Extension of AOs		O
Well potential		O
α value in reference atom		O

According to variational principle, total energy calculated by Eq. (6.1) or (6.7) is good measure of quality of MOs. This means that MOs of DV-Xα method can be tuned through total energy calculated with those MOs. The tuned MOs will be useful to calculate other chemical or physical quantities.

There are many kinds of parameter for DV-Xα calculation, which are explained in detail by Adachi's collaborator (Iwasawa and Adachi 1996). Those especially influencing total energies are listed in Table 6.1 and will be explained in the following sections.

6.4.2 Reference Atom

Because atoms in the different environments feel the different potentials, they should have different AOs. For a H atom bonded to the C atom and the H atom bonded to the O atom in a CH_3OH molecule, a user can prepare different reference atoms in the f01 file and the program ESPIN prepares slightly different AOs appropriate for each reference atom. This function is inherited from SSPIN and will be one of the greatest advantages of DV-Xα method. One should utilize this function to obtain better results with relatively low calculation cost.

6.4.3 Spin Polarization

In ESPIN, the electron configuration of the molecule is described by specifying the occupation number f_k^σ of MO ϕ_k^σ $(\sigma = \alpha, \beta)$. In the case of a molecule with 2n electrons, the default f05 file makes

$$f_k^\sigma = 1(k = 1, 2, 3, \ldots, n),\ f_k^\sigma = 0,\quad (k = n + 1, n + 2, n + 3, \ldots,).$$

In the case of a molecule with 2n + 1 electrons, the default f05 file makes

$$f_k^\alpha = 1\ (k = 1,2,3,\ldots,n,n+1),\ f_k^\alpha = 0\ (k = n+2,n+3,\ldots),$$

$$f_k^\beta = 1\ (k = 1,2,3,\ldots,n),\ f_k^\beta = 0\ (k = n+1,n+2,\ldots).$$

Let's define spin polarization like $S_z = \frac{1}{2}\left(\sum_k f_k^\alpha - \sum_k f_k^\beta\right)$.

These settings make S_z of the molecule the minimum, which is usually the satisfactory result for ground state. But there are some exceptional cases, like O_2 molecule and atoms which has high S_z according to Hund's rule. ESPIN sometimes finds results with appropriate S_z by the default f05 file. But it often fails to find appropriate results.

For the case of O_2 molecule, one can edit the f05 file and change occupation numbers of MOs like, $f_9^\alpha = 0 \to 1, f_8^\beta = 1 \to 0$ to give the molecule appropriate spin polarization $S_z = 1$.

Another way to alter S_z is to edit the f05 file and adjust the initial electron configuration of AOs of reference atom(s). The program ESPIN starts calculation from this settings. For example, in default f05 file of O atom the initial electron configuration of AOs are described as follows,

Quantum number			Occupation number	
n	l	Eigen value	α-spin	β-spin
1.0	0.0	−19.72000	1.00000	1.00000
2.0	0.0	−1.07000	1.00000	1.00000
2.0	1.0	−0.52000	2.00000	2.00000

Here "n" means principal quantum number and "l" means orbital-angular-momentum quantum number of the atom. If above default parameters are used, ESPIN gives the result without spin polarization. The ground state of O atom is ^3P and only $S_z = 1$ can correspond to ^3P state. By adjusting initial electron configuration as follows,

Quantum number			Occupation number	
n	l	Eigen value	α-spin	β-spin
1.0	0.0	−19.72000	1.00000	1.00000
2.0	0.0	−1.07000	1.00000	1.00000
2.0	1.0	−0.52000	3.00000	1.00000,

ESPIN gives the result with $S_z = 1$ and with total energy 3.49 eV lower than that by default parameter. The initial electron configuration adjustment is indispensable for calculation of atoms, and often adaptable for molecules. But the program sometimes doesn't give the results as expected by the f05 file. A user has to make sure of final occupation numbers of MOs

6.4.4 Extension of Basis

If basis set is complete, MO method program will find the "genuine" MOs within the approximation of HF or HFS method. But in usual MO method, including DV-Xα method, basis set are limited and far from complete. For example, the default f05 file of O_2 molecule, the basis set comprises only 1s, 2s, 2p orbitals. This may be enough for calculation of O atom, but unsatisfactory to describe quantitatively bond formation in the molecule. Extension of basis set often improves the result. A user can edit the f05 file and add basis functions, 3s, 3p orbitals for example.

quantum number			occupation number	
n	l	eigen value	α-spin	β-spin
1.0	0.0	-19.72000	1.00000	1.00000
2.0	0.0	-1.07000	1.00000	1.00000
2.0	1.0	-0.52000	3.00000	1.00000
3.0	0.0	-0.20000	0.00000	0.00000
3.0	1.0	-0.10000	0.00000	0.00000

6.4.5 Well Potential and SCF Convergence

Radial components of AOs are obtained by solving the numerical differential equation which includes influence from the environment of the molecule. But as outer AOs are spreading broadly in the space, they are often difficult to solve consistently. To ease this situation, artificial "well potential" $\Phi_{well}(r_{1\nu})$ was introduced and overlaid on the potential around nucleus ν, represented by Eq. (6.3). Figure 6.15 shows the shape of "well" having radius r_{ido} and depth v_{ido}. Outside of r_{ido}, the well has exponential tail. In default f05 file, r_{ido} is set to be 1.5–3.0 atomic unit (depending on Z_ν) and v_{ido}. to be −2.0. One can edit the f05 file and adjust r_{ido} and v_{ido}. Therefore AOs shown in Fig. 6.1a, b are slightly different from genuine AOs of Zn atom.

If well potential is removed, what happens? Figure 6.16a demonstrates its effect on 4s and 4p orbitals of Zn atom, where orbitals without well potential draw tails beyond d = 6. But as can be seen in Fig. 6.2, almost all sample points for Zn atom are located inside of d = 6, which is insufficient space for numerical integration containing 4s or 4p orbitals. In such situation, Fock operator F_{rs} and overlap integral S_{rs} won't be calculated accurately. Though 4p orbitals are empty in Zn atom, they will be occupied partly and contribute to total energy in the molecules containing Zn atom. On the other hand, 4s and 4p orbitals with default well potential have short tails that don't cross d = 6 and they won't cause the numerical integration error.

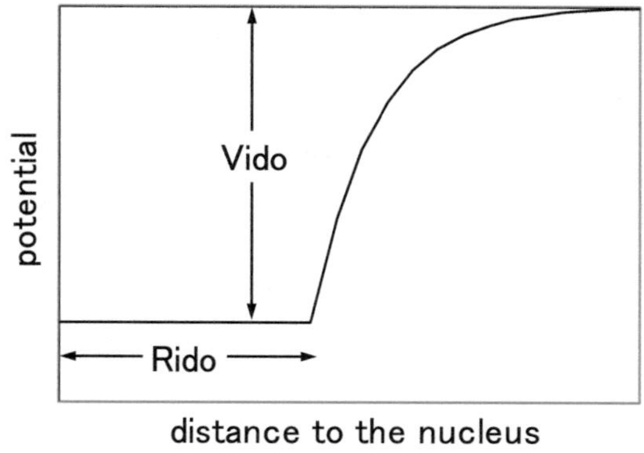

Fig. 6.15 Shape of well potential

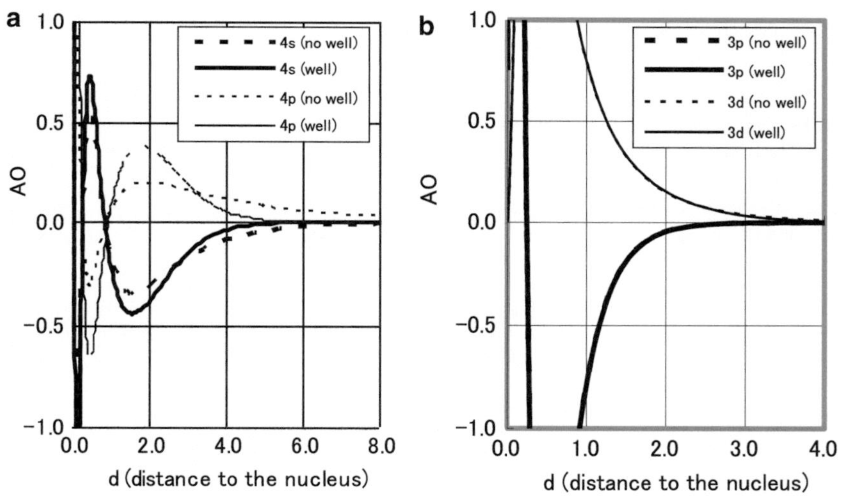

Fig. 6.16 (**a**) Radial components of 4s and 4p orbitals of Zn atom, with and without default well potential. (**b**) Radial components of 3p and 3d orbitals of Zn atom, with and without default well potential

Doesn't well potential affect inner AOs? Figure 6.16b demonstrates that 3p orbital with well potential and that without well potential can't be distinguished each other. And 3d orbitals with and without well potential has almost the same shapes too. Well dependence of AOs that reside inside of 3p are further smaller.

Of course, well potential has some influence on total energy. Figure 6.17 shows dependence of total energy of F atom on v_{ido}. But the influence of well seems to be far smaller than expected from v_{ido} and it may be strange that the total energy has minimum at $v_{ido} \approx 0$.

TESDA calculates total energy without well potential but ESPIN calculates MOs with well potential. From the viewpoint of TESDA, MOs calculated by ESPIN is

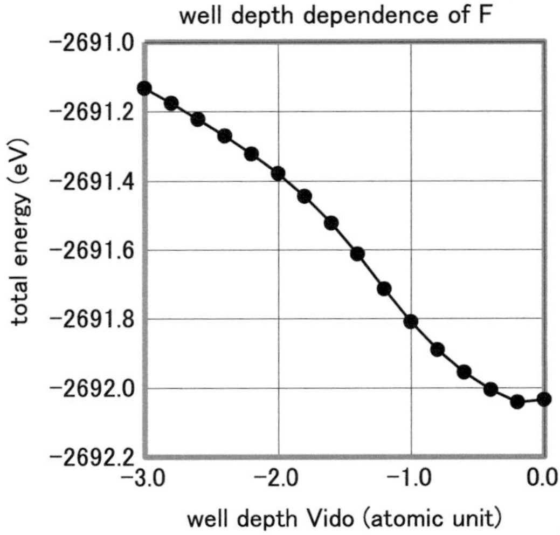

Fig. 6.17 Well depth V_{ido} dependence of total energy of F atom. Well radius R_{ido} is fixed to 2.5

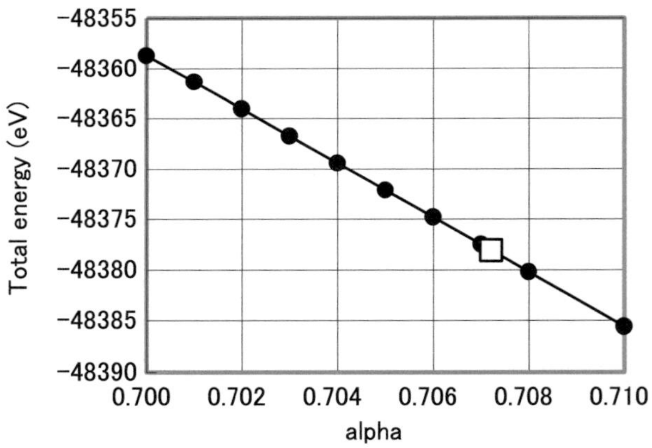

Fig. 6.18 The α value dependence of total energy of Zn. *Blank square* indicates optimized α that makes calculated total energy that by Hartree-Fock method

considered to be perturbed by well potential. The variational principle teaches that total energy calculated using genuine MOs has the extreme value (usually the minimum) and total energy calculated using MOs varied from genuine MOs has slightly larger value as demonstrated in Fig. 6.17.

6.4.6 α-Value of Xα Term

In default f05 file, α in Xα term is set to be 0.7 for every atom. But total energy depends on value of α and one can edit the f05 file to adjust α for each reference atom independently. Figure 6.18 shows α value dependence of total energy of Zn.

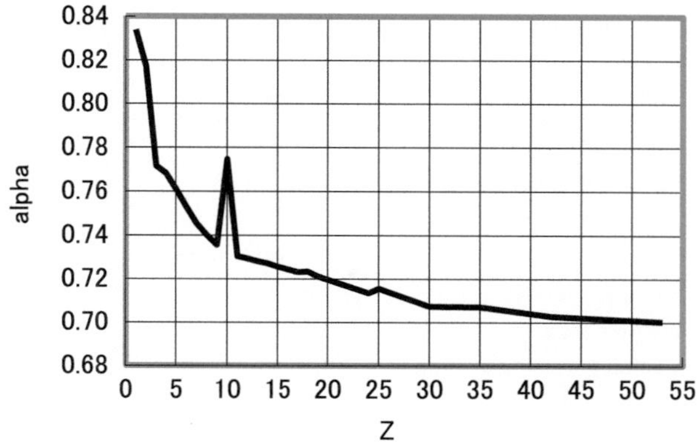

Fig. 6.19 Optimized α of atoms from H to I

One policy to optimize α for each atom is based on the fact that DV-Xα calculation without well potential is an approximation of Hartree-Fock method, for which strictly calculated total energies were published for all atoms (Fraga et al. 1976). Therefore one can optimize α for any element by adjusting α until total energy calculated by DV-Xα agrees with that calculated by Hartree-Fock method. In Fig. 6.18, the blank square indicates optimized α value that makes calculated total energy −483,378.12 eV or that by Hartree-Fock method.

Figure 6.19 shows optimized α value for elements obtained by this way from H to I. Here α is the highest for H and then decreases gradually to 0.7, some peaks exist though. These values are fairy smaller than those listed in other references. Such values may be optimized by using non-spin-DV-Xα method.

In f05 file of molecules, α value for inside of the influential area of each atom and α value for peripheral area of the molecule can be adjusted. The α value for peripheral area is an ambiguous concept. One plausible policy to decide such α is to average αs weighted by valence electrons of each atom in the molecule.

Policies described above are based on the assumption that SCF convergence can be obtained without well potential. But this isn't satisfied sometimes for actual calculation of molecules.

6.4.7 Default Parameters for Calculation of C_2H_n

In the preceding Sect. 6.3.4, parameters for calculation of C_2H_n were not described. Table 6.2 summarizes parameters used there. These are default settings prepared by MAKEF05.

Table 6.2 Parameters for calculations of C_2H_6, C_2H_4 and C_2H_2 in Sect. 6.3.4

Parameter	C_2H_6	C_2H_4	C_2H_2
Sample point number	32,000	24,000	16,000
Reference atoms	1(C), 1(H)	1(C), 1(H)	1(C), 1(H)
Spin polarization in AO	None	None	None
Spin polarization in MO	None	None	None
Default AOs	1s-2p(C),1s(H)	1s-2p(C),1s(H)	1s-2p(C),1s(H)
Extension of AOs	None	None	None
Well potential	Default	Default	Default
α value in reference atom	0.70	0.70	0.70
α value in peripheral area	0.70	0.70	0.70

6.5 Equilibrium Structure of Multi-atomic Molecule

6.5.1 *$SiCl_4$*

As **Td** symmetry was assumed, only Si-Cl bond length was the parameter to be optimized. Figure 6.20 shows Si-Cl bond length dependence calculated in default condition with 2,500, 5,000, 10,000 and 20,000 sample points. Δd(SiCl) denotes Si-Cl bond length deviation from experimental length. All four bond lengths were altered simultaneously. Calculated- and experimental equilibrium bond length is listed in Table 6.3. Excellent agreement with experimental bond length was obtained.

6.5.2 *XeF_4*

XeF_4, the molecule containing rare gas element, was investigated. As $\mathbf{D_{4h}}$ symmetry is assumed, calculated Xe-F bond length in default condition is listed in the column titled ***default*** of Table 6.3 and compared with experimental bond length.

All four bond lengths were altered simultaneously. It is fairy longer than experimental bond length. Default Basis functions of Xe are 1s-5p, which are occupied completely, and those of F are 1s-2p, which are almost completely occupied.

To increase freedom for bond formation, basis functions 5d, 6s 6p were included to Xe and 3s, 3p were included to F. Improved result with basis extension is described in the column titled ***extended basis*** in Table 6.3. Curves by "extended basis" condition with 10,000, 20,000 and 40,000 points are shown in Fig. 6.21. Total energy curves calculated with 2,500 or 5,000 sample points were scattered severely and couldn't be contained inside of this graph. Extension of basis functions improves optimized Xe-F bond length considerably.

Based on Figs. 6.20 and 6.21, 4,000 points per atom can be considered as the standard sample point number up to fifth row elements. In the examples introduced later, 4,000 points per atom is adopted in principle.

Fig. 6.20 Si-Cl bond length dependence of total energy of $SiCl_4$

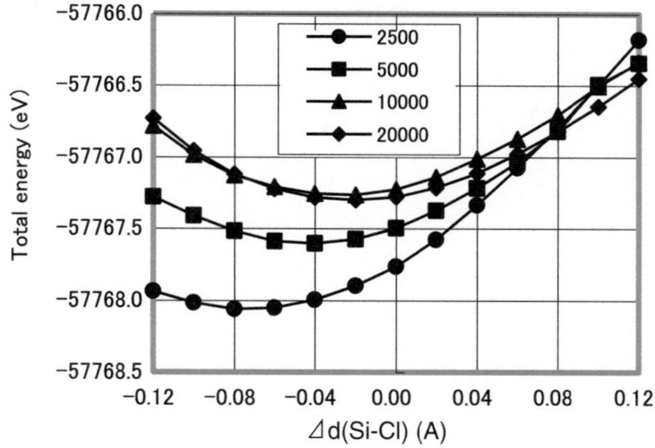

Table 6.3 Optimized Si-Cl bond length by default condition and Xe-F bond length by default or extended basis condition

	Parameter	Experimental	***Default***	***Extended basis***
$SiCl_4$	$d(Si-Cl)$	2.019 A	***1.9979 A***	–
XeF_4	$d(Xe-F)$	1.94 A	***2.0851 A***	***1.9729 A***

Italic font suggests parameters calculated by DV-Xα method

Fig. 6.21 Xe-F bond length dependence of total energy of XeF_4. Curves by 2,500 or 5,000 sample points were out of range in this graph

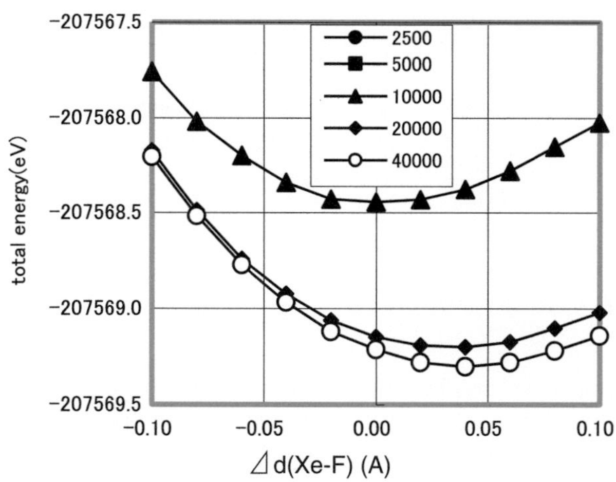

6.5.3 ClF_3

ClF_3 is a molecule with $\mathbf{C_{2v}}$ symmetry and Fig. 6.22 illustrates its structure. The parameters to be optimized are $d(Cl-F_a)$, $d(Cl-F_b)$ and $\theta(F_a-Cl-F_b)$. Optimized parameters calculated in ***default*** condition and those in ***extended basis*** condition (4s, 4p for Cl and 3s, 3p for F) are summarized in Table 6.4. All structural parameters were improved by ***extended basis*** condition, but the agreement of $d(Cl-F_b)$ is still poor.

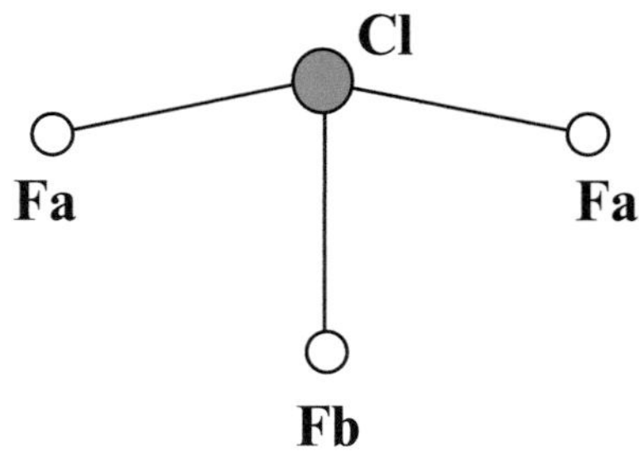

Fig. 6.22 Structure of ClF_3 molecule

Table 6.4 Structural parameters of ClF_3 by default condition and extended basis condition

Parameter	Experimental	*Default*	*Extended basis*
$d(Cl-F_a)$	1.698 A	***1.7645 A***	***1.7192 A***
$d(Cl-F_b)$	1.598 A	***1.7219 A***	***1.6712 A***
$\theta(F_a-Cl-F_b)$	87.5°	***91.37°***	***89.78°***

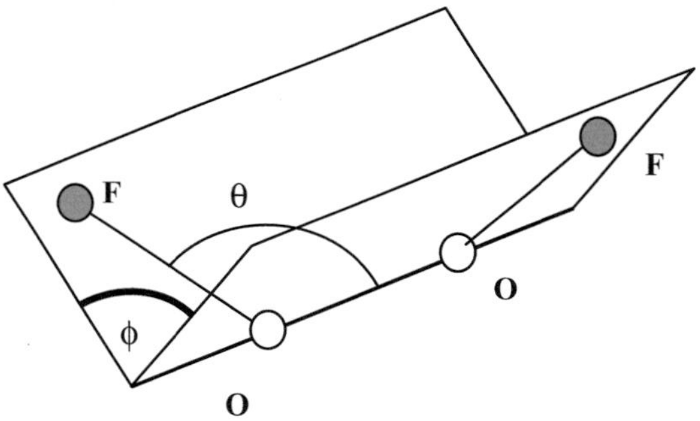

Fig. 6.23 Structure of FOOF molecule. ϕ is the dihedral angle

6.5.4 FOOF

Foresman and Frisch introduced the FOOF molecule as a "pathological case" in their handbook and listed calculated results by various ab-initio and DFT programs in Gaussian package (Foresman and Frisch 1996). C_2 symmetry structure of FOOF molecule is illustrated in Fig. 6.23. Some of their results and experimental ones are summarized in Table 6.5. Complicated basis set has to be used to obtain reasonable results. The counterparts calculated by DV-Xα in ***default*** condition and ***extended basis*** (3s,3p for O and 3s,3p for F) are also summarized in Table 6.5. Results by DV-Xα are superior to HF and seem to be equivalent to DFT with small basis set.

Table 6.5 Structural parameters of FOOF molecule calculated by ab-initio method, DFT and DV-Xα in default and extended basis condition (ext.)

Parameter	Experimental	HF	CCSD(T)	B3LYP		***DV-Xα***	
		(A)	(B)	(C)	(D)	***Default.***	***Ext. basis***
$d(O-O)$	1.217 A	1.296	1.216	1.266	1.241	***1.287***	***1.253***
$d(O-F)$	1.575 A	1.352	1.614	1.497	1.515	***1.599***	***1.619***
$\theta(F-O-O)$	109.5 A	106.3	108.9	108.3	108.7	***111.6***	***110.7***
$\phi(F-O-O-F)$	87.5°	85.1	87.8	86.8	87.2	***84.8***	***86.3***

Basis set: (A) = 6-311 + G(3d), (B) = 6-311 + G(3df), (C) = 6-31G(d) (D) = 6-311G(2d)
CCSD(T): "Coupled Cluster method including single, double and optionally triple terms"

Table 6.6 Optimized parameters of $Cr(CO)_6$ complex calculated by RHF with two kind of basis set and calculated by DV-Xα in default and extended basis conditions

		RHF		*DV-Xα*	
Parameter	Experimental	STO-3G	3-21G	***Default***	***Extended basis***
d(Cr-C)	1.92	1.79	1.93	***1.774***	***1.834***
d(C-O)	1.16	1.17	1.13	***1.256***	***1.170***

Table 6.7 Optimized parameters of $Fe(C_5H_5)_2$ calculated by DV-Xα in default condition

Parameter	Experimental	***Default.***
$d(Fe-C)$	2.064	***2.038***
$d(C-C)$	1.440	***1.465***
$d(C-H)$	1.098	***1.132***

6.5.5 $Cr(CO)_6$

As $\mathbf{O_h}$ symmetry (octahedron configuration) was assumed, d(Cr-C) and d(C-O) were optimized. Foresman and Frisch introduced these results calculated by RHF with two kinds of basis set, which are summarized in Table 6.6. The counterparts calculated by DV-Xα in default and extended conditions are listed in Table 6.6, too. Extended basis condition makes d(Cr-C) longer and d(C-O) shorter, but the improvement for d(Cr-C) is insufficient. Here basis set of C and O were extended to 3p but basis set of Cr couldn't be extended because of SCF convergence failure.

6.5.6 $Fe(C_5H_5)_2$

As $\mathbf{D_{5h}}$ symmetry (eclipsed conformation) was assumed, d(Fe-C), d(C-C) and d(C-H) were optimized. And all H atoms were assumed to be located on planes of C_5 ring. In the author's calculating environment, TESDA can't treat more than 52,000 sample points, $Fe(C_5H_5)_2$ was calculated in default condition with 52,000 points, which means 2,476 points per atom. All parameters were calculated with errors less than 3 %, as shown in Table 6.7.

6.6 Atoms

Total energies of isolated atoms are often necessary to evaluate bond dissociation energy (BDE) of molecules. Even though such data were already published, total energy actually required are not published ones but ones calculated by DV-Xα method in the equivalent conditions (the equivalent sample points and the same well potential, α-value, etc.) to calculation of the molecule.

Table 6.8 Occupation numbers of 2p orbitals of C atom by initial electron configuration (a) and (b)

Result	(a)		(b)	
Spin	α	β	α	β
#3(2p)	1	1	1	0
#4(2p)	0	0	1	0
#5(2p)	0	0	0	0

As isolated atoms are in their highest spin polarized state according to Hund's rule, one have to calculate total energy carefully. It is difficult to relate general electronic state of an atom to occupation numbers of MOs and the utilization of discrete variational multielectron (DV-ME) method is desired (Ogasawara et al. 2007). Fortunately, electronic state of the ground state is often related to spin polarizable occupation numbers of MOs.

6.6.1 *Typical Element Atoms*

Spin polarizations of typical element atoms are determined with the highest p-orbitals. Let's take up second row element atoms as example. For C atom, electronic states $^3P, ^1D, ^1S$ are actualized under electron configuration $2p^2$, and the ground state is known to be ^3P. But the preprocessor program MAKEF05 prepare default initial electron configuration for 2p orbital like (a) in the below list. Occupation number of each 2p orbital calculated with it is listed in Table 6.8. This state isn't spin polarized and doesn't correspond to the ground state ^3P.

Quantum number				Occupation number	
	n	l	Eigen value	α-spin	β-spin
(a)	2.0	1.0	−0.330000	1.00000	1.00000
(b)	2.0	1.0	−0.330000	2.00000	0.00000

If electron configuration (b) listed above is used, results (b) in Table 6.8 are obtained. In this case, because $S_z = 1$ can be given only by ^3P among $^3P, ^1D, ^1S$ states, we can say $S = 1$. Further as its α-electron configuration is that of a closed shell from which one electron is removed, its orbital angular momentum is the same as one electron and we can say $L = 1$. Therefore we can identify (b) with the ground state ^3P of a C atom.

For a N atom, electron states $^4S, ^2D, ^2P$ are actualized on electron configuration $2p^3$, and the ground state is ^4S.

Quantum number				Occupation number	
	n	l	Eigen value	α-spin	β-spin
(a)	2.0	1.0	−0.420000	1.50000	1.50000

Table 6.9 Occupation numbers of 2p orbitals of N atom

Result	(a)	
Spin	α	β
#3(2p)	0	1
#4(2p)	0	1
#5(2p)	0	1

Table 6.10 Occupation numbers of 2p orbitals of O atom by initial electron configuration (a) and (b)

Result	(a)		(b)	
Spin	α	β	α	β
#3(2p)	1	1	1	1
#4(2p)	1	1	1	0
#5(2p)	0	0	1	0

Default condition (a) in Table 6.9 gives $S_z = -\frac{3}{2}$. Because only $S = \frac{3}{2}$ state can give $S_z = -\frac{3}{2}$ and a closed shell of β electrons gives $L = 0$, we can identify (a) with the ground state ^{4}S of a N atom.

Quantum number				Occupation number	
	n	l	Eigen value	α-spin	β-spin
(a)	2.0	1.0	−0.520000	2.00000	2.00000
(b)	2.0	1.0	−0.520000	3.00000	1.00000

For an O atom, electronic states ^{3}P,^{1}D,^{1}S are actualized on electron configuration 2p^4, and the ground state is ^{3}P, just the same as a C atom.

Results (a) and (b) are correspond to condition (a) and (b) in above list respectively. Because $S_z = 1$ of (b) in Table 6.10 is included only in $S = 1$ and a closed shell of α spin electrons don't contribute to orbital angular momentum, (b) can be identified with ^{3}P. A user has to rewrite f05 to obtain the ground state of O atom.

As can be seen in above examples, the program sometimes automatically finds the ground state of atom, but often misses it. It is recommended to adjust initial condition in f05 file and make sure of occupation number of MOs.

6.6.2 Transition Metal Element Atoms

Spin polarizations of transition metal element atoms are determined with mostly the highest nd-orbitals. Let's take up 3d transition metal element atoms as example. Table 6.11 shows their electron configurations, ground state and other states having the same spin polarization as the ground state. Cr, Mn, Fe atoms have only one

Table 6.11 The ground states and other states having the same spin

Element	Configuration	Ground state	Other maximum spin state
Ti	$4s^2,3d^2$	^3F	^3P
V	$4s^2,3d^3$	^4F	^4P
Cr	$4s,3d^5$	^7S	None
Mn	$4s^2,3d^5$	^6S	None
Fe	$4s^2,3d^6$	^5D	None
Co	$4s^2,3d^7$	^4F	^4P
Ni	$4s^2,3d^8$	^3F	^3P

maximum spin state respectively, we can assign the maximum spin-polarized state for the ground state.

But for Ti, V, Co, Ni atoms, as there exist a maximum spin state other than the ground state listed it the Table 6.11, it adds an ambiguous factor to assignment of the maximum spin polarized states to the ground state.

6.7 Diatomic Molecules

Other than optimized structure, bond dissociation energy (BDE) and molecular vibration wavenumber (ω_e) were calculated and compared with experimental data. And molecules including elements picked up from whole the periodic table were evaluated. For such purposes diatomic molecules are suitable, because there are plenty of structural, thermodynamic or spectroscopic data for them and wide variety of elements can form homo- or hetero-diatomic molecules (Lide 2003 and Luo 2007).

6.7.1 Bond Dissociation Energy

Bond dissociation energy of a diatomic molecule is the total energy difference between a molecule in equilibrium and the molecule whose nuclei are separated far away. Let's take up a H_2 molecule. If two reference atoms were prepared and the spin polarizations of atom 1 and atom 2 are adjusted like below, DV-Xα method gives reasonable results like Fig. 6.24.

Quantum number				Occupation number	
	n	l	Eigen value	α-spin	β-spin
1)	1.0	0.0	−0.300000	1.00000	0.00000
2)	1.0	0.0	−0.300000	0.00000	1.00000

Here two atoms are reversely spin polarized each other and both electrons can be localized in separated H atoms. By applying Eq. 6.12 to total energy curve in

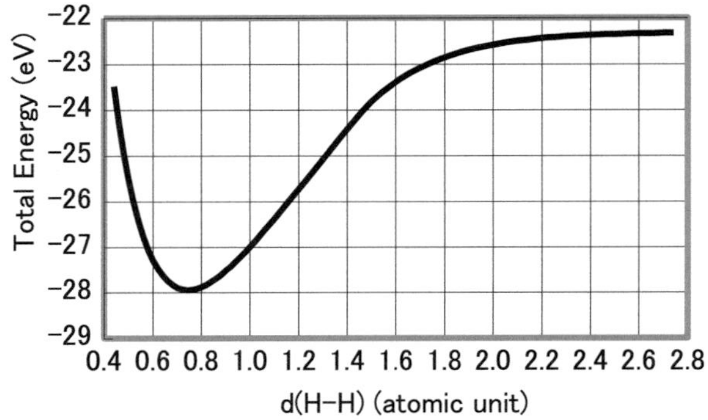

Fig. 6.24 Bond dissociation of H_2 molecule. Two reference atoms are utilized. One of them has a α spin and the other has β spin

Table 6.12 BDE of H_2 molecule by two kind of method. (a) Two reference atoms method. (b) Calculation of the difference between H_2 molecule and double H atoms

Calculation condition	*(a)* Fig. 6.24	*(b) H_2-2H*	Experimental
Bond length (A)	***0.7460***	***0.7460***	0.7414
Total energy at minimum (eV)	***−27.95***	***−27.90***	–
Reference total energy (eV)	***−22.33***	**2 × *−11.16***	–
Bond dissociation energy (eV)	***5.62***	***5.58***	4.5187

Fig. 6.24, the equilibrium bond length and the total energy minimum are evaluated. The difference between total energy minimum and that at the right edge of the curve indicates the BDE, which is listed in the column titled Fig. 6.24 of the Table 6.12. This procedure can be applied for other diatomic molecules. But even if two reference atoms are prepared, SCF convergence often fails for separated-atoms situation.

Generally executable way is to calculate total energy of a H_2 molecule near the equilibrium and total energy of an isolated H atom.

The difference between the total energy minimum of a H_2 molecule and the double of total energy of H atom is equivalent to the BDE. The calculated results are summarized in the column titled ***H_2-2H*** of the Table 6.12.

Calculated results by procedure(a) and (b) in Table 6.12 are almost the same. From now on, procedure (b) will be adopted. To obtain more accurate BDEs, total energies of atoms and the molecules should be calculated with the equivalent condition, like well potential, α-value, etc.

6.7.2 Molecular Vibration

The curvature of total energy curve as shown in Fig. 6.24 is related to the molecular vibration. In the case of diatomic molecule comprising atom 1 and atom 2, wavenumber ω_e (cm^{-1}) of the molecular vibration is represented as

$$\omega_e\left[cm^{-1}\right] = \frac{1}{2\pi c}\sqrt{\frac{k}{\mu}} = 737.48\sqrt{\frac{D_e\alpha^2\left[eV/A^2\right]}{\mu[amu]}}, \tag{6.14}$$

where μ^1 and μ^2 are mass of each atom, reduced mass is defined $\mu = \frac{\mu^1 \cdot \mu^2}{\mu^1 + \mu^2}$, k is the force constant and $D_e\alpha^2$ is the coefficient of a harmonic term in Eq. (6.11). By substituting $D_e\alpha^2$ from calculation of Table 6.12, $\omega_e =$ 5,071.23 (cm^{-1}) was obtained. Observed one is 4,401.21 (cm^{-1}).

6.7.2.1 Diatomic Molecules of Second Row Element

Equilibrium bond lengths, BDEs and molecular vibration wavenumbers ω_e of diatomic molecules consisting of the same second row element atoms were calculated by not only ***default basis*** but also ***extended basis*** (***default basis*** + 3s, 3p).

Equilibrium bond lengths by ***default*** condition were calculated fairly longer than experimental lengths. They were improved by ***extended basis*** condition to some extent. This tendency is often observed in other diatomic molecules and will be discussed in Sect. 6.9.

Molecular vibration wavenumber ω_es, especially by ***extended basis*** condition, were close to experimental ones, which suggests potential curves around equilibrium bond length are calculated well for these molecules.

Although calculated ω_es were agreed well with experimental ones, BDEs of O_2 and F_2 were more inaccurate in ***extended basis*** than in ***default condition***. As BDE of diatomic molecule is evaluated by the relation $E_{molucule} - E_{atom1} - E_{atom2}$, poor accuracy of total energies of isolated O and F atoms may degrade accuracy of BDE.

6.7.2.2 Halogen Molecules

The same parameters as Table 6.13 were calculated for halogen molecules other than F_2 (already investigated in the preceding section) in ***default*** condition and in ***extended basis*** conditions (***default basis*** + (n + 1)s, (n + 1)p orbitals) and summarized on Table 6.14.

Errors of bond length and wavenumber ω_e are relatively small and improved by ***extended basis*** to some extent. Despite of that, accuracies of BDEs are poor again.

6.7.2.3 Alkali Metal Molecules

The same parameters as Table 6.13 were calculated for alkali metal molecules in ***default*** conditions and were summarized on Table 6.15. Calculated results don't agree with experimental ones at all. Besides they couldn't be improved by ***extended*** basis. In the case of these molecules, elimination of well potential was found to drastically improve their accuracies, which will be discussed in Sect. 6.9.

Table 6.13 Equilibrium bond lengths, BDEs and molecular vibration wavenumbers of diatomic molecules comprising the same second row element atoms by two kinds of conditions

	Condition	Exp.	*Default*	*Extended basis*
C_2	Bond length (A)	1.2425	***1.3040***	***1.2663***
	BDE (eV)	6.173	***9.110***	***6.473***
	ω_e (cm^{-1})	1,854.7	***1,847.4***	***1,980.8***
N_2	Bond length (A)	1.0977	***1.2306***	***1.1446***
	BDE (eV)	9.793	***7.871***	***8.911***
	ω_e (cm^{-1})	2,358.6	***1,957.4***	***2,331.0***
O_2	Bond length (A)	1.2074	***1.3745***	***1.3168***
	BDE (eV)	5.155	***6.266***	***7.187***
	ω_e (cm^{-1})	1,580.2	***1,357.0***	***1,540.3***
F_2	Bond length (A)	1.4119	***1.5219***	***1.4873***
	BDE (eV)	1.645	***3.331***	***4.517***
	ω_e (cm^{-1})	916.6	***1,056.5***	***990.5***

Table 6.14 Equilibrium bond lengths, BDEs and molecular vibration wavenumbers of halogen diatomic molecules by two kinds of conditions

	Condition	Experimental	*Default*	*Extended basis*
Cl_2	Bond length (A)	1.9878	***2.0044***	***2.0064***
	BDE (eV)	2.513	***4.088***	***3.719***
	ω_e (cm^{-1})	559.7	***675.4***	***623.8***
Br_2	Bond length (A)	2.2811	***2.3665***	***2.3265***
	BDE (eV)	2.009	***3.022***	***3.919***
	ω_e (cm^{-1})	325.32	***396.7***	***378.6***
I_2	Bond length (A)	2.6663	***2.7149***	***2.6135***
	BDE (eV)	1.5780	***2.558***	***3.404***
	ω_e (cm^{-1})	214.5	***248.8***	***252.9***

Table 6.15 Equilibrium bond lengths, BDEs and molecular vibration wavenumbers of alkali metal molecules by default condition

	Condition	Experimental	*Default*
Li_2	Bond length (A)	2.6729	***2.3934***
	BDE (eV)	0.261	***1.180***
	ω_e (cm^{-1})	351.4	***561.5***
Na_2	Bond length (A)	3.0789	***2.4439***
	BDE (eV)	0.775	***1.506***
	ω_e (cm^{-1})	159.1	***472.2***
K_2	Bond length (A)	3.9051	***3.0730***
	BDE (eV)	0.590	***1.649***
	ω_e (cm^{-1})	92.0	***175.6***

Table 6.16 Equilibrium bond lengths, BDEs and molecular vibration wavenumbers of ionic molecules by default condition

	Condition	Experimental	***Default***
HF	Bond length (A)	0.9169	***0.9890***
	BDE (eV)	5.904	***7.054***
	ω_e (cm^{-1})	4,138.3	***3,715.2***
NaCl	Bond length (A)	2.3609	***2.0215***
	BDE (eV)	4.271	***6.665***
	ω_e (cm^{-1})	366	***479.1***
KBr	Bond length (A)	2.8208	***2.7325***
	BDE (eV)	3.929	***6.697***
	ω_e (cm^{-1})	213	***216.9***

Table 6.17 Equilibrium bond length, BDE and molecular vibration wavenumber w_e of MnH by DV-Xα method

Condition	Experimental	***Default***	***Default***	***Extended basis***
S_z In f05		***Default***	***Adjusted***	***Adjusted***
S_z in result		***2***	***3***	***3***
Bond length	1.7308	***1.5390***	***1.6396***	***1.6311***
BDE	2.6443	***3.227***	***3.628***	***1.881***
ω_e (cm^{-1})	1,548	***2,066***	***1,937***	***1,772***

Here the row titled "condition" describes just basis functions and, well potential and α. The row titled "S_z in f05" means initial status of spin polarization of AOs of the Mn reference atom

6.7.2.4 Ionic Molecules

Ionic molecules were investigated and results are summarized in Table 6.16.

Calculated results don't agree well with experimental ones, which may be caused partly by the same reason as in the case of alkali molecules.

6.7.2.5 MnH

At first MnH was calculated in ***default*** condition. The equilibrium bond length was far shorter than experimental one and its spin polarization S_z in calculated results was 2, as listed in Table 6.17. But because MnH is a small atom and Hund's law may hold here, ***the spin polarization*** S_z of Mn atom in f05 file was ***adjusted*** like list below, with other conditions fixed to default status. It improved bond length to some extent and its spin polarization became 3. Actually the ground state of MnH is known to be $^7\Sigma$. Further ***extended basis*** and no well and optimized α condition were tested.

Quantum number			Occupation number	
n	l	Eigen value	α-spin	β-spin
`3.0`	`2.0`	`−0.400000`	`5.00000`	`0.00000`
`4.0`	`0.0`	`−0.200000`	`1.00000`	`1.00000`
`4.0`	`1.0`	`−0.100000`	`0.00000`	`0.00000`

For molecule containing transition metal atom, calculating condition should be carefully prepared.

6.8 Atomization Energy, Enthalpy Change and Excitation Energy

Foresman and Frinsch introduced some calculations of atomization energy, enthalpy change accompanied with reaction and excitation energy in their book. The author traced their calculations by DV-Xα and compared both results.

6.8.1 The Atomization Energy of CO_2

Equilibrium bond length R(C-O) and atomization energy D_0 of CO_2 calculated by four kinds of ab-initio and DFT methods were cited in Table 6.18. Here "atomization energy" means the energy required to break a molecule into isolated atoms and the same energy as bond dissociation energy (BDE) for diatomic molecules.

Both of R(C-O) and D_0 were reproduced well by almost all methods except for D_0 by HF. The results by ***default*** DV-Xα calculation were not so good, but were improved by ***extended basis***

6.8.2 The Enthalpy Change of the Reaction $O_3 + Cl \rightarrow ClO + O_2$

The reaction $O_3 + Cl \rightarrow ClO + O_2$ is the first step of ozone destruction. Foresman and Frisch analyzed this reaction by 15 kinds of method and basis set, among which, 6 examples are listed in Table 6.19. Here atomization energies $D_0(O_3)$, $D_0(O_2)$ and $D_0(ClO)$ are listed and enthalpy change ΔH is calculated by using them.

Although ΔHs were reproduced relatively well, agreement of three kinds of calculated atomization energy with experimental ones in Table 6.19 were poor except for B3LYP. It may suggest that calculation of CO_2 in Table 6.18 is an exceptional case and quantitative total energy calculation of isolated atom is not easy by other methods.

Table 6.18 Equilibrium bond length R(C-O) and atomization energy D_0 of CO_2

	Experimental	HF	MP2	BLYP	B3PW91	***Default***	***Ext. basis***
R(C-O)	1.162	1.143	1.1780	1.183	1.168	***1.231***	***1.196***
D_0 (eV)	16.55	10.56	16.42	16.38	16.52	***17.75***	***16.78***

Table 6.19 Atomization energies of O_3, O_2 and ClO, and enthalpy change accompanied with the reaction $O_3 + Cl \rightarrow ClO + O_2$. by various methods and basis sets. (a): HF/6-31G(d) (b): B3LYP/6-31G(d) (c): QCISD(T)/6-31G(d) (d): HF/6-311+G(3df) (e): B3LYP/6-311+G(3df) (f): QCISD (T)/6-311+G(3df)

	Exp.	(a)	(b)	(c)	(d)	(e)	(f)
$D_0(O_3)$	6.16	−0.62	6.02	3.89	−0.31	6.02	5.52
$D_0(O_2)$	5.12	1.17	5.29	4.38	1.35	5.27	4.92
D_0(ClO)	2.74	−0.08	2.51	2.02	0.21	2.83	2.54
ΔH	−1.69	−1.70	−1.79	−2.51	−1.87	−2.08	−1.94

Table 6.20 Atomization energies of O_3, O_2 and ClO, and enthalpy change accompanied with the reaction $O_3 + Cl \rightarrow ClO + O_2$ by DV-Xα in various conditions

	Experimental.	***Default***	***Extended basis***	***Ext. basis and*** $V_{ido} = -0.5$
$D_0(O_3)$	6.16	***4.99***	***7.41***	***5.43***
$D_0(O_2)$	5.12	***5.29***	***7.02***	***4.35***
D_0(ClO)	2.74	***4.13***	***4.38***	***2.34***
ΔH	−1.69	***−4.43***	***−3.99***	***−1.94***

Table 6.21 Ionization energy and electron affinity (E.A.) of PH_2 calculated by DFT and DV-Xα in various conditions

	Experimental	B3LYP	***Default***	***Extended basis***	***Ext. basis and*** $V_{ido} = -0.5$
Ionization	9.82	9.95	***7.34***	***8.24***	***9.53***
E.A.	1.26	1.24	***−3.63***	***−2.17***	***0.26***

Calculated results by DV-Xα are summarized in Table 6.20. Atomization energies calculated by ***default*** condition was not so bad, but enthalpy change was overestimated. By ***extended basis*** condition, atomization energies were rather overestimated. But shallow well $\boldsymbol{V_{ido} = -0.5}$ instead of $V_{ido} = -2.0$ in e***xtended basis*** condition improved the situation, which will be discussed in Sect. 6.9.

6.8.3 Ionization Energy and Electron Affinity of PH_2

Ionization energy and electron affinity (E.A.) of PH_2 were calculated in ***default*** and in ***extended basis*** conditions and compared with calculations of B3LYP/6-31G(d) by Foresman and Frisch. Results are summarized in Table 6.21. The agreement of ionization energies was not bad. But even with ***extended basis*** **and** $\boldsymbol{V_{ido} = -0.5}$ condition, the agreement of E.A.s is still poor, even though it became positive.

Table 6.22 Multiplicities, symmetries and corresponding transitions of four excited states of C_2H_2 calculated by Foresman and Frisch

State	Multiplicity	Symmetry	Transition
1	Triplet	B_{1u}	$\pi-\pi^*$
2	Triplet	B_{3u}	$\pi-3s$
3	Singlet	B_{3u}	$\pi-3s$
4	Singlet	B_{1u}	$\pi-\pi^*$

Table 6.23 Multiplicities, symmetries and corresponding transitions of four excited states of C_2H_2 calculated by Foresman and Frisch and by DV-Xα in extended basis and extended basis and $V_{ido}=-0.5$ conditions

State	Symmetry	Experimental	CIS/6-31 + G(d)	***Ext. basis***	***Ext. basis and $V_{ido}=-0.5$***
1	$^3B_{1u}$	4.36	3.78	***4.91***	***4.42***
2	$^3B_{3u}$	7.66	7.43	***10.84***	***7.73***
3	$^1B_{3u}$	6.98	7.83	***6.30***	***5.88***
4	$^1B_{1u}$	7.15	7.98	***11.12***	***7.77***

6.8.4 Excitation Energy of C_2H_2

Foresman and Frisch calculated four excited states of C_2H_2 in Table 6.22 using CI-single (CIS) method with 6-31 + G(d) basis set. Their results are listed in Table 6.23.

In the MOs of C_2H_2 ground state, 1st to 8th MOs with α-spin and β-spin were occupied. Eighth MO corresponds to π orbital, 9th MO to π* orbital and 10th orbital to 3s orbital. For example, $^3B_{1u}$ state is described by the transition (8th $\alpha-\mathrm{MO}\rightarrow$ 9th $\beta-\mathrm{MO}$) and $^1B_{3u}$ state is described by the transition (8th $\alpha-\mathrm{MO}\rightarrow$ 10th $\alpha-\mathrm{MO}$). Total energies of these states were calculated in ***extended basis*** condition and ***extended basis*** **and** $\mathbf{V_{ido}=-0.5}$ condition and then excitation energies were evaluated. Latter condition improves their accuracies up to the level of CIS/6-31 + G(d). All results were summarized in Table 6.23.

6.9 Discussion

As far as structure optimization of multi-atomic molecules is concerned, results by DV-Xα method were found to be superior at least to those by simple Hartree-Fock method and often close to those by higher level ab-initio or DFT method. Calculated structures of C_2H_n, in Figs. 6.9, 6.10, 6.11, 6.13, and 6.14, $SiCl_4$ and XeF_4 in Table 6.3, ClF_3 in Table 6.4 and $Fe(C_5H_5)_2$ in Table 6.7 agreed with experimental structures with errors less than 3 %, even in ***default*** condition or in ***extended basis*** conditions. Calculated results of FOOF molecule in Table 6.5, $Cr(CO)_6$ in Table 6.6

Table 6.24 Errors ratio of calculated C-C bond length of various molecules

	Error (%)
C_2H_6	−1.6
C_2H_4	−0.1
C_2H_2	−0.2
C_2	+4.9

and CO_2 in Table 6.18 were comparable with those by various methods in Gaussian package.

These are the astonishing results, judging from that DV-Xα method is just the approximation of Hartree-Fock method or a kind of DFT method having the most primitive functional. Accurate and adaptive basis functions of DV-Xα method must play the essential roles to obtain accurate equilibrium structures.

On the other hand, Table 6.24 shows the error ratios of calculated C-C bond lengths calculated by DV-Xα in ***default*** condition to experimental ones of various molecules. As mentioned above, C-C bond lengths of C_2H_n were reproduced very exactly, but that of C_2 was poorly reproduced. Generally structure optimization of diatomic molecules seems to be more difficult than that of multi-atomic molecules. But it will give us valuable suggestions to improve total energy calculation by DV-Xα method.

In the total energy calculation by DV-Xα method, five factors degrading accuracy of total energy are enumerated as follows.

1. **numerical integration error**
2. **limited number of basis functions**
3. **well potential**
4. **self consistent charge (SCC) approximation**
5. **Xα potential approximation**

Numerical integration error was already considered in Sect. 6.3.2. Based on it, almost all total energies were calculated by using 4,000 sample points per atom, which was assumed to be enough except for small total energy difference, like the low conformational barrier of C_2H_6. Let's investigate how factor (2) to (5) affect quality of MOs and the accuracy of total energy.

Overestimated bond lengths of C_2, N_2, O_2 and F_2 calculated by default condition in Table 6.13 were improved by **extension of basis functions** to some extent. The number of basis functions of default condition is too limited for electrons to rearrange enough and extension of basis gives them the freedom to form bonds strong enough. It is reasonable that this prescription is effective for XeF_4, molecule including rare-gas atom and halogen atom, in Table 6.3, because their default basis functions are almost occupied.

Extension of basis functions seems to be a recommendable method to increase accuracy.

In Table 6.15, bond lengths of alkali metal molecules are fairly shorter than experimental ones. But in this case, extension of basis set, which was effective for molecules in Tables 6.13 and 6.14, has no effect to improve the disagreement of

Table 6.25 Bond length, BDE and ω_e of Na_2 molecule with well depth altered from 0.0 eV to −2.0 eV. For well depth = 0.0, calculation of Na atom wasn't converged and BDE couldn't be evaluated

	Experimental	$V_{ido} = 0.0$	−0.5	−1.0	−1.5	−2.0
Bond length (A)	3.0789	3.1520	2.7228	2.5603	2.4910	2.4439
BDE (eV)	0.775	–	0.736	1.092	1.320	1.506
ω_e (cm^{-1})	159.1	175.0	328.9	388.3	432.9	472.2

structural parameters. Maybe it's because outer AOs of alkali atoms are almost empty and there remains enough freedom for bond formation. On the other hand, **well potential** was found to affect structural parameters as demonstrated for the case of Na_2 molecule in Table 6.25. Deeper well strikingly degrades bond length, BDE and ω_e. Introduction of well potential shrinks outer AOs in alkali metal molecules and may shorten their bond lengths.

Further, for enthalpy change in Table 6.20, ionization energy and electron affinity in Table 6.21 and excitation energy in Table 6.23, well depth $\boldsymbol{V_{ido}} = \mathbf{-0.5}$ condition was found to improve their agreements with experimental data. The reason to use $\boldsymbol{V_{ido}} = \mathbf{-0.5}$ instead of $V_{ido} = 0.0$ is that SCF convergence of MO calculation often fails or becomes unreliable in $V_{ido} = 0.0$.

Considering these results, especially for the situation that total energy itself is targeted, it is desirable to establish the failure-free procedure of SCF calculation without well potential.

Doesn't **SCC approximation** affect total energy accuracy? The potential that electrons feel in the program ESPIN and that in program TESDA is different in two ways. That is, ESPIN in default condition adopts well potential and SCC approximation and the TESDA doesn't adopt both of them. According to variational principle, MOs calculated by ESPIN will be seen from TESDA to be perturbed by these approximations. Therefore SCC approximation can affect the accuracy of total energy.

Figure 6.25 shows the well depth V_{ido} dependences of total energy of HF molecule. The total energy minimum resides at $V_{ido} = -1.2$. But the minimum of F atom resided at $V_{ido} \approx 0.0$ in Fig 6.17.

These different dependences may be explained by considering influence of SCC approximation. For an atom, the potential of which has the spherical symmetry, the potential by SCC approximation will be similar to actual potential of the atom. Just well potential is a perturbation and total energy curve will have the minimum at $V_{ido} = 0.0$, no perturbation condition. But for a molecule which doesn't have spherical symmetry, potential of SCC approximation apparently differs from actual potential of the molecule. In this case, addition of well potential may decrease the deviation of total potential (SCC potential + well potential) from actual potential of the molecule. In such case, minimum of total energy may reside at other than $V_{ido} = 0.0$ as shown in Fig. 6.25. In other word, this different V_{ido} dependence between atom and molecule suggests deviation of calculated MO from the accurate

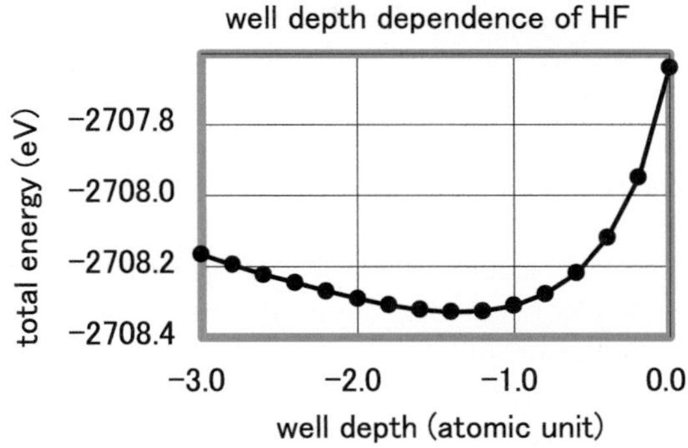

Fig. 6.25 Well depth dependence of total energy of a HF molecule. Well radius is fixed to 2.5

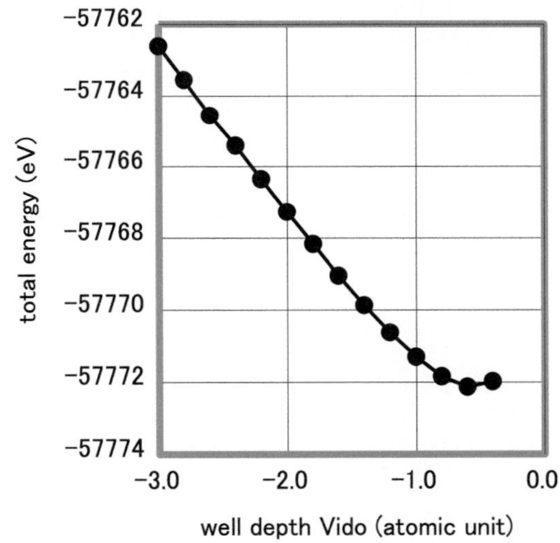

Fig. 6.26 Well depth dependence of total energy of $SiCl_4$ molecule. Well radious is fixed to 2.5

MO obtained without SCC approximation. Further it is one of possible reasons of poor reproductions of BDE in general.

Figure 6.26 shows V_{ido} dependence of total energy of SiF_4. Here total energy minimum resides nearer to $V_{ido} = 0.0$ than that of HF. As SiF_4 molecule has higher symmetry than HF molecule, the difference between actual potential and that by SCC approximation is relatively small. This will explain the minimum position in Fig. 6.26 and is one of the possible reasons for the tendency that calculated structures of diatomic molecules, whose structure is unsymmetrical, are often less accurate than those of multi-atomic molecules.

Adjustment of Xα potential has theoretical foundation explained in Sect. 6.4.6. In the case that absolute value of total energy itself is targeted, it will establish good calculation condition. But in usual chemical application, such cases are rather rare. Adjustment of α-value seems to affect structural parameters not so much. Further accurate total energy by strict Hartree-Fock method was calculated without well potential and is often difficult to reproduce by DV-Xα method because of SCF calculation failure. For these reasons the author doesn't consider adjustment of α-value as a standard procedure to increase accuracy.

From the practical viewpoint, one of the greatest problems of total energy calculation of multi-atomic molecules by DV-Xα was that it couldn't optimize molecular structures automatically. But the author has developed the method to calculate the force acting on each atom in the molecule (energy gradient) and demonstrated examples of molecules comprising of up to period three atoms (Nakagawa 2012) and of molecules comprising of up to period five atoms (Nakagawa 2013).

At the end of this discussion, the author wants to emphasize that research of MOs that generates accurate total energy will be very helpful to obtain accurate physical or chemical quantities other than total energy.

Conclusions

1. Total energies of molecules can be obtained quantitatively by using the MOs calculated by DV-Xα method with 2,000–4,000 points per atom from H to Xe, provided that the numerical error cancelling procedure is adopted. The accuracy of total energy depends mostly on quality of MOs, which depends on the calculation condition of DV-Xα.
2. Structures of some multi-atomic molecules were reproduced with errors less than 3 % even using MOs calculated by default condition. Such excellent results may be achieved by accurate and adaptive atomic orbitals, which are utilized as basis functions in DV-Xα method.
3. Strictly speaking, MOs of DV-Xα method are slightly different from genuine MOs on the point that they are affected by "well potential", which has outstanding effect to prevent SCF convergence failure, and SCC (self consistent charge) approximation, which makes computational cost far lower.
4. To obtain accurate results for wider variety of molecules or to treat energy itself, DV-Xα calculation condition should be devised. Considerable improvements were achieved by extension of basis functions or shallowing of well potential, or their hybrid condition. Their effects are reasonably explained from the viewpoint of calculation mechanisms of MOs and total energy.
5. DV-Xα method couldn't optimize structures of multi-atomic molecules and it was inconvenient for practical applications. But the author has developed the method to calculate energy gradient and to optimize molecular structure automatically. Details were published elsewhere.

Acknowledgment The author wishes to thank the members of "strategic program developing committee" of the Society for Discrete Variational Xα for fruitful discussions on my research and Dr. AEleen Frisch for allowing me to cite valuable calculated results by programs in Gaussian package.

References

Adachi H, Tsukada M, Satoko C (1978) Discrete variational Xα cluster calculations. I. Applications to metal clusters. J Phys Soc Jpn 45:875–883

Foresman JB, Frisch A (1996) Exploring chemistry with electronic structure methods, 2nd edn. Gaussian, Pittsburgh

Fraga S, Karwowski J, Sanena KMS (1976) Handbook of atomic data. Elsevier, Amsterdam

Iwasawa M, Adachi H (1996) Calculations of electronic state by DV-Xα method. Sankyo Shuppan, Tokyo (in Japanese)

Kowada Y, Nishihara W, Ogasawara K (2009) Total cluster energy calculation of lithium ion conductors by the DV-Xalpha method. Int J Quantum Chem 109:2658–2663

Nakagawa K (2003) Total energy calculation of molecules using spin-DV-Xα molecular orbitals. Bull Soc Discret Var Xα 16:93–97 (in Japanese)

Ogasawara K, Watanabe S, Toyoshima H, Brik MG (2007) Handbook on physics and chemistry of rare earths, vol 37. Elsevier, North-Holland, Chapter 231

Ziegler T, Rauk A (1977) On the calculation of bonding energies by the Hartree Fock Slater method. Theor Chim Acta 46:1–10

Throughout this article, experimental structural parameters molecules and molecular vibration wavenumbers were cited from

Lide DR (ed) (2003) Handbook of chemistry and physics, 84th edn. CRC Press, Boca Raton

Experimental bond dissociation energies of diatomic molecules were cited from

Luo YR (2007) Comprehensive handbook of chemical bond energies. CRC Press, Boca Raton

Nakagawa K (2012) Total energy calculation and structure optimization of molecules by DV-Xα method. J Comput Chem Jpn 11:194–201. doi:10.2477/jccj.2012-0011

Nakagawa K (2013) The novel numerical integration method superior to Monte-Carlo method and its application for DV-Xa method. Bull Soc Discret Var Xa 26(1&2):98–103

Chapter 7
Energy Expression of the Chemical Bond Between Atoms in Hydrides and Oxides and Its Application to Materials Design

Masahiko Morinaga, Hiroshi Yukawa, and Hiromi Nakai

7.1 Introduction

Nowadays, total energy calculation is so common in every field of materials science. However, information of the chemical bond between atoms is still limited along the total energy calculation alone. In 1955 Wigner and Seitz (1955) predicted that if one had a great calculating machine, one might apply it to the problem of solving the Schrödinger equation for each metal and obtain thereby the interesting physical quantities, such as the cohesive energy, the lattice constant, and similar parameters. It is not clear, however, that a great deal would be gained by this. Presumably the results would agree with experimentally determined quantities and nothing vastly new would be learned from the calculation. This prediction made by two pioneers of solid state physics appears still true in these days, even though the computational science has made great progress.

In addition, the Mulliken population analysis (Mulliken 1955) is so common in the field of molecular orbital calculation, and the nature of the chemical bond between atoms has been defined well using the standard concept of covalent and

---

M. Morinaga (✉)
Toyota Physical and Chemical Research Institute, Nagakute, Aichi 480-1192, Japan

Department of Materials Science and Engineering, Graduate School of Engineering, Nagoya University, Furo-cho, Chikusa-ku, Nagoya 464-8603, Japan
e-mail: morinaga@toyotariken.jp

H. Yukawa
Department of Materials Science and Engineering, Graduate School of Engineering, Nagoya University, Furo-cho, Chikusa-ku, Nagoya 464-8603, Japan

H. Nakai
Department of Chemistry and Biochemistry, School of Advanced Science and Engineering, Waseda University, Okubo, Shinjuku-ku, Tokyo 169-8555, Japan

T. Ishii et al. (eds.), *The DV-Xα Molecular-Orbital Calculation Method*,
DOI 10.1007/978-3-319-11185-8_7

ionic bonds. However, with this analysis it is still difficult to compare quantitatively the chemical bond strengths among a variety of materials, because both covalent and ionic interactions are operating in most materials. To solve this problem, the chemical bond should be estimated quantitatively in an energy scale.

Recently, Nakai has proposed a new analyzing technique called energy density analysis (EDA) (Nakai 2002; Nakai et al. 2007). In this method, the total energy of a system, computed by Kohn-Sham type density functional theory (DFT) (Kohn and Sham 1965), is partitioned into atomic energy densities, and the characteristics of the chemical bond are understood in terms of each atomic energy density instead of the total energy. Using this energy density analysis, we have proposed a unified approach to the chemical bond in hydrides and hydrocarbons (Shinzato et al. 2007a). Special attention is directed toward hydrogen storage materials such as complex hydrides (e.g., $NaAlH_4$), metal hydrides (e.g., $TiFeH_2$) and hydrocarbons (e.g., cyclohexane, C_6H_{12}) (Shinzato et al. 2007a, b). The nature of the chemical bonds between atoms is very different among these materials, and hence it is needed to analyze the chemical bond in an energy scale to discuss them on a common ground. Recently, this approach has been applied to the quantitative evaluation of some oxide catalysts (e.g., Nb_2O_5) on the hydrogen desorption reaction of MgH_2 (Hirate et al. 2009), and also of some chloride catalysts (e.g., $TiCl_3$) on the decomposition reaction of $NaAlH_4$ (Hirate et al. 2011a).

Subsequently, attention is then directed toward metal oxides, since they possess a variety of chemical and physical properties such as superconductivity, ferroelectricity, magnetism and proton conductivity (Yoshino et al. 2003). For example, a perovskite-type oxide, $BaTiO_3$, shows ferroelectricity in the tetragonal phase and paraelectricity in the cubic phase. The emergence of various properties is originated from the diverse nature of the chemical bond in metal oxides. In order to understand the bond formation in various metal oxides in a fundamental manner, energy density analyses are performed focusing mainly on the structural stability problems. Local structure, average structure and phase transition are investigated in binary metal oxides such as zirconia (ZrO_2), Cr oxides, Fe oxides and Ti oxides, and in ternary perovskite-type oxides (Shinzato et al. 2011).

It is well known that the phase transition is one of the fundamental problems in the perovskite-type oxides (Kingery et al. 1976). For example, phase transition occurs either by the cooperative displacements of both Ti and O atoms along <100> direction in $BaTiO_3$ (Miyake and Ueda 1947), or by tilting or rotation of TiO_6 (or ZrO_6) octahedra around O-Ti (or Zr)-O crystal axes in $CaTiO_3$, $SrTiO_3$, $CaZrO_3$ and $SrZrO_3$ (Ahtee et al. 1976, 1978). A series of phase transitions is examined to draw a common feature with the aid of the atomization energy concept.

This unified approach based on the atomization energy concept has recently made great progress. It has indeed a great potential of being widely used in the field of materials science. In this paper, this unique approach is reviewed in a consistent way, aiming to understand the chemical bond between atoms in hydrides and oxides without using the standard concept of covalent and ionic bonds.

7.2 Calculation Method

7.2.1 Geometry Optimization

The crystal structures of hydrides and oxides are optimized by the total energy minimization using the plane-wave pseudopotentital method. For this purpose, the first-principle calculations based the density functional theory (DFT) are performed with a generalized gradient approximation (GGA) by Perdew et al. (1996). The implementation of DFT employed here combines a plane-wave basis set with the total energy pseudopotential method, as is embodied in the CASTEP code (Milman et al. 2000). The present calculations used are based upon the ultrasoft pseudopotentials proposed by Vanderbilt (1990). The plane-wave cutoff energy is chosen to be 380 eV, because this cutoff energy is found to achieve the convergence of the total energies within 0.03 eV, as compared to the results with the cutoff energies up to 600 eV. The sampling in the reciprocal space is done with the *k*-points grids. For example, $5 \times 5 \times 2$ for tetragonal $NaAlH_4$, $6 \times 6 \times 6$ for cubic $KMgH_3$, $5 \times 5 \times 9$ for monoclinic $TiFeH_2$ (Shinzato et al. 2007a, b), $6 \times 6 \times 6$ for Al_2O_3 and $8 \times 8 \times 8$ for cubic $CaTiO_3$ (Shinzato et al. 2007c, 2011). Also, the molecule structure of hydrocarbons is optimized using the AM1 method (Anders et al. 1993).

7.2.2 Energy Density Analysis (EDA)

The electronic structures for optimized crystal lattice of hydrides are obtained by the DFT calculations under the periodic boundary condition (PBC) using Gaussian03 program package (Frisch et al. 2004). The adopted functional is the BLYP functional, which consists of the Slater exchange (Slater 1951), the Becke (B88) exchange (Becke 1988), the Vosko-Wilk-Nusair (VWN) correlation (Vosko et al. 1980) and the Lee-Yang-Parr (LYP) correlation functionals (Lee et al. 1988). The following modified Gaussian basis sets are adopted: the correlation-consistent polarization plus the valence double zeta (cc-pVDZ) basis sets of Dunning (1989; Woon and Dunning 1993) without d-type functions for H, Li, B, N Na, Mg and Al, the Ahlrichs TZV basis set (Schafer et al. 1994) without the outer s function for K and Si, the Huzinaga basis sets (Huzinaga et al. 1984) without the outer s function and constructed to be double zeta class for Na, Ba, V, Cr, Fe, Co, Ni, Rb, Zr, Nb, Mo and Pd, the 6-31G basis sets (Rassolo et al. 1998; Hehre et al. 1972) without outer s function for O, Al, Ti and Ca, the 3-21G basis set (Dobbs and Hehre 1986) without outer s function for Sr, the Ahlrichs VDZ basis set (Schafer et al. 1992) without d-type function for C. In this study, the EDA calculations under PBC (Nakai et al. 2007) are performed by linking the original code for the EDA with Gaussian03.

Following the EDA (Nakai 2002), the atomic energy density of A atom is evaluated by,

$$E^A = E_{NN}^A + T_S^A + E_{Ne}^A + E_{CLB}^A + E_{XC}^A, \tag{7.1}$$

where E_{NN}^A is the nuclear-nuclear repulsion energy density, T_S^A is the non-interacting kinetic energy density, E_{Ne}^A is the nuclear-electron attraction energy density, E_{CLB}^A is the Coulomb energy density, E_{XC}^A is the exchange-correlation energy density.

In Eq. (7.1), for example, E_{XC}^A is evaluated by the partial sum for the numerical quadrature technique,

$$E_{XC}^A = \sum_g^{grid} \omega_g p_A(r_g) F_{XC}(r_g), \tag{7.2}$$

where ω_g is the weighting factor, $p_A(r_g)$ is the partition function, and $F_{XC}(r_g)$ is the exchange-correlation functional. The other terms in Eq. (7.1), which are evaluated by the analytical integration with the Kohn-Sham orbitals, are partitioned into their energy densities on the analogy of Mulliken population analysis (Frisch et al. 2004). Further detailed explanation of the calculation is given elsewhere (Nakai 2002; Nakai et al. 2007).

It is noted here that the Mulliken population analysis (Mulliken 1955) is the analysis for the electron density, namely, the electron density distribution over a molecule or crystal. Inevitably it is a convenient method for treating the covalent or ionic character of the chemical bond between atoms in the molecule or crystal. On the other hand, EDA is the analysis for the energy density, and the bond nature is expressed in an energy scale. So, it gives us a tool to compare the bond strengths quantitatively among a variety of hydrides or oxides.

7.2.3 Atomization Energy

For binary hydrides, MH, the respective atomic energy densities of M and H are related closely to the nature of the chemical bond relevant to M and H atoms in MH. When the energy of the isolated neutral atom, E_M^{atom} (or E_H^{atom}), is taken as a reference, the atomization energy, ΔE_M (or ΔE_H) is defined as,

$$\Delta E_M = E_M^{atom} - E_M^{hydride}, \tag{7.3}$$

$$\Delta E_H = E_H^{atom} - E_H^{hydride}, \tag{7.4}$$

where $E_M^{hydride}$ and $E_H^{hydride}$ are the atomic energy densities for M and H in MH, respectively. In case of ternary hydrides, $(M1M2)H_n$, ΔE_M is defined as $(\Delta E_{M1} + \Delta E_{M2})/n$, that is the average atomization energy of M1 and M2 to be counted per hydrogen atom. Even in the other type of ternary hydrides, ΔE_M is defined in a similar way.

Then, the cohesive energy, E_{coh}, of the hydride per hydrogen atom is defined as,

$$\Delta E_M + \Delta E_H = E_{coh}. \tag{7.5}$$

Thus, ΔE_M and ΔE_H are the components of E_{coh}. The advantage of using the atomization energy is to know the contribution of each constituent element to the hydride formation. Such a quantity is never able to be obtained from the total energy calculation alone.

By setting that $y = \Delta E_H$ and $x = \Delta E_M$, we obtain a relation, $y = -x + E_{coh}$. So, E_{coh} is expressed as a point of intersection of this line and y-axis at $x = 0$.

For hydrocarbons expressed as a chemical formula, C_mH_n, ΔE_H and ΔE_C are defined as,

$$\Delta E_H = E_H^{atom} - E_H^{hydrocarbon}, \tag{7.6}$$

$$\Delta E_C = \left(E_C^{atom} - E_C^{hydrocarbon}\right) \times (m/n). \tag{7.7}$$

Here, ΔE_C is the average atomization energy of carbon per hydrogen atom in C_mH_n. For simplicity, the sum of ΔE_H and ΔE_C is hereafter called cohesive energy, E_{coh}, of the hydrocarbon per hydrogen atom, although it exists in either solid or liquid or gas.

For oxides, the atomization energy is defined in a similar way. For example, for binary monoxide, MO, by using the energy of the isolated neutral atom, E_M^{atom} (or E_O^{atom}) as a reference, the atomic energy density, ΔE_M for M (or ΔE_O for O) is defined as,

$$\Delta E_M = E_M^{atom} - E_M^{oxide}, \tag{7.8}$$

$$\Delta E_O = E_O^{atom} - E_O^{oxide}. \tag{7.9}$$

In case of binary oxides, $M1_xO_y$, ΔE_M is defined as $x\Delta E_{M1}/y$ that is the atomization energy to be counted per O atom. Also, in case of ternary oxides, $(M1_xM2_y)O_z$, ΔE_M is defined as $(x\Delta E_{M1} + y\Delta E_{M2})/z$ that is the average atomization energy of M1 and M2 atoms to be counted per O atom.

7.3 Accuracy of Calculation

First, to show the reliability of the present calculation, the heat of formation, ΔH, for binary metal hydrides, MH_n, is calculated assuming that $M + (n/2)H_2 \rightarrow MH_n$, and compared with the experiment (Fukai 1994). As shown in Fig. 7.1, there is good agreement between the calculated and the experimental values.

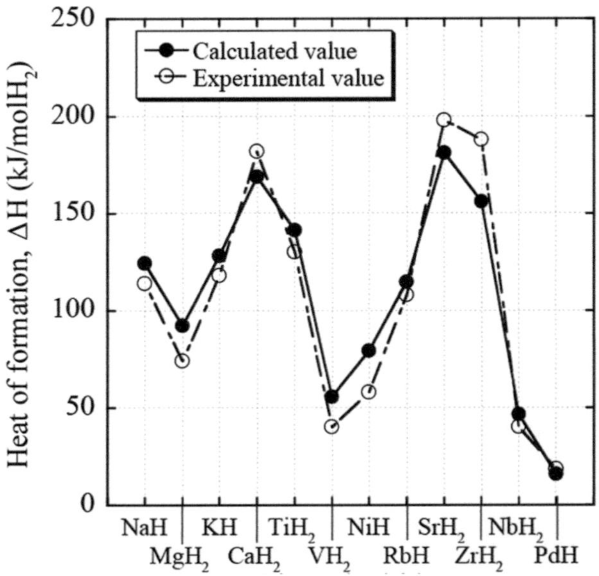

Fig. 7.1 Calculated and experimental heat of formation for binary hydrides

The cohesive energy, E_{coh}, is a measure of the electronic stability of materials. As shown in Table 7.1 (Shinzato et al. 2007a), the difference between the calculated and the experimental values for E_{coh} lies within 0.5 eV for binary metal hydrides and 0.2 eV for hydrocarbons (Weast et al. 2003). In case of oxides, as shown in Table 7.2, the difference between the calculated and the experimental values for E_{coh} lies within 0.5 eV except for FeO (Shinzato et al. 2011). A large discrepancy in FeO is probably attributable to the non-stoichiometry and the defect structure, since the present calculation is performed assuming a perfect crystal. Thus, the present calculation is performed in a reasonable manner.

7.4 Hydrides and Hydrocarbons

7.4.1 Crystalline Hydrides

7.4.1.1 Metal Hydrides

The ΔE_H vs. ΔE_M diagram for metal hydrides is shown in Fig. 7.2 (Shinzato et al. 2007a). This ΔE_H vs. ΔE_M diagram is hereafter called "atomization energy diagram". When the hydrides have a resemblance in the chemical bonding state, their locations are close to each other on the diagram. For example, binary hydrides of transition elements (e.g., NiH and PdH, both shown by solid square in Fig. 7.2) appear in the higher ΔE_H region than those of typical elements (e.g., NaH and MgH_2, both shown by open square in Fig. 7.2). This indicates that transition elements could stabilize the hydrogen state remarkably in the binary hydrides. In

Table 7.1 Comparison between the calculated and the experimental values for cohesive energy, E_{coh}, for binary metal hydrides and hydrocarbons (unit: eV)

(a)	E_{coh} (cal.)	E_{coh} (exp.)		E_{coh} (cal.)	E_{coh} (exp.)
NaH	4.39	3.96	TiH_2	5.52	5.38
MgH_2	3.34	3.41	VH_2	5.08	5.13
AlH_3	3.22	3.56	NiH	7.20	7.32
KH	3.38	3.78	ZrH_2	6.13	6.29
CaH_2	4.27	4.12	NbH_2	6.48	6.20
RbH	3.34	3.64	PdH	6.17	6.58
SrH_2	3.76	4.04			
(b)	E_{coh} (cal.)	E_{coh} (exp.)		E_{coh} (cal.)	E_{coh} (exp.)
C_2H_6	4.95	4.86	C_6H_6	9.68	9.49
C_2H_4	5.93	5.81	C_7H_8	8.80	8.64
C_2H_2	8.45	8.54	C_8H_{10}	8.26	8.14

Table 7.2 Comparison between the calculated and the experimental values for cohesive energy, E_{coh}, for oxides (unit: eV)

	E_{coh} (cal.)	E_{coh} (exp.)		E_{coh} (cal.)	E_{coh} (exp.)
Na_2O	9.16	9.10	Cr_2O_3	9.19	9.26
MgO	9.91	10.34	CrO_2	8.13	7.74
Al_2O_3	10.67	10.65	FeO	8.85	9.68
SiO_2	9.55	9.62	Fe_3O_4	8.41	8.69
K_2O	7.90	8.17	Fe_2O_3	8.28	8.28
CaO	11.08	11.01	SrO	10.27	10.41
TiO_2	10.23	9.93	Nb_2O_5	9.31	9.51
TiO	12.83	12.88	ZrO_2	11.13	11.40
Ti_2O_3	11.44	11.11	MoO_2	9.12	9.05
V_2O_3	10.60	10.35	BaO	10.50	10.18

other words, hydrogen atom interacts more strongly with transition elements than typical elements.

For ternary metal hydrides, $M1M2H_n$, where M2 is a transition metal (e.g., $TiFeH_2$ shown by open circle in Fig. 7.2), the M2-H interaction is important, since H is located in the neighborhood of M2 (e.g., M2 = Fe) in the crystal lattice. In case of Mg_2NiH_4, hydrogen atom is located in the neighborhood of Ni, and the strong interaction is operating between Ni and H in it. In agreement with this, Mg_2NiH_4 is located near NiH rather than MgH_2 on the atomization energy diagram shown in Fig. 7.2. However, the ΔE_H value is smaller in Mg_2NiH_4 than in NiH, indicating that the H state is destabilized by the presence of Mg in Mg_2NiH_4. On the other hand, $TiFeH_2$ is located well above TiH_2, so that the ΔE_H value is larger in $TiFeH_2$ than in TiH_2, indicating that H state is further stabilized by the presence of Fe in the neighborhood instead of neighboring Ti. It is very interesting that $TiFeH_2$ which desorbs hydrogen at room temperature has the largest cohesive energy, 8.3 eV, among the metal hydrides.

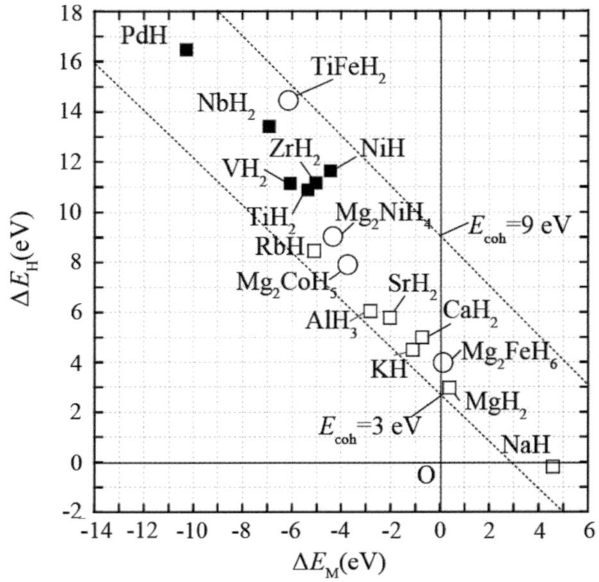

Fig. 7.2 Atomization energy diagram for metal hydrides

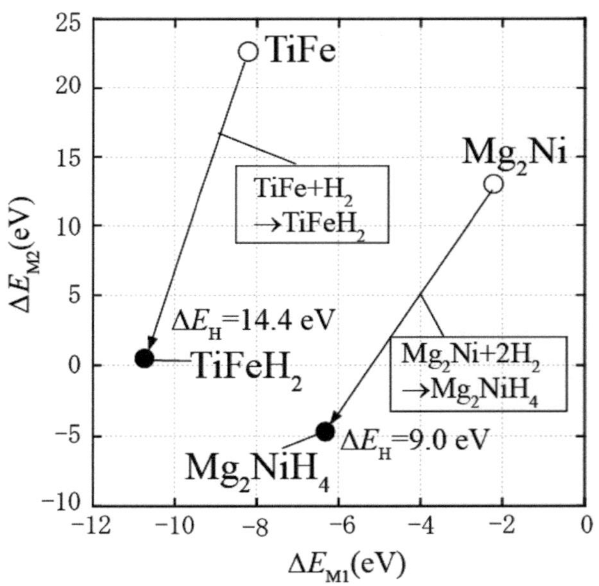

Fig. 7.3 ΔE_{M1} vs. ΔE_{M2} diagram for metal hydrides

Mg_2NiH_4 and $TiFeH_2$ are formed by the hydrogenation of intermetallic compounds, Mg_2Ni and TiFe, respectively. Therefore, the further analyses are carried out with these compounds. To understand the role of metal atoms in the bond formation, the values of ΔE_{M1} (M1 = Ti or Mg) and ΔE_{M2} (M2 = Fe or Ni) are plotted in Fig. 7.3 for TiFe and Mg_2Ni, together with the values for $TiFeH_2$ and

Mg_2NiH_4. ΔE_{Ti} for TiFe and ΔE_{Mg} for Mg_2Ni are negative, whereas ΔE_{Fe} for TiFe and ΔE_{Ni} for Mg_2Ni are positive. By hydrogenation, the atomization energy decreases in both M1 and M2. However, the decrement is more remarkable in ΔE_{M2} than in ΔE_{M1}. This is because, hydrogen exists in the neighborhood of M2 and interacts strongly with M2 (M2 = Fe or Ni) in the hydrides. As a result, most of the energy of M2 transfers to the hydrogen and the H state becomes very stable.

It is interesting to note that in either system of TiFe and Mg_2Ni there is a large difference between ΔE_{M1} and ΔE_{M2}, indicating that both TiFe and Mg_2Ni intermetallic compounds have the strong chemical bond between M1 and M2 atoms. This is a reason why the hydrogenation and dehydrogenation reactions take place smoothly without changing alloy compositions in them. Thus, this approach can clarify the role of constituent element in the formation of metal hydrides (Shinzato et al. 2007a).

7.4.1.2 Complex Hydrides

As shown in Fig. 7.4, complex hydrides are located in the narrow and small ΔE_M range (-2.2 eV $< \Delta E_M < -0.3$ eV), but in the wide ΔE_H range of 4.2–8.5 eV. ΔE_H changes in the order, $Mg(AlH_4)_2 < M1AlH_4$ (M1 = Li,Na,K) $< M1BH_4$ (M1 = Li, Na,K) $< LiNH_2$ (Shinzato et al. 2007a). The cohesive energy changes in the same order. Thus, the H state is stabilized well in compensation for the destabilized M state.

In case of alanate (Al) -type complex hydrides, the magnitude of ΔE_H changes in the order, $Mg(AlH_4)_2 < LiAlH_4 < NaAlH_4 < KAlH_4$ as is shown in Fig. 7.4. Hence,

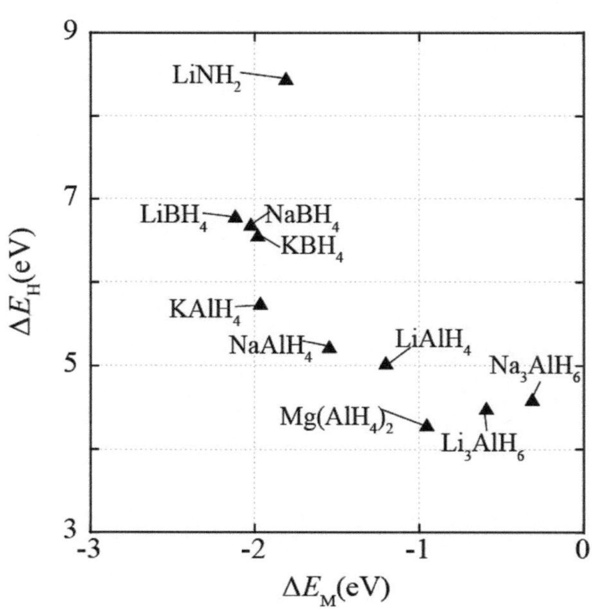

Fig. 7.4 Atomization energy diagram for complex hydrides

the H state in $Mg(AlH_4)_2$ is least stable among them, as is found experimentally (Komiya et al. 2007).

In addition, the coordination number effect can be understood by using the atomization energy diagram. In Na_3AlH_6 (or Li_3AlH_6), an Al atom is surrounded by six hydrogen atoms, whereas in $NaAlH_4$ (or $LiAlH_4$) an Al atom is surrounded by four hydrogen atoms. As shown in Fig. 7.4, ΔE_H is lower in Na_3AlH_6 (or Li_3AlH_6) than in $NaAlH_4$ (or $LiAlH_4$). Thus, the H state is less stable in Na_3AlH_6 (or Li_3AlH_6) than in $NaAlH_4$ (or $LiAlH_4$), as might be expected from the coordination number.

7.4.1.3 Atomization Energy Diagram for Crystalline Hydrides

For a variety of crystalline hydrides such as binary hydrides, ternary metal hydrides, complex hydrides and perovskite-type hydrides, atomization energy diagram is constructed in an energy scale as shown in Fig. 7.5 (Shinzato et al. 2007a, b). It is stressed here that every hydride can be shown on one figure, although there are

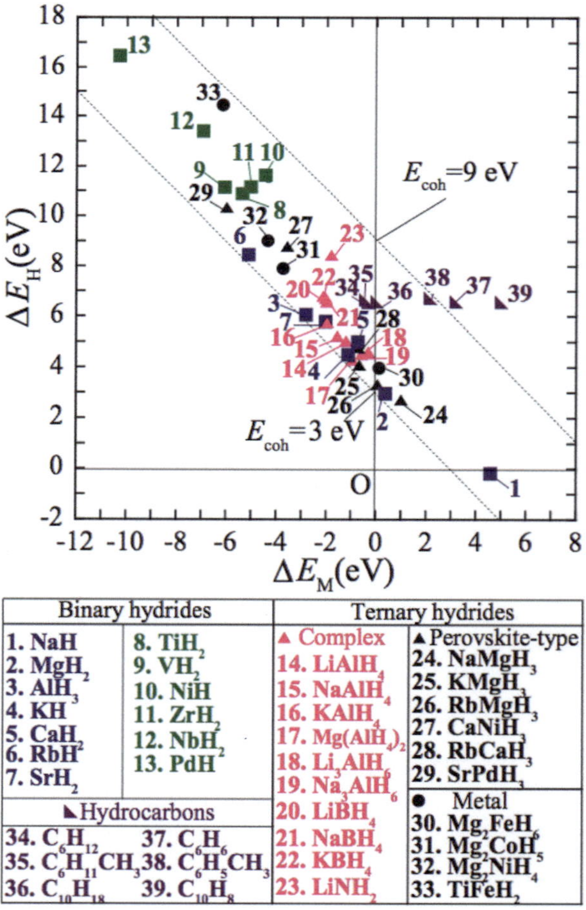

Binary hydrides		Ternary hydrides	
		▲ Complex	▲ Perovskite-type
1. NaH	8. TiH_2	14. $LiAlH_4$	24. $NaMgH_3$
2. MgH_2	9. VH_2	15. $NaAlH_4$	25. $KMgH_3$
3. AlH_3	10. NiH	16. $KAlH_4$	26. $RbMgH_3$
4. KH	11. ZrH_2	17. $Mg(AlH_4)_2$	27. $CaNiH_3$
5. CaH_2	12. NbH_2	18. Li_3AlH_6	28. $RbCaH_3$
6. RbH	13. PdH	19. Na_3AlH_6	29. $SrPdH_3$
7. SrH_2		20. $LiBH_4$	● Metal
◣ Hydrocarbons		21. $NaBH_4$	30. Mg_2FeH_6
34. C_6H_{12}	37. C_6H_6	22. KBH_4	31. Mg_2CoH_5
35. $C_6H_{11}CH_3$	38. $C_6H_5CH_3$	23. $LiNH_2$	32. Mg_2NiH_4
36. $C_{10}H_{18}$	39. $C_{10}H_8$		33. $TiFeH_2$

Fig. 7.5 Atomization energy diagram for crystalline hydrides and hydrocarbons

significant differences in the nature of the chemical bond among the hydrides. E_{coh} given by Eq. (7.5), lies in the range of 3–9 eV. As a whole, this range of E_{coh} ($=\Delta E_H + \Delta E_M$) is much smaller compared to the range of the atomization energy, $0\ \text{eV} < \Delta E_H < 18\ \text{eV}$ and $-11\ \text{eV} < \Delta E_M < 5\ \text{eV}$. Thus, each atomic state is well controlled or balanced in the hydrides so as to make E_{coh} as large as possible. Among the hydrides, a binary hydride, AlH_3, shows the smallest cohesive energy, 3.2 eV, so it releases hydrogen at low temperature (Graetz and Reily 2006), but the re-hydrogenation reaction never occurs in the moderate condition because of the low stability. This is also the case in a complex hydrides, $Mg(AlH_4)_2$ where both the E_{coh} and the ΔE_H values are small (Komiya et al. 2007).

It is evident from Fig. 7.5 that the region of complex hydrides is overlapped with that of the hydrides of typical elements (e.g., MgH_2 (No. 2)) on this diagram. Also, the locations of ternary metal hydrides are distributed over the wide range of ΔE_H, and some of them are located near the region of complex hydrides. In fact, Mg_2NiH_4 (No. 32), Mg_2CoH_5 (No. 31) and Mg_2FeH_6 (No. 30) are sometimes treated as complex hydrides instead of metal hydrides (Yvon 1998). In Fig. 7.5, Mg_2FeH_6 and Mg_2CoH_5 are located near the lower and upper region of complex hydrides, respectively. However, Mg_2NiH_4 is located in the region between the hydrides of transition elements and the complex hydrides, indicating the coexistence of the characteristics of both metal and complex hydrides in Mg_2NiH_4. Thus, the atomization energy diagram reflects important characteristics of the chemical bonds in hydrides.

7.4.2 Hydrocarbons

7.4.2.1 ΔE_C and ΔE_H for C_2H_6, C_2H_4, C_2H_2 and Arene

The present approach is extended further to the hydrocarbon, C_mH_n (Shinzato et al. 2007a). A simple analysis is first given in ethane (C_2H_6), ethylene (C_2H_4) and acetylene (C_2H_2). The chemical bond between C atoms in these hydrocarbons has been treated using the concept of the single bond for C_2H_6, the double bond for C_2H_4 and the triple bond for C_2H_2. The atomization energies for carbon, ΔE_C, and for hydrogen, ΔE_H, are shown in Fig. 7.6a. Here, ΔE_C is an atomization energy weighted by a factor, m/n, following Eq. (7.7). As is evident from Fig. 7.6a, ΔE_C increases linearly with the ratio of carbon number (m) to hydrogen number (n), m/n, whereas ΔE_H is nearly constant. For comparison, the bond energies between C atoms, E_{bond} (Pimentel and Spratley 1969) and its normalized values per hydrogen atom, E_{bond}' ($=E_{bond}/n$), are plotted in Fig. 7.6a. E_{bond} changes in the order, $C_2H_6 < C_2H_4 < C_2H_2$, in agreement with the order of ΔE_C. Also, the E_{bond}' line correlates well with the ΔE_C line, while showing a similar slope between them. Thus, the atomization energy for carbon, ΔE_C, reflects the bond strength between C atoms in these hydrocarbons.

Next shown is the arene such as benzene (C_6H_6), toluene ($C_6H_5CH_3$), napthalene ($C_{10}H_8$), p-xylen (C_8H_{10}), fluorene ($C_{13}H_{10}$), phenanthrene ($C_{14}H_{10}$), pyrene ($C_{16}H_{10}$), coronene ($C_{24}H_{12}$) (Shinzato et al. 2007a). Arene has a larger ratio,

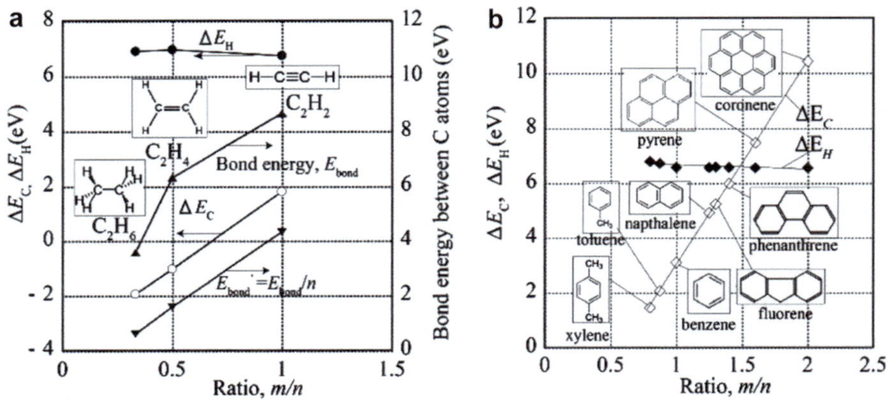

Fig. 7.6 Atomization energy for carbon and hydrogen in (**a**) C_2H_6, C_2H_4,C_2H_2 and (**b**) arene

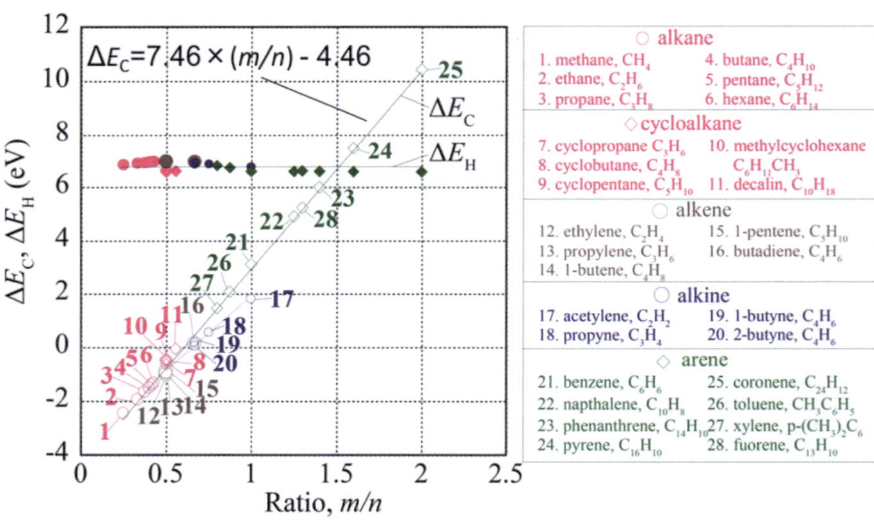

Fig. 7.7 Atomization energy for carbon and hydrogen for alkane, cycloalkane, alkene, alkine and arene

m/*n*, as compared to alkane, alkene and alkine. As shown in Fig. 7.6b, ΔE_H values are kept nearly constant and only the ΔE_C values vary linearly with *m*/*n*. So, the sum of them defined as E_{coh} changes with the *m*/*n* ratio in arene.

7.4.2.2 Correlation of ΔE_C and ΔE_H with the *m*/*n* Ratio in Hydrocarbon, C_mH_n

For a variety of hydrocarbons including alkane, cycloalkane, alkene, alkine and arene, ΔE_C and ΔE_H are plotted against *m*/*n*, as shown in Fig. 7.7 (Shinzato et al. 2007a). There is indeed a linear relationship between ΔE_C and *m*/*n*, and ΔE_C is expressed as,

$$\Delta E_C \text{ (eV)} = 7.46 \times (m/n) - 4.46. \tag{7.10}$$

The correlation factor is better than 0.998. According to Eq. (7.7), the slope is $(E_{\mathrm{C}}^{\mathrm{atom}} - E_{\mathrm{C}}^{\mathrm{hydrocarbon}})$ and it is nearly constant, 7.46 eV, regardless of the type of hydrocarbons. The alkine such as acetylene, propyne, 1-butyne and 2-butyne deviates slightly from this relation. On the other hand, the atomization energy for hydrogen, ΔE_H is a nearly constant, 6.76 eV, irrespective of the types of hydrocarbons. So, the cohesive energy per one hydrogen atom in hydrocarbons, E_{coh}, is expressed as,

$$E_{coh} \text{ (eV)} = 7.46 \times (m/n) + 2.30. \tag{7.11}$$

This beautiful relation is satisfied in the hydrocarbons, because the chemical stability of C_mH_n is dependent only on the *m*/*n* ratio. Thus, by using the atomization energies, ΔE_C and ΔE_H, the nature of the chemical bond in the hydrocarbons is understood without using the concept of single or multiple C-C bond.

For comparison, the results of some hydrocarbons used for hydrogen storage are plotted in Fig. 7.5. They are located near $LiNH_2$ (No. 23), or in a higher ΔE_H position than other complex hydrides.

7.4.3 Summary

An atomization energy diagram, which is obtained by the EDA is constructed for the first time to understand the chemical interaction in hydrides in an energy scale. All the crystalline hydrides can be located on this diagram, although there are significant differences in the nature of the chemical bond among the hydrides. Also, for hydrocarbon, ΔE_C, increases linearly with the ratio of carbon number to hydrogen number, *m*/*n*, while keeping ΔE_H constant. The chemical stability of C_mH_n is dependent only on the *m*/*n* ratio.

7.5 Metal Oxides

7.5.1 Binary Oxides

7.5.1.1 Atomization Energy Diagram and Local Structure

In binary metal oxides, a main part of the local structure is expressed by using the M-O interatomic distances, $d_{\text{M-O}}$, and the coordination number, *CN*, both of which affect the atomization energies, ΔE_O and ΔE_M. For example, a ΔE_O vs. ΔE_M diagram is shown in Fig. 7.8a for binary Cr oxides (Shinzato et al. 2011). There is a clear trend that ΔE_O increases with decreasing ΔE_M. The magnitude of ΔE_O changes in the order, $CrO_3 > CrO_2 > Cr_2O_3 > CrO$. This is the reverse order of the

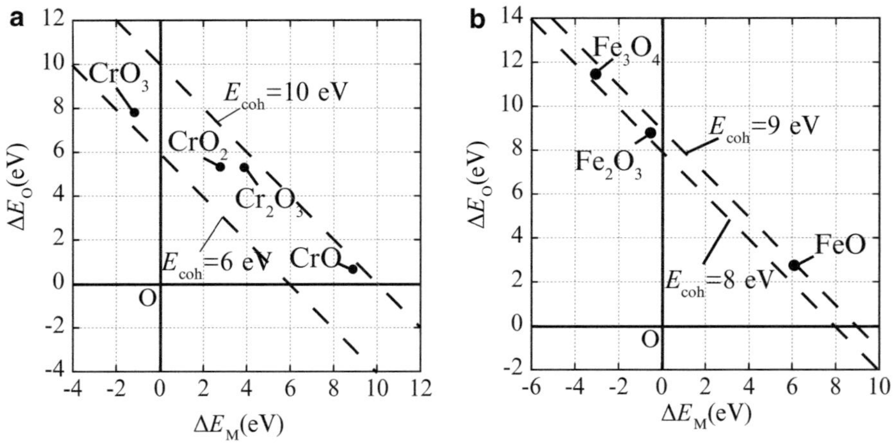

Fig. 7.8 Atomization energy diagram (**a**) for binary Cr oxides and (**b**) for binary Fe oxides

Table 7.3 Local structure in (a) Cr oxides, (b) Fe oxides and (c) Ti oxides

(a)	CrO_3	CrO_2	Cr_2O_3	CrO
$d_{Cr\text{-}O}$ (nm)	0.166	0.188	0.195	0.206
CN (Cr)	4	6	6	6
CN (O)	1,2	3	4	6

(b)	Fe_3O_4	Fe_2O_3	FeO
$d_{Fe\text{-}O}$ (nm)	0.185	0.188	0.204
CN (Fe)	4,6	6	6
CN (O)	4	4	6

(c)	TiO_2 (anatase)	TiO_2 (rutile)	Ti_2O_3	TiO
$d_{Ti\text{-}O}$ (nm)	0.196	0.197	0.204	0.214
CN (Ti)	6	6	6	6
CN (O)	3	3	4	6

average Cr-O interatomic distance, $d_{Cr\text{-}O}$, $CrO_3 < CrO_2 < Cr_2O_3 < CrO$ as shown in Table 7.3 (a). This indicates that the O atom state in binary Cr oxides is stabilized owing to the shorter Cr-O interatomic distance, but instead the Cr atom state is destabilized in some ways to optimize the Cr-O chemical bond. In addition, the coordination number of O around Cr, *CN* (Cr), is 4 for CrO_3 and 6 for CrO_2, Cr_2O_3 and CrO as listed in Table 7.3 (a). Thus, the O atom state is more stable in CrO_3 than in CrO_2, Cr_2O_3 and CrO, as might be expected from the coordination number effect (Nakai 2002) that the chemical bond becomes stronger as the coordination number decreases. However, the cohesive energy per O atom is lowest in CrO_3 among the binary Cr oxides.

Another atomization energy diagram is shown in Fig. 7.8b for binary Fe oxides. The magnitude of ΔE_O changes in the order, $Fe_3O_4 > Fe_2O_3 > FeO$. As shown in Table 7.3 (b), this order of ΔE_O is understood well by using the Fe-O interatomic

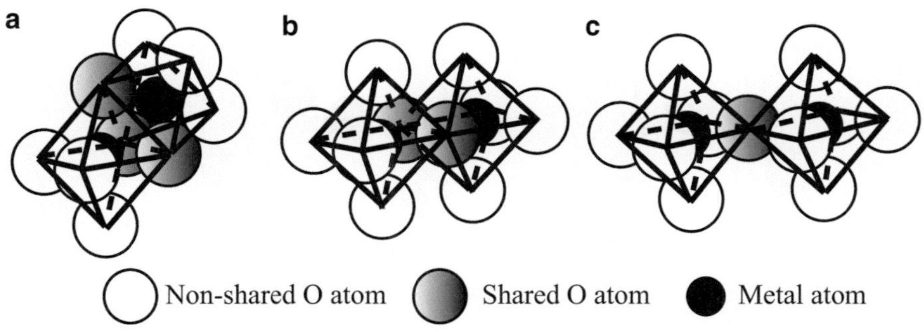

Fig. 7.9 Octahedra of (**a**) face, (**b**) edge and (**c**) vertex sharing

distance, $d_{\text{Fe-O}}$, and the coordination number of O around Fe, *CN* (Fe), as is similar to the Cr oxides. Thus, the atomization energies are sensitive to the local structure through the short-range chemical interaction operating between atoms.

7.5.1.2 Linkage of Polyhedra Over Crystal Space

Any crystal structures of metal oxides are composed of polyhedra such as the MO_6 octahedron and the MO_4 tetrahedron. As shown in Fig. 7.9, each polyhedron shares vertex, edge or face with the neighboring polyhedra (Yoshino et al. 2003, 2004). The chemical interactions operating between atoms in such polyhedra vary with the composition and density of the oxides. For example, from a geometrical point of view, the density increases in the order, vertex $<$ edge $<$ face sharing. Both the vertex and edge sharing are seen in most crystal structures of binary oxides, for example, in rutile-type structure and NaCl-type structure (Kingery et al. 1976). On the other hand, there are a few binary oxides (e.g., corundum-type oxides) in which polyhedra link together by the face sharing as well as the vertex and edge sharing.

A series of calculations is first carried out using four cluster models to investigate the atomization energy change with the sharing of polyhedra (Shinzato et al. 2011). As shown in Fig. 7.10a–d, the cluster models are made on the basis of the corundum-type crystal structure of Fe_2O_3, where all the sharing types coexist. Shown in the figure are (a) a single octahedron, (b) face-shared octahedra, (c) edge-shared octahedra and (d) vertex-shared octahedra. The calculated atomization energies for Fe, ΔE_{Fe}, in the clusters (a), (b), (c) and (d), are 0.1, -1.7, -1.7 and 1.2 eV, respectively. The atomization energies for O atom (or shared O atom), ΔE_O, in the clusters (a), (b), (c) and (d), are 3.5, 5.9, 6.7 and 6.2 eV, respectively. It is evident that the O atom state is stabilized by the sharing, but the face shared one is least stabilized among the shared atoms. This may be one of the reasons why the face-sharing is rare in the crystal structures of metal oxides.

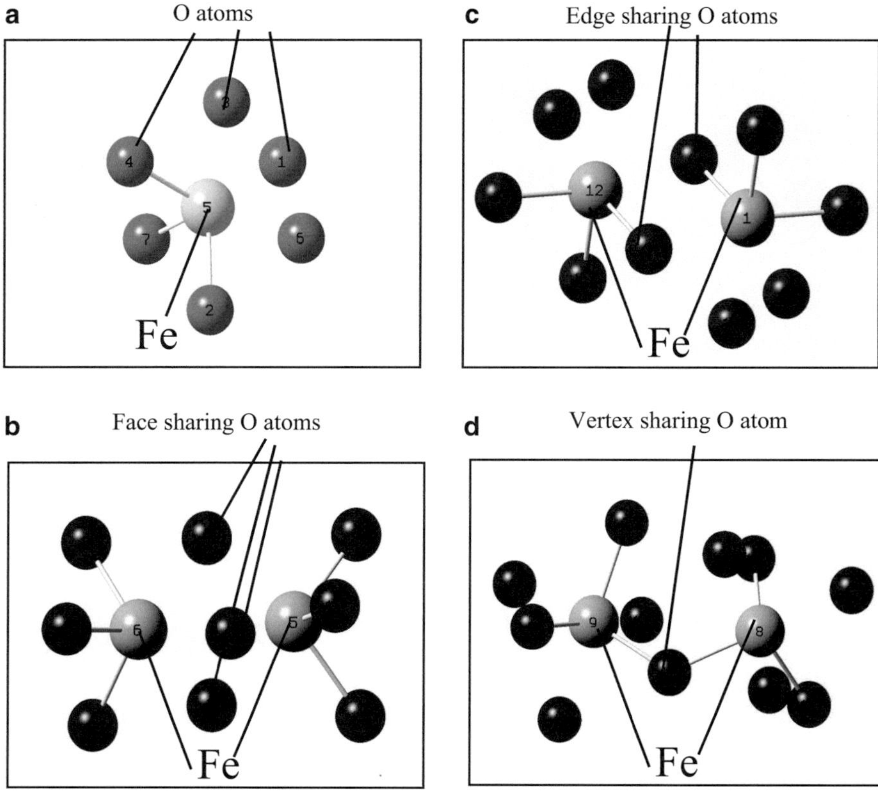

Fig. 7.10 Calculated Fe_2O_3 clusters of (**a**) one octahedron, and three clusters of octahedra shared by (**b**) face, (**c**) edge and (**d**) vertex

7.5.1.3 Density

An atomization energy diagram for binary Ti oxides is shown in Fig. 7.11. For TiO_2 such as anatase and rutile, Ti and O atoms form a distorted octahedron. The value of $d_{\text{Ti-O}}$ listed in Table 7.3 (c) is the average value of six interatomic distances between Ti and O. ΔE_O is larger in anatase than in rutile, although the local structure is almost similar between them as shown in Table 7.3 (c). This may be attributable to the difference in the density between anatase (3.84 Mg/m^3) and rutile (4.24 Mg/m^3). In other words, the density is another key parameter for describing oxide structures.

A correlation between the atomization energy and the density is shown in Fig. 7.12 for several binary metal oxides. There is a clear trend that ΔE_M increases and ΔE_O decreases with increasing density, as shown in (a) for Ti oxides, (b) for Cr oxides, and (c) for Fe oxides. A three-dimensional arrangement of polyhedra in crystal space determines the density. Such long-range interactions operating between atoms also reflect the values of the atomization energy.

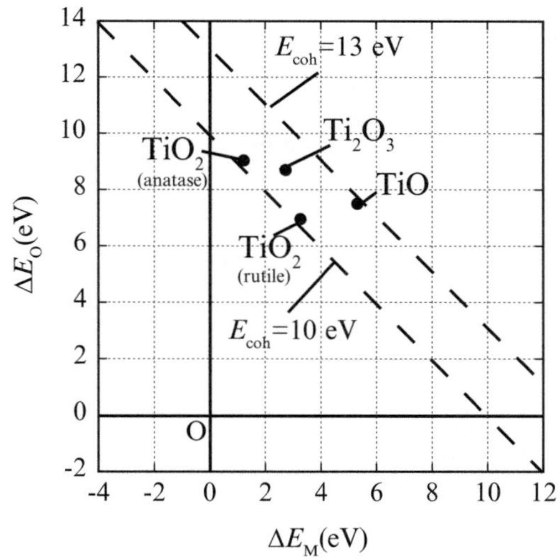

Fig. 7.11 Atomization energy diagram for binary Ti oxides

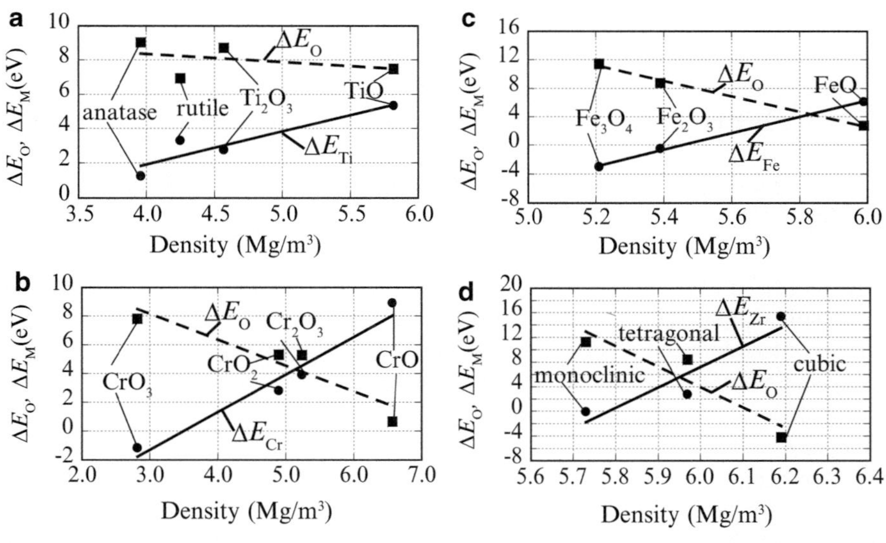

Fig. 7.12 The correlation between atomization energy and density of (**a**) binary Ti oxides, (**b**) binary Cr oxides (**c**) binary Fe oxides and (**d**) zirconia (ZrO_2)

As shown in Fig. 7.12d, this is the case even in zirconia (ZrO_2), in which there are three polymorphs; monoclinic, tetragonal and cubic phases. The phase transition occurs from monoclinic to tetragonal phase at about 1,300 K, and from tetragonal to cubic phase at about 2,600 K. As shown in Fig. 7.12d, the density increases monotonously with these phase transitions. This clearly indicates that the atomization energies vary significantly with the phase transition.

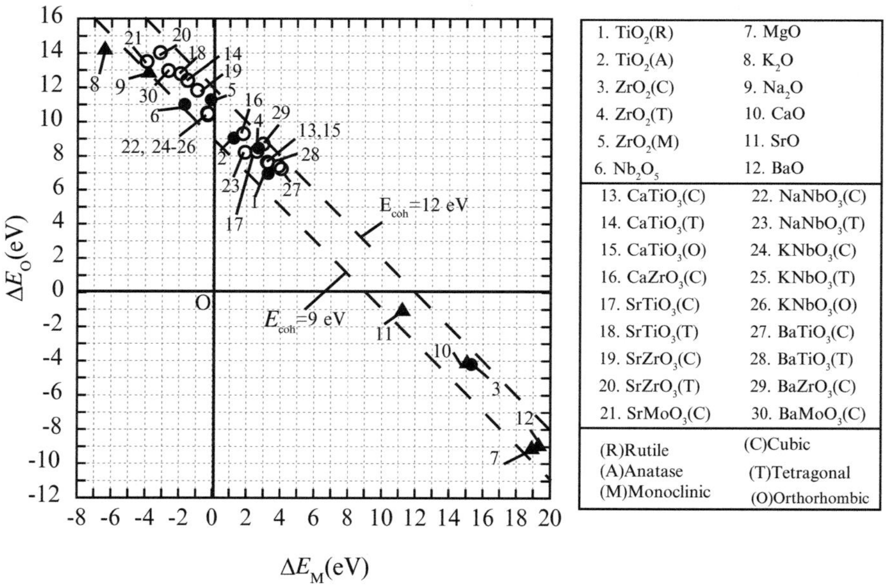

Fig. 7.13 Atomization energy diagram for perovskite-type oxides

7.5.2 *Perovskite-Type Oxides*

7.5.2.1 Atomization Energy Diagram for Perovskite-Type Oxides

The atomization energy diagram for perovskite-type oxides, $M1M2O_3$, is shown in Fig. 7.13 (Shinzato et al. 2007c, 2011). The perovskite-type oxides are shown by open circles, and for comparison some binary metal oxides are shown by solid circles or triangles in the figure. In every perovskite-type oxide, ΔE_O is larger than ΔE_M, indicating that the O atoms contribute to the cohesive energy more significantly than the metal atoms. The cohesive energy, E_{coh}, per O atom lies in the range of 9–12 eV.

As mentioned before, when there is a resemblance in the chemical bond between oxides, their locations must be close to each other on the diagram. A perovskite-type oxide, $M1M2O_3$, is formed by the reaction of either $M1O + M2O_2$ or $M1_2O + M2_2O_5$. It is evident from Fig. 7.13 that $M1M2O_3$ is located closer to the binary oxide, $M2O_2$ or $M2_2O_5$ than $M1O$ or $M1_2O$ on the diagram. For example, $CaTiO_3$ (No. 13 (cubic phase), No. 14 (tetragonal phase) or No. 15 (orthorhombic phase)) is located closer to TiO_2 (rutile (No. 1) and anatase (No. 2)) than CaO (No. 10). Also, $NaNbO_3$ (No. 22 (cubic phase) or No. 23 (tetragonal phase)) is located closer to Nb_2O_5 (No. 6) than Na_2O (No. 9). Thus, the perovskite-type oxide, $M1M2O_3$, inherits the nature of the chemical bond mainly from $M2O_2$ or $M2_2O_5$. However, M1 atoms still contribute to the cohesive energy to some extent.

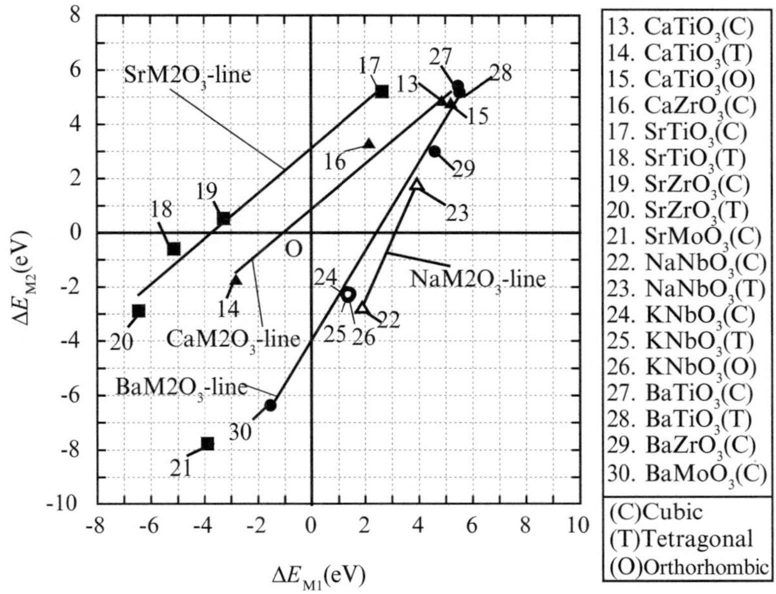

Fig. 7.14 ΔE_{M1} vs. ΔE_{M2} diagram for perovskite-type oxides

7.5.2.2 ΔE_{M1} vs. ΔE_{M2} Diagram for Perovskite-Type Oxides

As shown in Fig. 7.14, a ΔE_{M1} vs. ΔE_{M2} diagram is made in order to understand the role of each metal element in the bond formation of the perovskite-type oxides. The positions of $SrM2O_3$ (M1 = Sr, M2 = Ti, Zr) are aligned along a line drawn in the figure, indicating that there is a certain balance between ΔE_{Sr} and ΔE_{M2}. This trend is also seen in $CaM2O_3$ (M1 = Ca, M2 = Ti, Zr), $BaM2O_3$ (M1 = Ba, M2 = Ti, Zr) and $NaM2O_3$ (M1 = Na, M2 = Nb). Exception is $SrMoO_3$ (No. 21), which is not located along a $SrM2O_3$-line, but near $BaMoO_3$ (No. 30).

It is apparent from Fig. 7.14 that the values of ΔE_{M1} and ΔE_{M2} change with the crystal structure. For example, in case of $CaTiO_3$ the position moves from No. 13 for cubic structure to No. 14 for tetragonal structure. Similarly, in case of $SrTiO_3$ the position moves from No. 17 for cubic structure to No. 18 for tetragonal structure. As shown in Fig. 7.13, ΔE_O also varies with the crystal structure as does ΔE_{M1} or ΔE_{M2}.

In the perovskite-type $M1M2O_3$, not only M2 atom but also M1 atom plays a role in the oxide formation. In a geometrical point of view, M2 is surrounded by six O atoms to form a $M2O_6$ octahedron, whereas M1 is surrounded by twelve O atoms without forming any polyhedra. So, the bond nature is usually different between M1-O and M2-O. Namely, M1 atom weakly-bound to the surrounding O atoms, probably adjusts the coordinates readily and works to make the cohesive energy as large as possible, resulting in the energy balance between ΔE_{M1} and ΔE_{M2}. The existence of only $M2O_6$ octahedron in the structure is characteristic of the

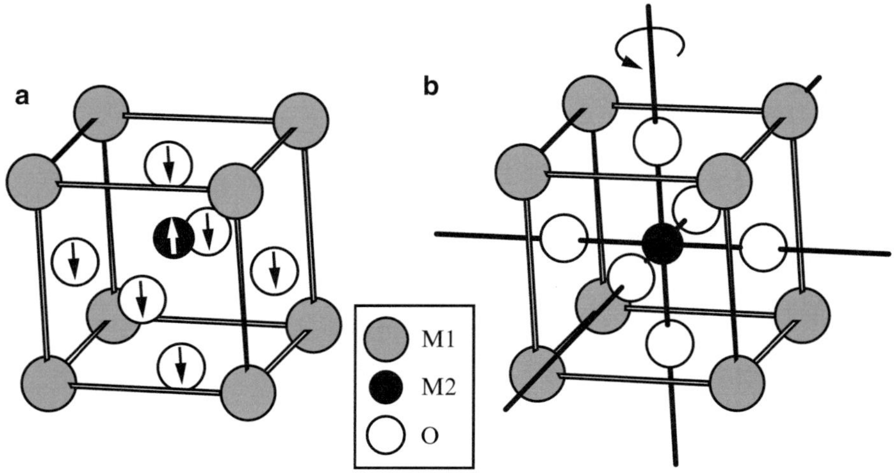

Fig. 7.15 Schematic illustration of (**a**) displacement-type and (**b**) tilting-type phase transitions

perovskite-type oxides. This unique structure will undergo the phase transition by the geometrical change inside the $M2O_6$ octahedron and by the attendant change of the weakly bound M1 atoms. As the result, both ΔE_{M1} and ΔE_{M2} change together with ΔE_O with the phase transition.

7.5.3 Phase Transition of Perovskite-Type Oxides

As illustrated in Fig. 7.15, there are two types of phase transition in the perovskite-type oxides. One type transition occurs by the cooperative displacements of both M2(=Ti) and O atoms along the <100> direction in $BaTiO_3$ (Miyake and Ueda 1947). The other type transition occurs by tilting (or rotation) of $M2O_6$ (=TiO_6 or ZrO_6) octahedra around O-M2(=Ti or Zr)-O crystal axes in $CaTiO_3$, $SrTiO_3$, $CaZrO_3$ and $SrZrO_3$ (Ahtee et al. 1976).

The atomization-energy changes attendant on the phase transition are simulated in $CaTiO_3$ and $BaTiO_3$. Crystal structure data used are the experimental values at room temperature, 1,273, 1,473, 1,523 and 1,633 K for $CaTiO_3$ (Kennedy et al. 1999; Liu and Lieberman 1993), and at room temperature, 1,387 and 1,645 K for $BaTiO_3$ (Kwei et al. 1993; Edwards et al. 1951). No structural optimization is performed in these calculations. In Fig. 7.16, the changes of the atomization energy, ΔE_i, are shown as a function of the density in $CaTiO_3$ and $BaTiO_3$. Here, the density is used in the horizontal coordinate instead of temperature. The density decreases as temperature increases, so it is lower in the cubic phase than in the tetragonal phase than in the orthorhombic phase. As shown in Fig. 7.16a for $CaTiO_3$, the atomization energy varies significantly with the tilting (or rotation)-type phase transition, although the cohesive energy scarcely changes

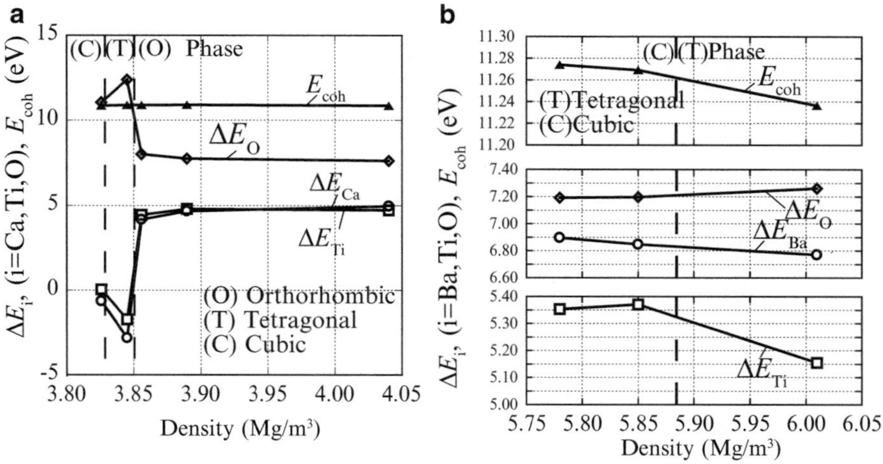

Fig. 7.16 Change of the atomization energy as a function of density for (**a**) $CaTiO_3$ and (**b**) $BaTiO_3$

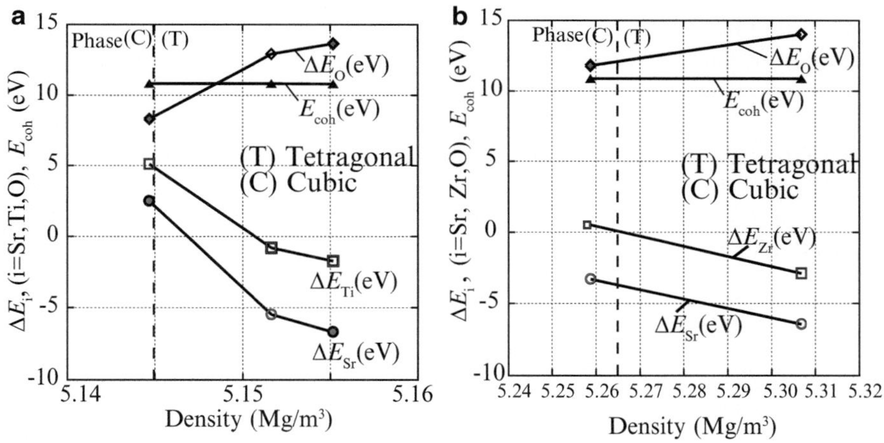

Fig. 7.17 Change of the atomization energy as a function of density for (**a**) $SrTiO_3$ and (**b**) $SrZrO_3$

with the phase transition. Namely, the ΔE_O value increases by the transition from cubic to tetragonal phase, and then it decreases by the transition from tetragonal to orthorhombic phase, followed by the nearly constant value in the orthorhombic phase. Both ΔE_{Ti} and ΔE_{Ca} change in a reverse way as does ΔE_O during the successive phase transition.

Also, as shown in Fig. 7.16b, the atomization energy change with the phase transition is much lower in $BaTiO_3$ than in $CaTiO_3$.

However, in either oxide, ΔE_O increases by the transition from cubic phase to tetragonal phase. The other two examples, $SrTiO_3$ and $SrZrO_3$, are shown in Fig. 7.17a and b, respectively.

The transition from cubic to tetragonal phase makes the O atom state stable in both $SrTiO_3$ and $SrZrO_3$. This is very similar to the result shown in Fig. 7.16a.

According to our previous study on $SrZrO_3$ (Yoshino et al. 2004), the average Zr-O interatomic distance increases by tilting or rotation of ZrO_6 octahedron, and the average Zr-O bond strength is recovered to some extent by forming a distorted ZrO_6 octahedron by the phase transition (Yoshino et al. 2004). The minimum Zr-O interatomic distance in cubic $SrZrO_3$ at 1,443 K is close to the shortest Zr-O interatomic distance in pure ZrO_2, so that the phase transition (i.e., the tilting of ZrO_6 octahedron) occurs around 1,443 K, and the attendant increase in the Zr-O interatomic distance works to retain the Zr-O bond strength. Unless such a tilting does take place, the Zr-O bond will become very weak according to a hypothetical calculation (Yoshino et al. 2004). Thus, a gain of ΔE_O by tilting or rotation is attributable to the recovery in the Zr-O bond strength with the increase in the Zr-O interatomic distance. Instead, ΔE_M decreases, resulting in nearly null change in the cohesive energy with the transition. It is stressed that the tilting (or rotation) makes a distorted ZrO_6 octahedron in which four Zr-O interatomic distances increase, and other two distances keep nearly unchanged with tilting (or rotation). Such a structural change reflects well the values of ΔE_O, ΔE_{Zr} and ΔE_{Sr} as shown in Fig. 7.17b.

On the other hand, as shown in Fig. 7.15, all the O atoms in $BaTiO_3$ are displaced in the same way and the same magnitude along a <100> direction by the transition from cubic to tetragonal phase. In case of tetragonal $BaTiO_3$ the Ti-O bond is strengthen for a longer Ti-O pair and weakened for a shorter Ti-O pair (Yoshino et al. 2004). But still the average Ti-O bond strength scarcely changes with the phase transition (Yoshino et al. 2004). As a result, the attendant change of ΔE_O is very limited as shown in Fig. 7.16b. This trend is also seen in $KNbO_3$.

7.5.4 Summary

Following the energy density analyses, the nature of the chemical bond in binary metal oxides and perovskite-type oxides is investigated by focusing on the local structure, average structure and phase transition. The atomization energies, ΔE_O and ΔE_M, are sensitive to the local and average atomic arrangements, and both ΔE_O and ΔE_M correlate well with the overall density of binary metal oxides.

In the perovskite-type oxides, $M1M2O_3$, there is an energy balance between ΔE_{M1} and ΔE_{M2}. As for the phase transition, the stable state of O atoms is formed in $CaTiO_3$ or $BaTiO_3$ by the tilting-type or the <100> displacement-type phase transition from cubic to tetragonal phase, although the increment of ΔE_O is much smaller in $BaTiO_3$ than in $CaTiO_3$.

7.6 Application of Atomization Energy Approach to Catalytic Problems

7.6.1 Atomization Energy Diagram for Binary Metal Oxides

Magnesium hydride (MgH_2) has been considered as one of the promising candidates for hydrogen storage, because it possesses a high hydrogen capacity of 7.6 wt%. However, the enthalpy change, ΔH, of the formation of MgH_2 is −74 kJ/mol H_2 (Stampfer et al. 1960; Cummings and Powers 1974), indicating that MgH_2 is a relatively stable hydride. As a result, the rate of the desorption reaction, i.e., $MgH_2 \rightarrow Mg + H_2$, is slow in the moderate conditions.

It is found experimentally that some transition metal oxides (e.g., Nb_2O_5) have a large catalytic effect on the hydrogen desorption reaction of MgH_2 (Oelerich et al. 2001; Barkhordarian et al. 2003, 2004, 2006). However, no quantitative method has been proposed for understanding this catalytic problem yet. Recently, catalytic activities of metal oxides have been evaluated quantitatively using the atomization energy concept (Hirate et al. 2009).

For metal oxides expressed as M_xO_y, the energy of the isolated neutral atom, E_M^{atom} (or E_O^{atom}), is taken as a reference, and the atomization energy, ΔE_M (or ΔE_O), is defined as,

$$\Delta E_M = \left(E_M^{atom} - E_M^{oxide}\right) \times (x/y), \tag{7.12}$$

$$\Delta E_O = E_O^{atom} - E_O^{oxide}. \tag{7.13}$$

Here, E_M^{oxide} and E_O^{oxide} are the atomic energy densities for M and O in M_xO_y, respectively. Thus, ΔE_M is the average energy of M to be counted per O atom.

The plots of ΔE_M vs. ΔE_O, are shown in Fig. 7.18 for binary metal oxides, M_xO_y (Hirate et al. 2009). Those oxides which are located in the upper right region in Fig. 7.18, have large cohesive energies. Also, the contribution of each element in the oxide to the cohesive energy is understood from this figure. For example, the cohesive energy is nearly same between NbO_2 and Al_2O_3 (NbO_2: $E_{coh} = 10.1$ eV, Al_2O_3: $E_{coh} = 10.7$ eV). However, they are located in the very different positions in Fig. 7.18. NbO_2 has a large ΔE_O value, but a very small ΔE_M value, indicating that the O atoms in NbO_2 make a significant contribution to the cohesive energy. On the other hand, in case of Al_2O_3, the value of ΔE_O is almost zero. Instead, its ΔE_M value is very large. Thus, the metal atoms in Al_2O_3 contribute mainly to the cohesive energy.

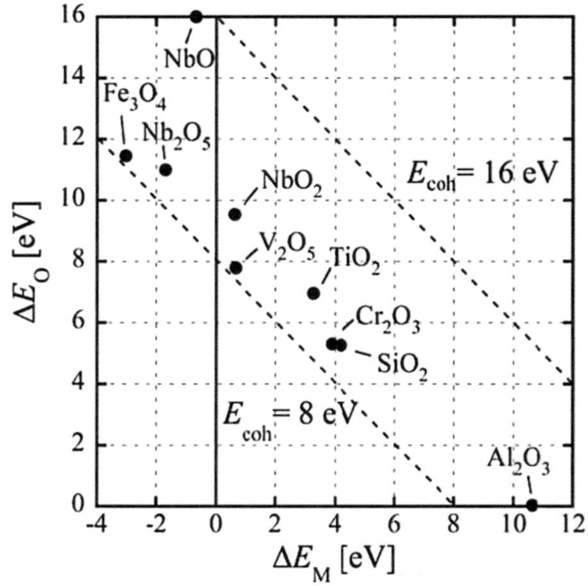

Fig. 7.18 ΔE_M vs. ΔE_O, for binary metal oxides, M_xO_y

Table 7.4 Hydrogen desorption rate measured at 573 K

	Desorption rate (10^{-2} wt%/s)		Desorption rate (10^{-2} wt%/s)
Nb_2O_5	10.2	Cr_2O_3	1.9
V_2O_5	6.0	TiO_2	1.9
NbO_2	3.5	Al_2O_3	0.7
NbO	2.0	SiO_2	0.2

7.6.2 *Quantitative Analysis of Catalytic Activities of Metal Oxides for MgH_2*

The hydrogen desorption rate measured at 573 K is adopted as a measure of the catalytic activity of each metal oxide (Hirate et al. 2009). The values reported by Barkhordarian et al. (2006) are listed in Table 7.4.

To evaluate the catalytic activity quantitatively, the atomization energies of constituent elements in metal oxides are used. As is evident from Eqs. (7.12) and (7.13), atomization energies are defined as the values per O atom. However, metal oxides, M_xO_y, are mixed by a mole unit in the experiment. So, the values of the atomization energies are needed to be converted into the values per mole unit. Therefore, each of ΔE_O and ΔE_M is multiplied by the number of O atoms, y, in M_xO_y, that is expressed as y × ΔE_O ($=(E_O^{\text{atom}} - E_O^{\text{oxide}}) \times y$), and y × ΔE_M ($=(E_M^{\text{atom}} - E_M^{\text{oxide}}) \times x$).

The measured desorption rate is plotted against the atomization energy for O atoms, y × ΔE_O, or metal atoms, y × ΔE_M, as shown in Fig. 7.19a and b (Hirate

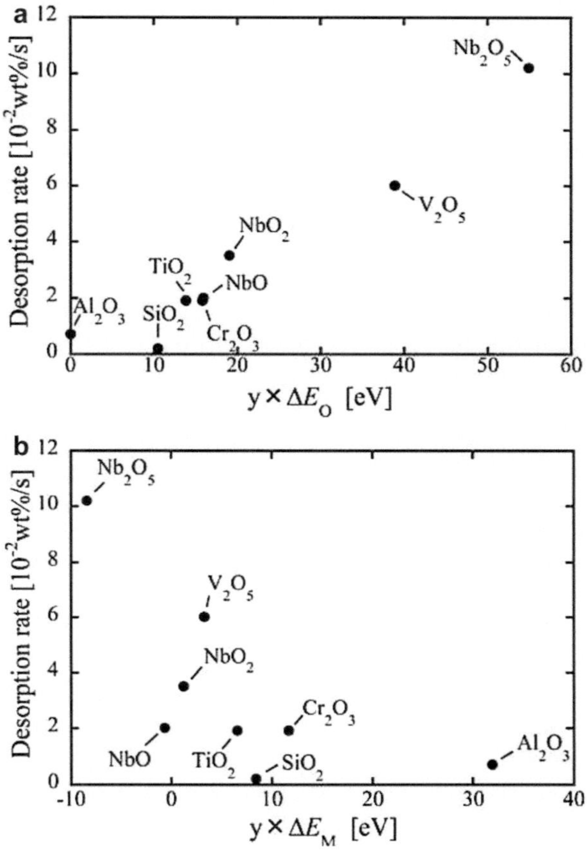

Fig. 7.19 The measured desorption rate plotted against the atomization energy for (**a**) O atoms, y × ΔE_O, or (**b**) metal atoms, y × ΔE_M

et al. 2009). The result shown in Fig. 7.19a indicates that metal oxides which have large y × ΔE_O values tend to accelerate the hydrogen desorption rate of MgH_2. On the other hand, the result shown in Fig. 7.19b is the reverse of this trend, since metal oxides which have small y × ΔE_M values tend to accelerate hydrogen desorption rate.

As mentioned before in Sect. 7.4 about the hydrogenation process of TiFe or Mg_2Ni, the element (e.g., Fe or Ni) with the higher atomization energy interacts more strongly with hydrogen atom than the element (e.g., Ti or Mg) with the lower atomization energy (Shinzato et al. 2007a). As a result, hydrogen atoms are located near Fe in $TiFeH_2$, and near Ni in Mg_2NiH_4. Thus, the element with the higher atomization energy in the metal oxide must interact more strongly with hydrogen atom in MgH_2. In other words, the O atoms in the metal oxides with larger y × ΔE_O values are supposed to interact more strongly with hydrogen atoms in MgH_2, as is shown in Fig. 7.19a. The stronger O-H interaction probably leads to the higher catalytic activities of oxides catalysts.

7.6.3 O-H Bonding During the Course of Dehydrogenation Reaction of Nb_2O_5-Catalyzed MgH_2

Using the atomization energy concept, it is predicted that the O atoms in metal oxides interact with hydrogen atoms in MgH_2 during the dehydrogenation of MgH_2 (Hirate et al. 2009). So, one experiment is performed to confirm the presence of the O-H interaction during dehydrogenation by using FT-IR spectroscopy (Hirate et al. 2011b).

For this experiment, four specimens are made from milled MgH_2 with 1 mol% Nb_2O_5 and used for the FT-IR measurements; (a) specimen released no hydrogen (called non-released specimen), (b) released hydrogen about 2.3 wt% (2.3 wt% released specimen), (c) released hydrogen about 5.6 wt% (5.6 wt% released specimen), and (d) released hydrogen completely (all-released specimen). Here, hydrogen is released from the specimen by heating at 573 K in vacuum for an appropriate time in the Sievert-type apparatus following a standard volumetric method.

A standard technique is employed for the measurement of FT-IR spectroscopy. With the specimens prepared as potassium bromide (KBr) pellets, FT-IR spectra are measured at room temperature in vacuum, using JASCO FT/IR-610 instrument (Hirate et al. 2011b).

According to previous experiments, the O-H bonding in the absorbed hydroxyl group appears in the region of 3,000 ~ 3,800 cm^{-1} due to the O-H stretching mode (Busca 1998; Burcham et al. 1999; Armaroli et al. 2000; Braga et al. 2005). So, the FT-IR spectra are measured in the region of 2,600 ~ 4,000 cm^{-1}, as shown in Fig. 7.20.

As shown in Fig. 7.20a, there are two bands at 3,420 and 3,500 cm^{-1}, and a broad band from 2,800 to 3,300 cm^{-1} in the spectrum from the non-released specimen. The absorption intensities decrease gradually with decreasing amount of hydrogen retained in the specimen, as shown in Fig. 7.20a–c. However, as shown in Fig. 7.20d, two small bands at 2,880 and 3,450 cm^{-1} are still observed in the all-released specimen, despite that no MgH_2 remains in it, judging from the non-appearance of any peaks from MgH_2 in the XRD profile. In addition, as shown in Fig. 7.20e, several bands appear in the milled MgH_2, although the intensities are weaker compared to those of the non-released specimen shown in Fig. 7.20a. The band at 3,450 cm^{-1} is also observed in the all-released specimen as shown in Fig. 7.20d. So, this may be related partially to the existence of MgO, since it is observed in both the milled MgH_2 and all-released specimens. Some O-H vibrations might retain on the surface of MgO, but the detail still remains unknown.

It is confirmed from the experiment that the O-H interaction between Nb_2O_5 and MgH_2 is operating during the course of dehydrogenation reaction of Nb_2O_5-catalyzed MgH_2. Therefore, the atomization energy of O atom, ΔE_O, in metal oxide catalysts is indeed a good measure of the catalytic activities in the dehydrogenation reaction of MgH_2.

Barkhordarian et al. (2006) investigated experimentally a catalytic mechanism of transition-metal oxides on the Mg hydrogen desorption reaction. Four factors

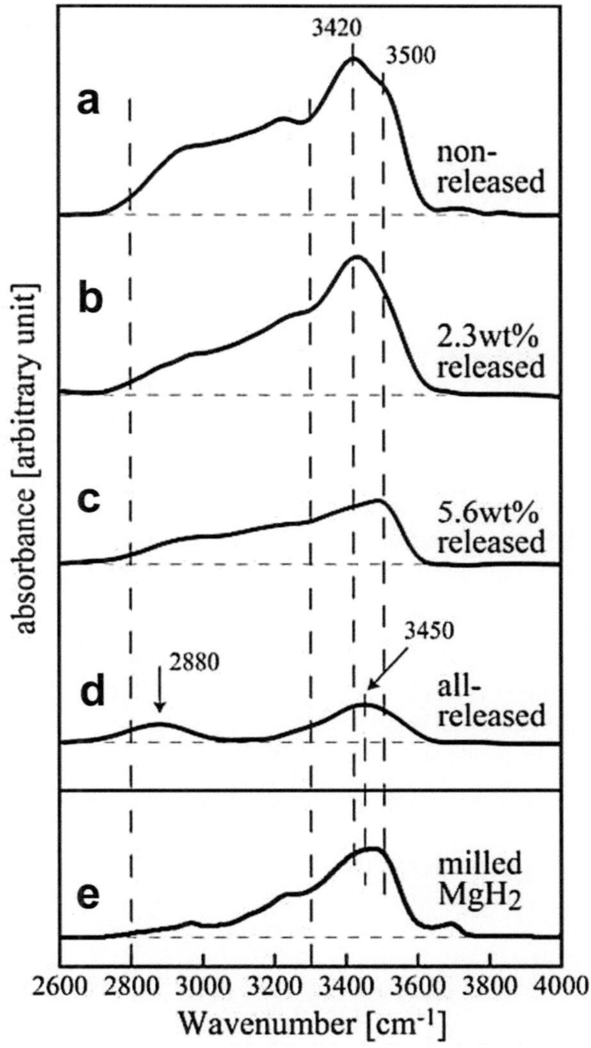

Fig. 7.20 The FT-IR spectra measured during the course of dehydrogenation of MgH_2 with 1 mol% Nb_2O_5

were proposed for the transition-metal oxide to be an effective catalyst. One of the factors is concerned with a high affinity of the transition-metal atom to hydrogen. Their model appears to be made on the basis of the idea that the transition-metal atom rather than the O atom in the oxide interacts more strongly with hydrogen atom in MgH_2. However, this contradicts the experimental result that Fe_3O_4 has a relatively high catalytic effect (Barkhordarian et al. 2003). This is because, the interaction between Fe and hydrogen atom is weak in the Fe-H system. On the other hand, the value of y × ΔE_O for Fe_3O_4 (45.8 eV) is intermediate between V_2O_5 (38.9 eV) and Nb_2O_5 (55.0 eV), in agreement with the experiment (Barkhordarian et al. 2003). Another contradiction is in our direct observation of the O-H bonding using FT-IR spectroscopy as is explained in the preceding paragraph. Further discussion is given elsewhere (Hirate et al. 2009).

In addition, it is known that the decomposition reaction of $NaAlH_4$ expressed as, $NaAlH_4 \rightarrow 1/3\ Na_3AlH_6 + 2/3\ Al + H_2$, is accelerated by mixing metal chloride catalysts (e.g., $TiCl_3$) (Bogdanovic and Schwickardi 1997). The catalytic activities of chlorides are evaluated quantitatively with the aid of the atomization energy (Hirate et al. 2011a).

7.6.4 Summary

The hydrogen desorption reaction of magnesium hydride (MgH_2), $MgH_2 \rightarrow Mg + H_2$, is accelerated by mixing catalytic metal oxides (e.g., Nb_2O_5). This catalytic effect is evaluated quantitatively using the atomization energy concept. The measured hydrogen desorption rate increases monotonously with increasing y × ΔE_O values of metal oxides, M_xO_y. Here, ΔE_O, is the atomization energy for the O atom in M_xO_y. This indicates that the O atom interacts mainly with hydrogen atom in MgH_2, in agreement with the observation of the O-H stretching mode in the FT-IR spectra during the dehydrogenation of the Nb_2O_5-catalyzed MgH_2.

Conclusion

A unified approach is reviewed for understanding the chemical bond between atoms in hydrides and oxides on the basis of the atomization energy concept. The results are summarized as follows.

1. The atomization energy diagram is constructed for various hydrides. All the hydrides can be located on this diagram, although there are significant differences in the nature of the chemical bond among the hydrides. It is found that there is very subtle but simple bond formation in hydrocarbons, C_mH_n, and the cohesive energy per hydrogen atom varies depending only on the m/n ratio.
2. The atomization energy diagram is also made for metal oxides, and used for comparing the bond characters among them. Also, a series of phase transitions of the perovskite-type oxides is explained in a consistent manner with the aid of the atomization energy.
3. The atomization energy is a good measure of the catalytic activities of metal oxides or metal chlorides on the dehydrogenation or decomposition reaction of hydrides. This is because, the atomization energy correlates well with the dehydrogenation reaction rate of hydrides.
4. Thus, this new approach allows us to understand the role of constituent elements in the bond formation of hydrides and oxides, and to get a new clue to materials design (e.g., catalyst design).

Acknowledgments The authors would like to express sincere thanks to the staff of the Computer Center, Institute for Molecular Science, Okazaki National Institute for the use of their supercomputers. This study is supported by a Grant-in-Aid for Scientific Research from the Ministry of Education, Culture, Sports, Science and Technology of Japan and by the Japan Society for the Promotion of Science.

References

Ahtee A, Ahtee M, Glazer AM, Hewat AW (1976) Structure of orthorhombic $SrZrO_3$ by neutron powder diffraction. Acta Cryst B32:3243–3246

Ahtee M, Glazer AM, Hewat AW (1978) High-temperature phases of $SrZrO_3$. Acta Cryst B 34:752–758

Anders E, Koch R, Freunscht P (1993) Optimization and application of lithium parameters for PM3. J Comp Chem 14:1301–1312

Armaroli T, Busca G, Carlini C, Giuttari M, Galletti AMR, Sbrana G (2000) Acid sites characterization of niobium phosphate catalysts and their activity in fructose dehydration to 5-hydroxymethyl-2-furaldehyde. J Mol Catal A Chem 151:233–243

Barkhordarian G, Klassen T, Bormann R (2003) Fast hydrogen sorption kinetics of nanocrystalline Mg using Nb_2O_5 as catalyst. Scr Mater 49:213–217

Barkhordarian G, Klassen T, Bormann R (2004) Effect of Nb_2O_5 content on hydrogen reaction kinetics of Mg. J Alloy Compd 364:242–246

Barkhordarian G, Klassen T, Bormann R (2006) Catalytic mechanism of transition metal compounds on Mg hydrogen sorption reaction. J Phys Chem B110:11020–11024

Becke AD (1988) Density-functional exchange-energy approximation with correct asymptotic-behavior. Phys Rev A 38:3098–3100

Bogdanovic B, Schwickardi MJ (1997) Ti-doped alkali metal aluminium hydrides as potential novel reversible hydrogen storage materials. J Alloy Compd 253/254:1–9

Braga VS, Dias JA, Dias SCL, de Macedo JL (2005) Catalyst materials based on Nb_2O_5 supported on SiO_2-Al_2O_3: preparation and structural characterization. Chem Mater 17:690–695

Burcham LJ, Datka J, Wachs IE (1999) In situ vibrational spectroscopy studies of supported niobium oxide catalysts. J Phys Chem B103:6015–6024

Busca G (1998) Spectroscopic characterization of the acid properties of metal oxide catalysts. Catal Today 41:191–206

Cummings DL, Powers GJ (1974) Storage of hydrogen as metal-hydrides. Ind Eng Chem Process Des Dev 13:182–192

Dobbs KD, Hehre WJ (1986) Molecular-orbital theory of the properties of inorganic and organometallic compounds 4–extended basis sets for 3rd-row and 4th-row, main-group elements. J Comput Chem 7:359–378

Dunning TH (1989) Gaussian basis sets for use in correlated molecular calculations. I. The atoms boron through neon and hydrogen. J Chem Phys 9:1007–1023

Edwards JW, Speiser R, Johnston HL (1951) Structure of barium titanate at elevated temperatures. J Am Chem Soc 73:2934–2935

Frisch MJ, Trucks GW, Schlegel HB, Scuseria GE, Robb MA, Cheeseman JR, Montgomery JA Jr, Vreven T, Kudin KN, Burant JC, Millam JM, Iyengar SS, Tomasi J, Barone V, Mennucci B, Cossi M, Scalmani G, Rega N, Petersson GA, Nakatsuji H, Hada M, Ehara M, Toyota K, Fukuda R, Hasegawa J, Ishida M, Nakajima T, Honda Y, Kitao O, Nakai H, Klene M, Li X, Knox JE, Hratchian HP, Cross JB, Bakken V, Adamo C, Jaramillo J, Gomperts R, Stratmann RE, Yazyev O, Austin AJ, Cammi R, Pomelli C, Ochterski JW, Ayala PY, Morokuma K, Voth GA, Salvador P, Dannenberg JJ, Zakrzewski VG, Dapprich S, Daniels AD, Strain MC, Farkas O, Malick DK, Rabuck AD, Raghavachari K, Foresman JB, Ortiz JV, Cui Q, Baboul

AG, Clifford S, Cioslowski J, Stefanov BB, Liu G, Liashenko A, Piskorz P, Komaromi I, Martin RL, Fox DJ, Keith T, Al-Laham MA, Peng CY, Nanayakkara A, Challacombe M, Gill PMW, Johnson B, Chen W, Wong MW, Gonzalez C, Pople JA (2004). Revision C.02. Gaussian, Inc., Wallingford CT

Fukai Y (1994) The metal-hydrogen system. Springer, Berlin

Graetz J, Reily JJ (2006) Thermodynamics of the alpha, beta and gamma-polymorphs of AlH_3. J Alloy Compd 424:262–265

Hehre WJ, Ditchfield R, Pople JA (1972) Self-consistent molecular-orbital methods 12 further extensions of Gaussian-type basis sets for use in molecular orbital studies of organic-molecules. J Chem Phys 56:2257–2261

Hirate H, Saito Y, Nakaya I, Sawai H, Shinzato Y, Yukawa H, Morinaga M, Baba T, Nakai H (2009) Quantitative approach to the understanding of catalytic effect of metal oxides on the desorption reaction of MgH_2. Int J Quant Chem 109:2793–2800

Hirate H, Saito Y, Nakaya I, Sawai H, Yukawa H, Morinaga M, Nakai H (2011a) Quantitative evaluation of catalytic effect of metal chlorides on the decomposition reaction of $NaAlH_4$. Int J Quantum Chem 111:950–960

Hirate H, Sawai H, Yukawa H, Morinaga M (2011b) Role of O-H bonding in catalytic activity of Nb_2O_5 during the course of dehydrogenation of MgH_2. Int J Quantum Chem 111:2251–2257

Huzinaga S, Andzelm J, Klobukowski M, Radzioandzelm E, Sakai Y, Tatewaki H (1984) Gaussian basis sets for molecular calculation. Elsevier, New York

Kennedy BJ, Howard CJ, Chakoumakos BC (1999) Phase transitions in perovskite at elevated temperatures – a powder neutron diffraction study. J Phys Condens Matter 11:1479–1488

Kingery WD, Bowen HK, Uhlman DR (1976) Introduction to ceramics. Wiley, New York/London

Kohn W, Sham LJ (1965) Self-consistent equation including exchange and correlation effects. Phys Rev 140:A1133–A1138

Komiya K, Morisaku N, Shinzato Y, Orimo S, Ohki Y, Tatsumi K, Yukawa H, Morinaga M (2007) Synthesis and dehydrogenation of $M(AlH_4)_2$ ($M = Mg, Ca$). J Alloy Compd 446–447:237–241

Kwei GH, Lawson AC, Billinge SJL, Cheong SW (1993) Structures of the ferroelectric phases of barium-titanate. J Phys Chem 97:2368–2377

Lee C, Yang W, Parr RG (1988) Development of the Colle-Salvetti correlation-energy formula into a functional of the electron-density. Phys Rev B 37:785–789

Liu X, Lieberman RC (1993) X-ray powder diffraction study of $CaTiO_3$ perovskite at high temperatures. Phys Chem Miner 20:171–175

Milman V, Winkler B, White JA, Pickard CJ, Payne MC, Akhmatskaya EV, Nobes RH (2000) Electronic structure, properties, and phase stability of inorganic crystals: a pseudopotential plane-wave study. Int J Quantum Chem 77:895–910

Miyake S, Ueda R (1947) On phase transformation of $BaTiO_3$. J Phys Soc Jpn 2:93–97

Mulliken RS (1955) Electronic population analysis of LCAO-MO molecular wave functions. 1. J Chem Phys 23:1833–1840

Nakai H (2002) Energy density analysis with Kohn-Sham orbitals. Chem Phys Lett 363:73–79

Nakai H, Kurabayashi Y, Katouda M, Atsumi T (2007) Extension of energy density analysis to periodic boundary condition calculation: evaluation of locality in extended systems. Chem Phys Lett 438:132–138

Oelerich W, Klassem T, Bormann R (2001) Metal oxides as catalysts for improved hydrogen sorption in nanocrystalline Mg-based materials. J Alloy Compd 315:237–242

Perdew JP, Burke K, Wang Y (1996) Generalized gradient approximation for the exchange-correlation hole of a many-electron system. Phys Rev B 54:16533–16539

Pimentel GC, Spratley RD (1969) Chemical bonding clarified through quantum chemistry. Holden-day, San Francisco

Rassolo VA, Pople JA, Ratner M, Windus TL (1998) 6-31G* basis set for atoms K through Zn. J Chem Phys 109:1223–1229

Schafer A, Horn H, Ahlrichs R (1992) Fully optimized contracted Gaussian-Basis sets for atoms Li to Kr. J Chem Phys 97:2571–2577

Schafer A, Huber C, Ahlrichs R (1994) Fully optimized contracted Gaussian-basis sets of triple zeta valence quality for atoms Li to Kr. J Chem Phys 100:5829–5835

Shinzato Y, Yukawa H, Morinaga M, Baba T, Nakai H (2007a) A unified approach to the analysis of the chemical bond in hydrides and hydrocarbons. Acta Mater 55:6673–6680

Shinzato Y, Yukawa H, Morinaga M, Baba T, Nakai H (2007b) Energy density analysis of the chemical bond between atoms in perovskite-type hydrides. J Alloy Compd 446–447:96–100

Shinzato Y, Saito Y, Yukawa H, Morinaga M, Baba T, Nakai H (2007c) New expression of the chemical bond in perovskite-type oxides. Mater Sci Forum 561–565:1823–1826

Shinzato Y, Saito Y, Yoshino M, Yukawa H, Morinaga M, Baba T, Nakai H (2011) Energy expression of the chemical bond between atoms in metal oxides. J Phys Chem Solids 72:853–861

Slater JC (1951) A simplification of the Hartree-Fock method. Phys Rev 81:385–390

Stampfer JF Jr, Holley CE Jr, Suttle JF (1960) The magnesium-hydrogen system. J Am Chem Soc 82:3504–3508

Vanderbilt D (1990) Soft self-consistent pseudopotentials in a generalized Eigen value formalism. Phys Rev B 41:7892–7895

Vosko SH, Wilk L, Nusair M (1980) Accurate spin-dependent electron liquid correlation energies for local spin-density calculation – a critical analysis. Can J Phys 58:1200–1211

Weast RC, Astle MJ, Beyer WH (2003) CRC handbook of chemistry and physics: a ready-reference book of chemical and physical data, 84th edn. CRC Press, Boca Raton

Wigner EP, Seitz F (1955) Qualitative analysis of the cohesion in metals. Solid State Phys 1:97–126

Woon DE, Dunning TH (1993) Gaussian-Basis sets for use in correlated molecular calculations. 3. The atoms aluminum through argon. J Chem Phys 98:1358–1371

Yoshino M, Liu Y, Tatsumi K, Tanaka I, Morinaga M, Adachi H (2003) Local geometries and energetics of hydrogen in acceptor-doped $SrZrO_3$. Solid State Ion 162–163:127–133

Yoshino M, Yukawa H, Morinaga M (2004) Modification of local electronic structures due to phase transition in perovskite-type oxides, $SrBO_3$ ($B = Zr$, Ru, Hf). Mater Trans 45(7):2056–2061

Yvon K (1998) Complex transition-metal hydrides. Chimia 52:613–619

Part III
Applications

Chapter 8
Comparative Study on Optical Properties of YPO_4: Mn, Zr Phosphor by Experiment and Calculation

Mamoru Kitaura, Shinta Watanabe, Kazuyoshi Ogasawara, Akimasa Ohnishi, and Minoru Sasaki

8.1 Introduction

Optical properties of Mn^{2+} impurities in solids have been studied from the viewpoint of the fundamentals in material science and the industrial applications for lighting and display. A large number of Mn^{2+} doped inorganic phosphors have been successfully synthesized so far (Tamatani 2007). Now, inorganic phosphors are generally activated using rare-earth ions such as Ce^{3+}, Eu^{2+}, Eu^{3+} and Tb^{3+}, because the quantum efficiency of emission is very high for these ions. Nevertheless, the Mn^{2+} ion is still utilized as the first choice in the development of new functional phosphors. The reason is because the emission due to Mn^{2+} ions exhibits high quantum efficiency and good color purity. Therefore, the Mn^{2+} ion is one of plausible candidates likely to replace the rare-earth ions. Since it is anticipated that ensuring rare-earth ions tends to be difficult, the phosphors including this ion are worthwhile noting.

Recently, Kaneyoshi and Nakazawa have developed a new Mn doped phosphors using yttrium phosphate (YPO_4) as a host material (Kaneyoshi and Nakazawa 2005). Figure 8.1 shows emission spectra of YPO_4 doped with Mn^{2+} ions (a) and Mn^{2+} and Zr^{4+} ions (b), which were measured at room temperature under excitation with vacuum ultraviolet photons at 176 nm. The YPO_4: Mn phosphor exhibits a very weak blue-green (BG) emission band around 477 nm. From analogy with some metal oxides doped with Mn^{2+} ions, the BG band has been assigned to intra-3*d* transitions of Mn^{2+} ions. This band is drastically enhanced as Zr^{4+} ions are added. For the physical mechanism of such a striking feature, the resonant energy transfer

M. Kitaura (✉) • A. Ohnishi • M. Sasaki
Department of Physics, Faculty of Science, Yamagata University 1-4-12 Kojirakawa, Yamagata 990-8560, Japan
e-mail: kitaura@sci.kj.yamagata-u.ac.jp

S. Watanabe • K. Ogasawara
Department of Chemistry, School of Science and Technology, Kwansei Gakuin University 2-1 Gakuen, Sanda, Hyogo 669-1337, Japan

T. Ishii et al. (eds.), *The DV-Xα Molecular-Orbital Calculation Method*,
DOI 10.1007/978-3-319-11185-8_8

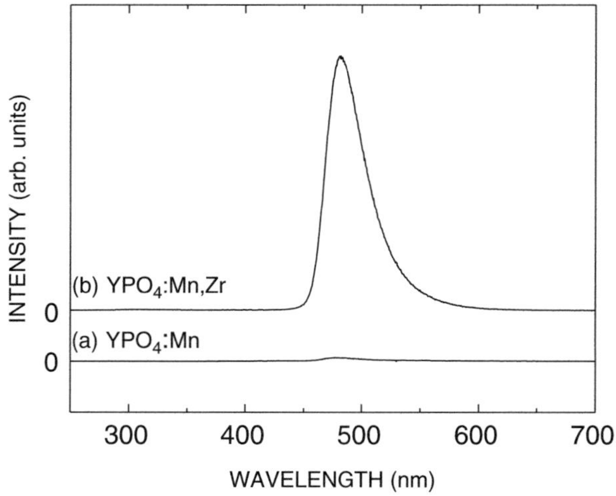

Fig. 8.1 Emission spectra of YPO_4: Mn (*a*) and YPO_4: Mn, Zr (*b*) measured at room temperature under excitation with vacuum ultraviolet photons at 176 nm. The 477 nm band due to the intra-3*d* transitions of Mn^{2+} ions becomes larger when Mn^{2+} ions are incorporated with Zr^{4+} ions

between Zr^{4+} and Mn^{2+} ions seems likely at first glance. However, the emission band due to the charge transfer transition from O-2*p* orbitals to Zr-4*d* orbitals and the absorption bands due to intera-3*d* transitions of Mn^{2+} ions are very weak. The resonant energy transfer from Zr^{4+} ions to Mn^{2+} ions is not as active as the BG band is strengthened.

Kitaura et al. have measured electron spin resonance (ESR) spectra of YPO_4: Mn and YPO_4: Mn, Zr, and investigated the correlation between the BG band and the local structure of Mn^{2+} ions (Kitaura et al. 2007). It was found that ESR signals due to Mn^{2+} ions were considerably changed by the presence of Zr^{4+} ions. From the spectral analysis using the spin-Hamiltonian for an axial symmetry, it turned out that the number of Mn^{2+} ions occupying Y^{3+} lattice sites is increased by the addition of Zr^{4+} ions. On this basis, the enhancement of the BG band was interpreted as an increase in the number of Mn^{2+} ions substituting Y^{3+} ions. It is more likely that Zr^{4+} ions play the role as a compensator to remove the charge inconsistency between Mn^{2+} and Y^{3+} ions.

Generally, energy levels of impurities such as transition-metals (TMs) and rare-earths (REs) in solids are sensitive to the chemical environment surrounding them. Since instructive information on the energy levels is included in optical spectra, optical experiment is of great importance. From this viewpoint, absorption spectra have been measured for various materials doped with TM and RE ions. However, it is difficult to see the local electronic structures of impurities by absorption spectra only. On the other hand, theoretical approaches based on the molecular-orbital (MO) theory have also been utilized in investigations of the local electronic structures of TMs and REs in solids. The discrete variational Xα (DV-Xα) method (Adachi et al. 1978) is also one of such methods. Accordingly, the combination of optical experiment and DV-Xα calculation is expected to be a more powerful tool to probe into TM and RE impurities directly.

The aim of this chapter is to demonstrate the validity of theoretical calculations for the analysis of optical properties due to TM impurities in phosphors. To do so, optical properties of Mn^{2+} ions in YPO_4: Mn, Zr were studied in detail by experiment and calculation. Theoretical calculations based on the relativistic DV-Xα and discrete variational multi-electron (DVME) method were performed in order to obtain the information on many-electron energy levels of the Mn^{2+} ions. The effect of cluster size on the impurity MO levels of a Mn^{2+} ion was evaluated through the parameters on crystal field splitting and orbital population. The theoretical spectrum of absorption was compared with the experimental spectrum. Agreement between the two spectra was good. The origins of distinctive structures were identified according to the combination of one-electron MO wave functions in Slater determinants. The energy position and lifetime of a Mn^{2+} emission were also determined using the results of calculation. The theoretical values thus obtained were checked with the experimental values. They were generally acceptable values.

8.2 Energy Levels of Mn^{2+} Ions in Inorganic Compounds

A Mn^{2+} ion has five electrons in the outermost 3*d* orbitals. The 3*d* electrons interact with each other through electron–electron repulsion. In liquids and solids, they are strongly influenced by properties of their surrounding anions, namely, "ligands". By the presence of such ligands, the 3*d* orbitals of Mn^{2+} are split into three-fold degenerate t_2 and two-fold degenerate *e*. When a Mn^{2+} ion is placed at the center of an octahedral coordination by six anions, the t_2 orbital has a lower energy than the *e* orbital. According to the Hund's rule, the ground state in an octahedral coordination is represented by the lowest-energy $(t_2)^3(e)^2$ electron configuration, where all of five electrons are received in up-spin states. This configuration is the high-spin state with $S = 5/2$, in contrast to the low-spin state with $S = 1/2$ of the $(t_2)^5$ ground state configuration. As depicted in Fig. 8.2, the lowest-energy excited state is represented by the $(t_2)^4(e)^1$ configuration generated by a spin-flip transition from the $(t_2)^3(e)^2$ configuration. On the other hand, the relationship between the t_2 and

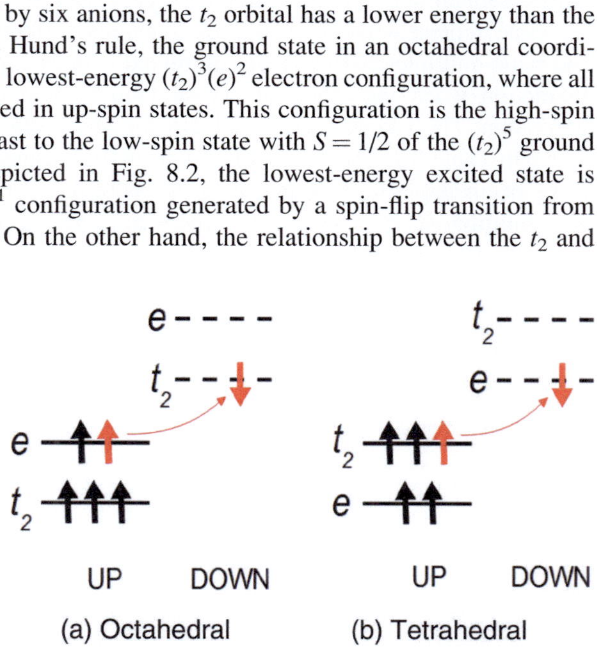

Fig. 8.2 Schematic illustration of the spin-flip transition of a Mn^{2+} ion in octahedral (**a**) and tetrahedral coordination (**b**). The spin-flip transitions indicated by *red arrows* connect the ground state configurations $(t_2)^3(e)^2$ and $(e)^2(t_2)^3$ with the lowest excited state configurations $(t_2)^4(e)^1$ and $(e)^3(t_2)^2$, respectively

e orbitals is inverted in the tetrahedral coordination formed by four anions. The ground state configuration $(e)^2(t_2)^3$ is basically same as that in the octahedral coordination, but the lowest-energy excited state is represented as the $(e)^3(t_2)^2$ configuration caused by a spin-flip transition. This configuration is different from the lowest-energy excited state configuration of $(t_2)^4(e)^1$ in the octahedral coordination. The five configurations have a common feature on spin multiplicity between the octahedral and tetrahedral coordination. The ground state is the spin-sextet state, while the lowest-energy excited state is a spin-quartet state. The optical transition between these states having different spin numbers is basically forbidden due to the spin-selection rule. This feature well-explains the fact that Mn^{2+} emissions usually have the long lifetime of millisecond or sub-millisecond range.

Optical properties due to intra-3*d* transitions of Mn^{2+} ions cannot be described accurately with only single use of t_2 and *e* orbitals, because the representations of these orbitals are based on the one-electron approximation. For the description of impurity energy levels, many-electron correlation effects among five electrons occupying t_2 and *e* orbitals have to be considered. This is generally called as "multiplets". From the electron configurations of $(t_2)^5$, $(t_2)^4(\mathrm{e})^1$, $(t_2)^3(\mathrm{e})^2$, $(t_2)^2(\mathrm{e})^3$, $(t_2)^1(\mathrm{e})^4$, the representations of multiplets shown in Table 8.1 are obtained.

The representations of multiplets have been used in the Tanabe-Sugano diagram for the $(3d)^5$ electron system, in order to describe many-electron energy levels (Tanabe and Sugano 1954a, b). In this diagram, multiplet energy levels are mainly dominated in term of two parameters, Δ and B. The parameter Δ is the crystal field splitting parameter expressed as the energy difference between the t_2 and *e* orbitals, and the parameter B an electron–electron repulsion parameter influencing the energy separations among multiplet energy levels. In Table 8.1, one can see the same representations of multiplets among different electron configurations. The same multiplet energy levels are mixed with each other through configuration interaction when the effect of crystal field is near to many-electron coulomb interaction. The values of Δ, B and Δ/B for Mn^{2+} ions in some host materials are shown in Table 8.2. The Δ/B of Mn^{2+} ions in the tetrahedral site is smaller than that in the octahedral site. This fact is in line with the well-known relation on crystal field splitting, $|\Delta_{\mathrm{tetra}}| = 4|\Delta_{\mathrm{octa}}|/9$, where Δ_{tetra} and Δ_{octa} indicate the crystal field splitting in tetrahedral and octahedral sites, respectively. Generally, the lowest-energy transition of ${}^4T_1 \rightarrow {}^6A_1$ is regarded as the origin of the Mn^{2+} emission in

Table 8.1 Representations of the multiplets originating from the $(3d)^5$ electron configuration in the cubic crystal field

Configurations	Representations of multiplets
$(t_2)^5$	${}^2T_2$
$(t_2)^4(e)^1$	${}^2T_1$, ${}^2T_2$, ${}^2E$, ${}^2A_1$, ${}^2A_2$, ${}^4T_1$, ${}^4T_2$
$(t_2)^3(e)^2$	${}^2T_1$, ${}^2T_2$, ${}^2E$, ${}^2A_1$, ${}^2A_2$, ${}^4T_1$, ${}^4T_2$, ${}^4E$, ${}^4A_1$, ${}^4A_2$, ${}^6A_1$
$(t_2)^2(e)^3$	${}^2T_1$, ${}^2T_2$, ${}^2E$, ${}^2A_1$, ${}^2A_2$, ${}^4T_1$, ${}^4T_2$
$(t_2)^1(e)^4$	${}^2T_2$

Table 8.2 Crystal field splitting Δ and Racah parameter B for Mn^{2+} ions in various oxides

Host	Coordination number	Δ (eV)	B (eV)	Δ/B
Zn_2SiO_4[a]	4	0.697	0.077	9.1
$MgGa_2O_4$[b]	4	0.645	0.077	8.4
$MgAl_2O_4$[c]	4	0.626	0.077	8.1
MnO[d]	6	1.25	0.093	13
$MnSiO_3$[e]	6	0.929	0.077	12
$ZrTiO_4$[f]	6	1.29	0.093	13

[a]Su et al. (2006)
[b]Costa et al. (2009)
[c]Mohler and White (1994)
[d]Mehra (1971)
[e]Lakshman and Reddy (1973)
[f]Dondi et al. (2006)

compounds with cubic structure. The energy separation between the $^6A_1$ and $^4T_1$ states is increase with Δ/B, as shown in the Tanabe-Sugano diagram (Tanabe and Sugano 1954a, b). The peak energy of the Mn^{2+} emission becomes lower when Mn^{2+} ions occupy octahedral sites; hence, the emission color of Mn^{2+} ions in the octahedral and tetrahedral coordination is quite different.

8.3 Calculation Methods

The cluster calculation based on the DV-Xα method allows us to obtain one-electron MO energy levels of materials without using empirical parameters. The merit of this method is to generate realistic potentials and wave functions with relatively small basis set. This feature makes it possible to complete computational processes within small iterations even for large size molecules. Accordingly, the DV-Xα method is very useful for better understanding of structural, electrical, optical and magnetic properties of materials. Actually, there are many examples on applications to material design and spectroscopic analysis.

The DV-Xα method further advanced including the relativistic effect. The relativistic one-electron Hamiltonian for solids is expressed as

$$h(\mathbf{r}) = c\tilde{\boldsymbol{\alpha}}\mathbf{p} + \tilde{\beta}c^2 - \sum_{\nu} \frac{Z_\nu}{|\mathbf{r}_i - \mathbf{R}_\nu|} + V_0(\mathbf{r}) + \sum_{\mu} \frac{Z_\mu^{eff}}{|\mathbf{r}_i - \mathbf{R}_\mu|}, \tag{8.1}$$

where c is the velocity of light, $\mathbf{p}$ the momentum operator, $\tilde{\boldsymbol{\alpha}}, \tilde{\beta}$ the Dirac matrices, Z_ν and $\mathbf{R}_\nu$ the charge and position of the νth nucleus, and Z_μ^{eff} and $\mathbf{R}_\mu$ the effective charge and position of the μth ion outside the model cluster. The potential term $V_0(\mathbf{r})$ of (8.1) is expressed as

$$V_0(\mathbf{r}) = \int \frac{\rho(\mathbf{r}')}{|\mathbf{r} - \mathbf{r}'|} d\mathbf{r}' + V_{xc}(\mathbf{r}), \tag{8.2}$$

where ρ the electron density, and $\mathbf{r}$ the position of an electron. The first and second terms of (8.2) represent the electron repulsive interaction and the exchange-correlation interaction, respectively. The exchange-correlation interaction term is expressed using Slater's Xα potential (Slater 1974). Dirac equation is solved in all of relativistic MO calculation, from which the eigenvalues (one-electron orbital energies) and corresponding eigenvectors were obtained. Using the eigenvectors, relativistic MO wave functions are also constructed from the linear combination of atomic orbital spinors.

The results calculated by the relativistic DV-Xα method are of ground state in a mean-field approximation. They are not sufficient for an analysis of the optical transitions between ground and excited states. On the other hand, the relativistic DVME method is a hybrid method of the one-electron MO calculation and the many-electron configuration interaction calculation. When using this method, excited states are dealt simultaneously with the ground state. The many-electron Hamiltonian is constructed using the one-electron operator $h(\mathbf{r})$ and two-electron operator $V_{ee}(\mathbf{r}, \mathbf{r}')$,

$$H = \sum_{i=1}^{n} h(\mathbf{r}_i) + \sum_{i}^{n} \sum_{i>j}^{n} V_{ee}(\mathbf{r}_i, \mathbf{r}_j), \tag{8.3}$$

where n is the number of selected electrons, $\mathbf{r}_i$ is the position of the ith electron. The one-electron operator $h(\mathbf{r})$ is almost the same as (8.1), but the potential $V_0(\mathbf{r})$ has to be replaced by the following explicit formula for the potential proposed by Watanabe and Kamimura (1989) using Dirac-Fock-Slater approximation, which is expressed as

$$V_0(\mathbf{r}) = \int \frac{\rho_0(\mathbf{r}')}{|\mathbf{r} - \mathbf{r}'|} d\mathbf{r}' + \frac{3}{4} \left[\frac{\rho(\mathbf{r}) V_{\mathrm{xc}}\{\rho(\mathbf{r})\} - \rho_0(\mathbf{r}) V_{\mathrm{xc}}\{\rho_0(\mathbf{r})\}}{\rho_{\mathrm{imp}}(\mathbf{r})} - V_{\mathrm{xc}}\{\rho_{\mathrm{imp}}(\mathbf{r})\} \right], \tag{8.4}$$

where ρ_{imp}, ρ_0, and ρ is the electron density of impurity MOs, the core and valence MOs, and all the occupied MOs, respectively, and V_{xc} is the Slater's Xα potential (Slater 1974). The two-electron operator $V_{ee}(\mathbf{r}_i,\mathbf{r}_j)$ represents the electron–electron repulsion between electrons received in impurity states,

$$V_{ee}(\mathbf{r}_i, \mathbf{r}_j) = \frac{1}{|\mathbf{r}_i - \mathbf{r}_j|}. \tag{8.5}$$

Since the two-electron operators does not include the relativistic effect, the frequency-independent Breit operator $B(\mathbf{r}_i, \mathbf{r}_j)$ was introduced as a correction term for the deviation of Hamiltonian for the electron–electron repulsion. The term of $B(\mathbf{r}_i, \mathbf{r}_j)$ is expressed as

$$B(\mathbf{r}_i, \mathbf{r}_j) = -\frac{1}{2r_{ij}}\left\{\boldsymbol{\alpha}_i \cdot \boldsymbol{\alpha}_j + \frac{(\boldsymbol{\alpha}_i \cdot \mathbf{r}_{ij})(\boldsymbol{\alpha}_j \cdot \mathbf{r}_{ij})}{r_{ij}^2}\right\}, \tag{8.6}$$

where $\boldsymbol{\alpha}_i$ and $\boldsymbol{\alpha}_j$ are Dirac matrices. The eigenvalues and corresponding eigenvectors are obtained by the diagonalization of the many-electron Hamiltonian of (8.3). Using the eigenvectors, the wave functions of multiplet energy levels are also obtained explicitly as a linear combination of Slater determinants. The relativistic DVME method has been successfully applied to the investigation of $4f^{\mathrm{n}} - 4f^{\mathrm{n}-1}5d^1$ absorption spectra of rare-earth ions (Watanabe and Ogasawara 2008; Watanabe et al. 2010), and $L_{2,3}$-edge XANES and electron-energy-loss near-edge structure (ELNES) spectra of TM ions (Ogasawara et al. 2001).

As mentioned above, many-electron wave functions are derived through the computational procedures by the relativistic DVME methods. In this method, the oscillator strength for electric dipole transitions between multiplet energy levels is calculated using the following equations,

$$I_{if} = \frac{2}{3}(\mathrm{E}_f - \mathrm{E}_i)\left|\langle \Psi_f | \sum_{j=1}^{n} \mathbf{r}_j \cdot \mathbf{e} | \Psi_i \rangle^2\right|, \tag{8.7}$$

where Ψ_i and Ψ_f are many-electron wave functions for the initial and final states of transitions, and E_i and E_f are energy eigenvalues of the two states. $\mathbf{r}_j$ denotes the position of the jth electron, and $\mathbf{e}$ expresses the unit vector parallel to the direction of the electric field. For comparison with experimental spectra, multiplet energy levels were broadened by assuming the Gaussian function with full width of half maximum (FWHM) of 0.1 eV.

8.4 Comparison of Optical Properties of YPO_4: Mn, Zr Between Experiment and Calculation

8.4.1 Experimental

YPO_4: Mn, Zr phosphors were synthesized by the conventional solid state reaction. The commercial powers of Y_2O_3, $(NH_4)_2HPO_4$, $MnCO_3$, and ZrO_2 were used as starting materials. These powders were mixed using a ball-mill, and sintered at 1,300 °C in N_2 atmosphere for 3 h. The concentration of Mn^{2+} and Zr^{4+} ions was adjusted to be 1.0 mol%. The composition and crystallinity of phosphor powders were checked by X-ray diffraction measurements.

Optical experiment was mainly carried out at the beamline of BL7B at UVSOR of Institute for Molecular Science. The powder samples are not transparent, because of light scattering at the surface. The measurements of absorption spectra are not possible using the powder samples. In our experiment, the excitation spectrum for the 477 nm emission due to the intra-3*d* transitions of Mn^{2+} ions was adopted as the substitution of absorption spectrum, in order to compare with the theoretical absorption spectrum calculated by the relativistic DVME method. YPO_4: Mn, Zr phosphors were excited using monochromatic light dispersed by the 3-m normal incident monochromator. The substrates of quartz and pyrex glasses were inserted into the front of the sample chamber, in order to remove the high-order light. The emission intensity was monitored using the combination system of a grating monochromator and a photomultiplier tube during the measurements of excitation spectra.

Decay curves of the 477 nm emission were measured at various temperatures under excitation with third harmonic generation (THG) light pulses ($h\nu = 355$ nm) from a Nd^{3+}:YAG laser system. The pulse duration and repetition rate were 20 ns and 10 Hz, respectively. The lifetime was analyzed by assuming that the decay curve fits a single exponential function.

8.4.2 One-Electron MO Calculation

The crystal structure of YPO_4 belongs to the tetragonal system with the space group $I4_1/amd$ (Milligan et al. 1982). The lattice constants are $a = 0.68817$ nm and $c = 0.60177$ nm. Y^{3+} and P^{5+} ions are surrounded by eight and four O^{2-} ions, respectively. Two kinds of cluster models were employed to see the effect of cluster size on calculated results. The model clusters $(MnO_8)^{14-}$ (a) and $(MnY_4P_6O_{44})^{44-}$ (b) are shown in Fig. 8.3, which are named “cluster (a)” and “cluster (b)”, hereafter. The point symmetry around the central Mn^{2+} ion is D_{2d} symmetry in these clusters. Lattice relaxation is often induced when a Mn^{2+} ion is introduced at the Y^{3+} ion site, because the ionic radius of Mn^{2+} is smaller than that of Y^{3+} ion (Shannon 1976). However, the effect of lattice relaxation was not considered in our calculation. Effective Madelung potentials were constructed with point charges located at the

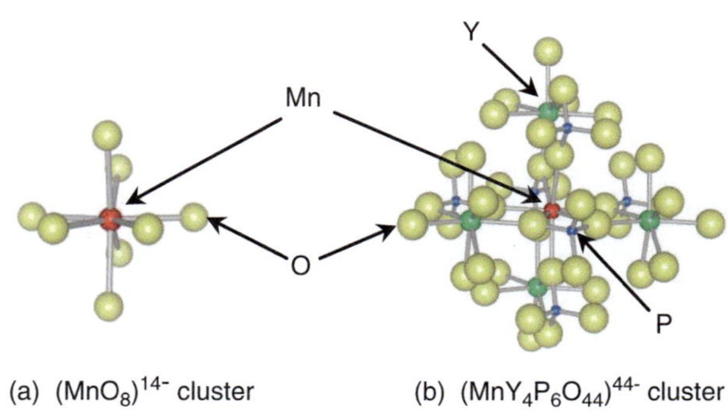

Fig. 8.3 Schematic diagrams of the model clusters, $(MnO_8)^{14-}$ (**a**) and $(MnY_4P_6O_{44})^{44-}$ (**b**), used in our calculation

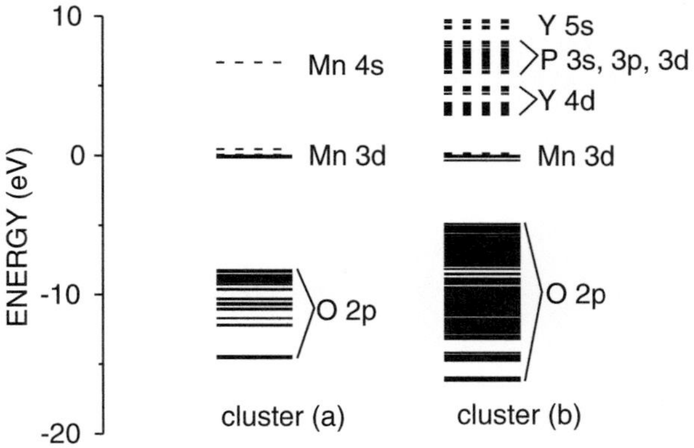

Fig. 8.4 One-electron MO energy level diagrams calculated using the clusters (**a**) and (**b**). *Solid* and *broken lines* indicate occupied and unoccupied MO levels. The highest-energy occupied MO levels were set as zero

atomic positions outside clusters. A partially extended basis set of 1*s*-4*p* for Mn, 1*s*-5*p* for Y, 1*s*-3*d* for P, and 1*s*-2*p* for O was used. The sampling points were set to be 100,000.

Figure 8.4 shows relativistic one-electron MO energy levels of the clusters (a) and (b). Solid and broken lines indicate the occupied and unoccupied levels, respectively. The energy of the highest-energy occupied level was taken as zero. The clusters (a) and (b) commonly have the valence band mainly of O-2*p* orbitals. On the other hand, the conduction band structure is different each other. For the cluster (a), only the Mn 4*s* orbital appears. It should be noted that the conduction band is not reproduced for the cluster (a), because of a lack of the surrounding cations in the cluster (b). For the cluster (b), the lower and upper parts are made up of Y-4*d* orbitals and P-3*s*, *p*, *d* orbitals, respectively. Higher-lying conduction states of Y-5*s* orbitals are also found. For the cluster (b), the forbidden gap between the valence and conduction bands was calculated to be 7.82 eV. The energy is comparable to the energy of the calculated forbidden gap of pure YPO_4 ($E_g = 7.89$ eV). The MO energy levels inside the forbidden gap are due to Mn-3*d* orbitals.

The MO energy levels of Mn-3*d* orbitals are composed of five levels, as shown by a magnified scale in Fig. 8.5. According to the order of their energies, the MO energy levels are described by $\Gamma_6(a_1)$, $\Gamma_7(b_1)$, $\Gamma_6(e)$, $\Gamma_7(e)$ and $\Gamma_7(b_2)$ for the cluster (a), and by $\Gamma_7(b_1)$, $\Gamma_6(a_1)$, $\Gamma_7(e)$, $\Gamma_6(e)$ and $\Gamma_7(b_2)$ for the cluster (b). Here, both the Mulliken notations and the Bethe symbols were used in order to make their origin clear. The $\Gamma_6(a_1)$ and $\Gamma_7(b_1)$ levels originate from the *e* level appearing under parent T_d symmetry, while the $\Gamma_7(e)$, $\Gamma_6(e)$ and $\Gamma_7(b_2)$ levels arise from the t_2 level. The effective crystal field splitting Δ_{eff} is defined as the energy difference between the barycenters of the *e* and t_2 levels. The value of Δ_{eff} was 0.249 eV for the cluster (a) and 0.270 eV for the cluster (b). These values are rather small compared to the crystal filed splitting in Table 8.2. The $\Gamma_6(e)$ and $\Gamma_7(e)$ levels originate from the *e* level by spin-orbit splitting. The energy splitting was 0.049 eV for the cluster (a) and 0.037 eV for the cluster (b). The spin-orbit coupling constant for the 3*d* electron of Mn can be calculated using the formula $\zeta_{3d} = 0.0355 + 0.0058 \cdot (Q_{Mn} - 1)$

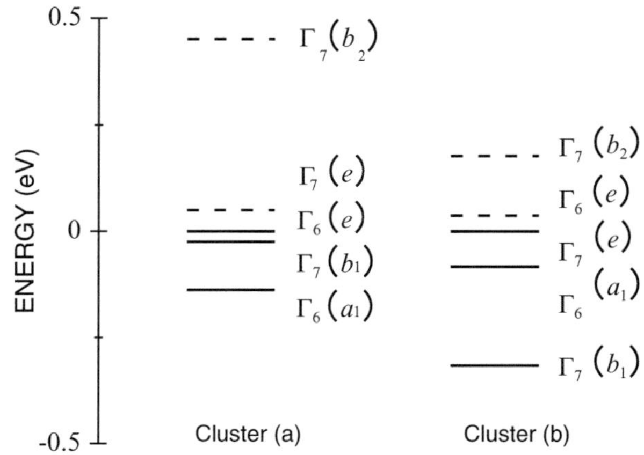

Fig. 8.5 Magnified one-electron MO energy level diagrams calculated using the clusters (**a**) and (**b**). These energy levels correspond to impurity energy levels of Mn-3*d* orbitals

Table 8.3 Proportions of atomic orbitals to the impurity MO levels in Fig. 8.5. The values were obtained from the results of the Mulliken's population analyses of the clusters (a) and (b)

		Mn		Ligand O		Y	P
Cluster	MO	3*d*	4*s* + 4*p*	2*s*	2*p*	4*d* + 5*s* + 5*p*	3*s* + 3*p* + 3*d*
(a)	Γ_6 (a_1)	0.9779	0.0008	0.0028	0.0185		
	Γ_7 (b_1)	0.9867	0.0000	0.0011	0.0122		
	Γ_6 (e)	0.9800	0.0000	0.0024	0.0173		
	Γ_7 (e)	0.9868	0.0000	0.0012	0.0120		
	Γ_7 (b_2)	0.9798	0.0072	0.0075	0.0126		
(b)	Γ_7 (b_1)	0.9768	0.0000	0.0002	0.0147	0.0000	0.0058
	Γ_6 (a_1)	0.9440	0.0002	0.0049	0.0371	0.0039	0.0099
	Γ_7 (e)	0.9490	0.0000	0.0042	0.0332	0.0004	0.0029
	Γ_6 (e)	0.9479	0.0004	0.0058	0.0304	0.0015	0.0079
	Γ_7 (b_2)	0.9400	0.0000	0.0138	0.0202	0.0000	0.0023

(Parrot and Boulanger 2001), where Q_{Mn} is the net charge of Mn obtained from the Mulliken population analysis. The value of Q_{Mn} was 2.031 for the cluster (a) and 1.421 for the cluster (b). The value of $\zeta_{3\mathrm{d}}$ was estimated to be 0.041 for the cluster (a) and 0.038 eV for the cluster (b). These values are agreement with the result of our calculation. From the Mulliken's population analysis of MOs, the contribution of atomic orbitals to impurity MO levels was evaluated qualitatively. The results are shown in Table 8.3, where the proportion of the contribution are listed on Mn-3*d*, Mn-4*s* + 4*p*, O-2*s*, O-2*p*, Y-4*d* + 5*s* + 5*p*, P-3*s* + 3*p* + 3*d* orbitals. As seeing the result of the cluster (a), the proportion of Mn-3*d* orbitals is more than 0.97. The MO energy levels are almost of Mn-3*d* orbitals, indicating that the Mn ion has the strong ionic character. The small contribution of O-2*s* and -2*p* orbitals is also found. The proportion of Mn-3*d* orbitals is slightly reduced in the result of the cluster (b), as compared to the value of the cluster (a). The population reduction is in contrast to the increase in the proportion of O-2*s*,2*p* orbitals. Since the atomic orbitals of cations cannot be neglected in the construction of one-electron MO wave functions, it is supposed that the cluster (a) is not suitable for the electronic state analysis of Mn^{2+} ions in YPO_4: Mn, Zr.

8.4.3 Effect of Ligand Field on Mn-3d Orbitals

When a Mn^{2+} ion is combined with other elements in solids or liquids, the covalent bond nature between them is generated more or less. The effect of ligand field modulates the electron wave functions of Mn-3*d* orbitals, which are spread out toward the direction of ligands. The expansion of Mn-3*d* wave functions causes not only the formation of Mn-O covalent bonds, but also the reduction of electron correlation among Mn-3*d* electrons. It is thus important to evaluate the effects of covalency and electron-correlation by the analysis of local electronic states of Mn^{2+} impurities doped in various compounds.

The effect of covalency is included in the two-electron integral, the formula of which is expressed by $J[ii] = <ii|g|ii>$, where i represents the orbitals of e or t_2 representations in the parent T_d symmetry. In order to analyze the effect of covalency qualitatively, it is useful to introduce the so-called orbital deformation parameters (Fazzio et al. 1984). The formulas are given by

$$\begin{aligned} \lambda_e &= J_{MO}[ee]/J_{AO}[ee] \\ \lambda_{t_2} &= J_{MO}[t_2t_2]/J_{AO}[t_2t_2], \end{aligned} \tag{8.8}$$

where J_{MO} and J_{AO} are the Coulomb integrals calculated using impurity MOs and pure Mn-3*d* atomic orbitals (AOs), respectively. The values of four Coulomb integrals $J_{MO}[a_1(e)a_1(e)]$, $J_{MO}[b_1(e)b_1(e)]$, $J_{MO}[e(t_2)e(t_2)]$, and $J_{MO}[b_2(t_2)b_2(t_2)]$ are directly calculated using the impurity MO energy levels. The values of four Coulomb integrals $J_{AO}[a_1(e)a_1(e)]$, $J_{AO}[b_1(e)b_1(e)]$, $J_{AO}[e(t_2)e(t_2)]$, and $J_{AO}[b_2(t_2)$ $b_2(t_2)]$ are approximately calculated using the AOs of a isolated Mn^{2+} ion in the YPO_4 lattice. In this calculation, the effective charge of Mn was set to be +1.427, same as that in the cluster (b). The Coulomb integrals and orbital deformation parameters are listed in Table 8.4.

Table 8.4 Calculated Coulomb integrals $J[ii]$ and orbital deformation parameters λ_i for 3*d*-MO energy levels

	$J[ii]$ (eV)			λ_i	
i	MOs (a)	MOs (b)	AOs	(a)	(b)
$b_1(e)$	22.9	22.4	22.6	1.01	0.991
$a_1(e)$	22.4	21.0	22.4	1.00	0.938
$e(t_2)$	22.4	21.0	21.9	1.02	0.959
$b_2(t_2)$	23.5	21.0	22.8	1.03	0.921

The values of $J[ii]$ were calculated using the MOs of the cluster (b) and the AOs of a Mn^{2+} ion

For the MOs of the cluster (a), the values of J_{MO} are larger than those of J_{AO}, and the values of λ are more than unity. This result seems not to be natural, because the universal relation $J_{AO} > J_{MO}$ is not satisfied. It is thus apparent that the values of J_{MO} are overestimated using the MOs of the cluster (a). For the MOs of the cluster (b), the values of J_{MO} are smaller than those of J_{AO}, and the values of λ are less than unity. These results follow the above-mentioned relation. The mean values of λ_e and λ_{t2} are estimated to be 0.965 and 0.938, respectively. These values are good agreement with $\lambda_e = 0.967$ and $\lambda_{t2} = 0.938$ in ZnS:Mn (Fazzio et al. 1984). The trend of $\lambda_e > \lambda_{t2}$ is also same, indicating that the spatial extension of e orbitals is slightly larger than that of t_2 orbitals. Therefore, the effect of covalency is comparable to that of ZnS:Mn.

8.4.4 Absorption Spectra

Figure 8.6a, b show experimental and theoretical absorption spectra due to intra-3d transitions of a Mn^{2+} ion in the YPO_4 lattice, respectively The experimental spectrum exhibits a prominent peak at 3.18 eV and a shoulder at 3.07 eV. Two weak peaks are also observed at 3.60 and 3.83 eV in the high-energy side of the

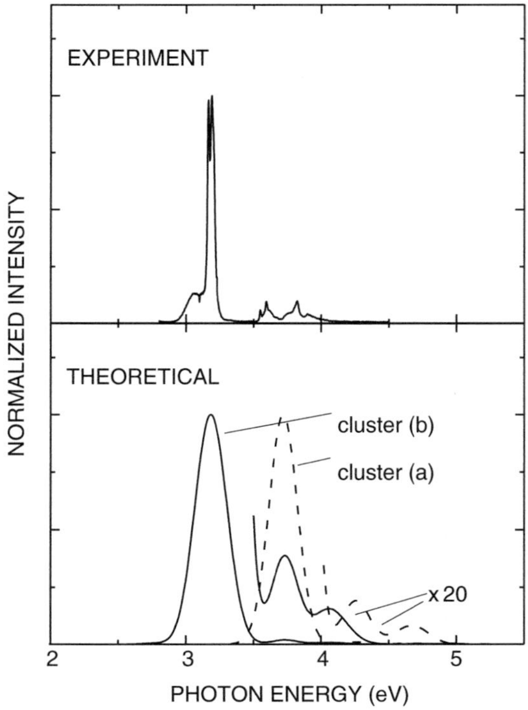

Fig. 8.6 Experimental and theoretical absorption spectrum of YPO_4: Mn, Zr. The excitation spectrum for the 477 nm emission was used as the experimental spectrum. The theoretical spectra calculated using the clusters (*a*) and (*b*) were represented by broken and solid lines, respectively

Table 8.5 Possible combinations of the electron configurations of Mn-3*d* orbitals under T_d and D_{2d} symmetries and the number of Slater determinants N_{SD}. The order of four sublevels under the D_{2d} symmetry was determined on the basis of the one-electron MO energy levels calculated using the cluster (b)

Electron configurations		
T_d	D_{2d}	N_{SD}
$(e)^4(t_2)^1$	$(b_1)^2\,(a_1)^2(e)^1$	4
	$(b_1)^2(a_1)^2(b_2)^1$	2
$(e)^3(t_2)^2$	$(b_1)^2\,(a_1)^1(e)^2$	12
	$(b_1)^2\,(a_1)^1(e)^1(b_2)^1$	16
	$(b_1)^2(a_1)^1(b_2)^2$	2
	$(b_1)^1(a_1)^2(e)^2$	12
	$(b_1)^1\,(a_1)^2(e)^1(b_2)^1$	16
	$(b_1)^1(a_1)^2(b_2)^2$	2
$(e)^2(t_2)^3$	$(b_1)^2(e)^3$	4
	$(b_1)^2(e)^2(b_2)^1$	12
	$(b_1)^2(e)^1(b_2)^2$	4
	$(b_1)^1(a_1)^1(e)^3$	16
	$(b_1)^1\,(a_1)^1(e)^2(b_2)^1$	48
	$(b_1)^1\,(a_1)^1(e)^1(b_2)^2$	16
	$(a_1)^2(e)^3$	4
	$(a_1)^2(e)^2(b_2)^1$	12
	$(a_1)^2(e)^1(b_2)^2$	4
$(e)^1(t_2)^4$	$(b_1)^1(e)^4$	2
	$(b_1)^1(e)^3(b_2)^1$	16
	$(b_1)^1(e)^2(b_2)^2$	12
	$(a_1)^1(e)^4$	2
	$(a_1)^1(e)^3(b_2)^1$	16
	$(a_1)^1(e)^2(b_2)^2$	12
$(t_2)^5$	$(e)^3(b_2)^2$	4
	$(e)^4(b_2)^1$	2
Total		252

3.18 eV peak. Although the data is not shown here, the intense band continuum due to the charge transfer transitions from O-2*p* orbitals to Mn-3*d* orbitals appears above 6.0 eV.

In order to calculate the theoretical spectrum, only MOs of the Mn-3*d* orbitals are taken into consideration in the construction of Slater determinants. Table 8.5 lists the combination of electron configurations under T_d and D_{2d} symmetry and the number of possible Slater determinants. The total number of Slater determinant is ${}_{10}C_5 = 252$. This is the same as the number of multiplet energy levels in the $(3d)^5$ electron system such as Mn^{2+} ions.

As shown in Fig. 8.6b, theoretical spectra for the clusters (a) and (b) exhibit a prominent peak at 3.71 and 3.20 eV, respectively. The peak energy is overestimated by 0.53 eV, as compared to the experimental spectrum of Fig. 8.6a. The discrepancy between experiment and calculation may come from the overestimation of the electron–electron repulsion energy among 3*d* electrons, because the effective crystal field splitting for the cluster (a) is almost the same as that for the cluster (b). On the other hand, the agreement of the peak energy is fairly good for the cluster (b). From this result, it is evident that a Mn^{2+} ion occupies the Y^{3+} ion site in YPO_4. The introduction of Mn^{2+} ions into the YPO_4 host lattice is promoted by the addition of Zr^{4+} ions, resulting in an enhancement of the Mn^{2+} emission band. Two weak peaks also appear in theoretical spectra. The peak energies do not coincide with those in the experimental spectrum. This result may imply that the electron–electron repulsion energy is slightly overestimated even in the cluster (b). The effect of electron-correlation has to be needed in order to analyze the absorption spectrum and local electronic state of a Mn^{2+} ion in YPO_4: Mn, Zr.

We discuss the features of experimental and theoretical spectra under D_{2d} symmetry. As given in Table 8.5, there are 25 electron configurations in the Mn^{2+} ion belonging to the $(3d)^5$ electron system. The contribution of each configuration to multiplet energy levels was evaluated by the configuration analysis, the results of which are shown in Fig. 8.7, together with the energy distribution of the calculated oscillator strength *f*. No contribution to the multiplet energy levels below 5 eV could be seen in ten electron configurations, and thus they were not indicated in Fig. 8.7.

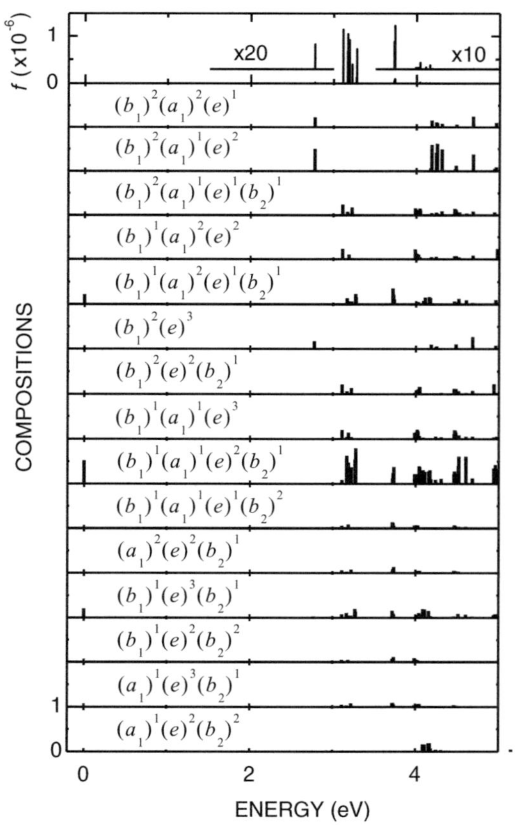

Fig. 8.7 Energy distribution of oscillator strength *f* and compositions of eight electron configurations to multiplet energy levels of a Mn^{2+} ion in YPO_4: Mn, Zr

The ground states around 0 eV is dominated by the $(b_1)^1(a_1)^1(e)^2(b_2)^1$ configuration. Since this feature is typical of the high-spin state of $S = 5/2$, the ground state can be expressed in term of the $^6A_1$ irreducible representation under D_{2d} symmetry. Admixture of $(b_1)^1(a_1)^2(e)^1(b_2)^1$ and $(b_1)^1(e)^3(b_2)^1$ configurations is also found at the ground state. The $^6A_1$ state slightly has the influence of level-mixing with the ^4E and $^4A_2$ states of $(b_1)^1(a_1)^1(e)^2(b_2)^1$, $(b_1)^1(a_1)^2(e)^1(b_2)^1$ and $(b_1)^1(e)^3(b_2)^1$ configurations through spin-orbit interaction, because the operator $\boldsymbol{t}(\mathbf{r})$ of the spin-orbit interaction Hamiltonian $H_{SO} = \boldsymbol{t}(\mathbf{r}) \cdot \boldsymbol{s}$ has the same symmetry as the basis of E and A_2 irreducible representations under D_{2d} symmetry. Judging from this fact, our calculated results seem to be reasonable.

The lowest-energy excited state appears around 2.77 eV. The major component is the $(b_1)^2(a_1)^1(e)^2$ configuration which results from the one-electron excitation from the highest-energy $b_2\uparrow$ state to the lowest-energy $b_1\downarrow$ state in the ground state configuration of $(b_1)^1(a_1)^1(e)^2(b_2)^1$. The $(b_1)^2(a_1)^1(e)^2$ configuration comes from the $(e)^3(t_2)^2$ configuration under T_d symmetry, and the direct product $b_1\uparrow \times b_1\downarrow \times a_1\uparrow \times e\uparrow \times e\uparrow$ contains a spin-quartet state of the A_2 irreducible representation. Therefore, it is evident that the $^4A_2$ state appears in the $(b_2)^2(a_1)^1(e)^2$ configuration. The $^4A_2$ state originates from the parent $^4T_1$ state by descending symmetry of $T_d \rightarrow D_{2d}$. The transition of $A_1 \rightarrow A_2$ is basically electric-dipole forbidden under D_{2d} symmetry, because the symmetry of the electric dipole moment is described by the irreducible representations of E and B_2. The value of f is expected to be small for this transition; hence, no peak is found at 2.77 eV in the theoretical spectrum. The $(b_1)^2(a_1)^2(e)^1$ and $(b_1)^2(e)^3$ configurations are included as minor components. Since these configurations yield ^2E states, they are miscible with the $^4A_2$ state through spin-orbit interaction.

The next excited state is located at 3.10 eV. This excited state originates in the $(b_1)^2(a_1)^1(e)^1(b_2)^1$ configuration realized by the one-electron excitation from the $e\uparrow$ state to the $b_1\downarrow$ state. This configuration causes the ^4E state of another $^4T_1$ origin. The $^6A_1 \rightarrow {}^4E$ transition is of an electric-dipole allowed type, and it is thus natural that the value of f is large for this transition. The $(b_1)^1(a_1)^2(e)^2$, $(b_1)^2(e)^2(b_2)^1$ and $(b_1)^1(a_1)^1(e)^3$ configurations also contribute to the excited state at 3.10 eV as the consequence of configuration interaction and spin-orbit coupling with the $^4E(^4T_1)$ excited state. These configurations are formed by one-electron excitation of $b_2\uparrow \rightarrow a_1\downarrow$, $a_1\uparrow \rightarrow b_1\downarrow$ and $b_2\uparrow \rightarrow e\downarrow$, respectively. The energies of them are slightly high compared to that of the $(b_1)^2(a_1)^1(e)^1(b_2)^1$ configuration, and thus the four configurations are remarkably mixed with each other.

A number of excited states are concentrated around 3.17 eV. These states mainly originate in the $(b_1)^1(a_1)^1(e)^2(b_2)^1$ configuration generated by one-electron transitions, e.g., the $a_1\uparrow \rightarrow a_1\downarrow$ transition. This configuration is apparently different from the ground state configuration giving the $^6A_1$ state. It produces $^4A_1$, $^4A_2$, $^4B_1$ and $^4B_2$ states under D_{2d} symmetry. There is also admixture of $(b_1)^2(a_1)^1(e)^1(b_2)^1$, $(b_1)^1(a_1)^2(e)^1(b_2)^1$ and $(b_1)^1(e)^3(b_2)^1$ configurations. These configurations are characterized by the irreducible representation E under D_{2d} symmetry. Since the B_2 and E states are open for the electric dipole transitions from the A_1 state, the large values of f are possible for the excited states with B_2 and E characters. Actually, one can see a prominent peak at 3.17 eV in the theoretical spectrum. Therefore, the 3.17 eV peak is mainly ascribed to one-electron transitions to the excited states with B_2 and E characters. However, the assignment of multiplet energy levels is not possible in their dense situation, because different configurations are strongly mixed in the calculated results based on the relativistic DVME method.

The largest contribution of the $(b_1)^1(a_1)^1(e)^2(b_2)^1$ configuration can be seen at 3.27 eV, where the value of f is also the largest. This feature may be an indication of the $^4A_1$ state, because this state is one of the orbital-singlet states among spin-quartet terms in the parent $(e)^2(t_2)^3$ configuration. The energy of an orbital-singlet state is higher than that of an orbital-doublet and -triplet states, if the spin-multiplicity is the same. Therefore, it is likely to connect the $^6A_1 \rightarrow {}^4A_1$ transition to the origin of the 3.27 eV peak. The $^6A_1 \rightarrow {}^4A_1$ transition is allowed by the level mixing of ^4E states due to spin-orbit interaction. In the energy matrices by Tanabe and Sugano (1954a), the matrix elements of Coulomb interaction for the $^4A_1$ state do not contain the term of crystal field splitting Δ, as same as the $^6A_1$ state. This means that the energy positions of $^6A_1$ and $^4A_1$ states are basically independent of the magnitude of Δ. In this case, since a narrow peak due to the $^6A_1 \rightarrow {}^4A_1$ transition is expected, the transition would be responsible for the 3.18 eV peak in the experimental spectrum.

The excited state at 3.72 eV is composed of $(b_1)^1(a_1)^2(e)^1(b_2)^1$ and $(b_1)^1(a_1)^1(e)^2(b_2)^1$ configurations. The former corresponds to the $^4B_1$ state coming from a ^4E state of the $(e)^2(t_2)^3$ configuration under parent T_d symmetry. The latter corresponds to the ^4E state coming from a $^4T_2$ state of the $(e)^3(t_2)^2$ configuration. Both configurations contribute to the excited state at 3.27 eV comparably, because they are mixed with each other by spin-orbit interaction. Although it is difficult to determine the origin of the 3.27 eV excited state, we tentatively assign this state to the ^4E state of $^4T_2$ origin by considering that it consists of neighboring four levels. Considerable admixture of the $^4B_1$ state would weaken the oscillator strength of the $^6A_1 \rightarrow {}^4E$ transition of electric dipole allowed type.

There are many multiplet energy levels above 4 eV, but the value of f is negligibly small. This is due to the appearance of spin-doublet states. The transitions from the spin-sextet ground state to the spin-doublet excited states are caused by two-electron excitation, e.g., the $e\uparrow e\uparrow \rightarrow e\downarrow e\downarrow$ transition in the $(e)^2(t_2)^3$ configuration. Such transitions are basically forbidden, because the matrix elements of the two-electron operator for Coulomb interaction are originally zero for them.

8.4.5 Energy Position and Lifetime of the Mn^{2+} Emission

Mn^{2+} doped inorganic materials have good properties as phosphors for lighting and displays. However, Mn^{2+} ions in such compounds usually emit phosphorescence with the lifetime longer than sub-millisecond order. This feature is a weakness in the application of Mn^{2+} doped phosphors, and thus the improvement of lifetime is strongly desirable. For elucidation of this subject, two approaches have been energetically performed so far. One is the development of phosphor nanocrystals doped with Mn^{2+} ions. In ZnS nanocrystals doped with Mn^{2+} ions, a fast decay component was certainly observed in the Mn^{2+} emission, but it was attributed to traps of ZnS nanocrystals (Smith et al. 2000). The other is the theoretical prediction of the lifetime of a Mn^{2+} emission. For single crystals of II-VI compounds doped with Mn^{2+} ions, the lifetime of a Mn^{2+} emission was accurately reproduced by cluster calculations (Boulanger et al. 2004). Accordingly, theoretical calculations are applicable to determine the lifetime of Mn^{2+} emissions in various compounds.

As depicted in Fig. 8.7, the relativistic DVME method enables us to calculate the oscillator strength for electron transitions among multiplet energy levels. Generally, the radiative transition probability of an emission is given by the Einstein's *A* coefficient which is proportional to the oscillator strength of electron transitions between initial and final states (Fowler and Dexter 1962). The formula is expressed as follows.

$$A = \frac{n(n^2+2)^2}{9} \cdot \frac{8\pi^2 e^2}{mc^3 h^2} \cdot \left|\langle E_i - E_f \rangle\right|^2 \cdot \frac{g_f}{g_i} \cdot \sum_{i,f} I_{fi}, \tag{8.9}$$

where n is the effective refractive index of the host material, e the elementary electric charge, m the electron rest mass, c the velocity of light in vacuum, h the Planck's constant, E_i and E_f energy eigenvalues of the initial and final states, g_i and g_f the number of the initial and final states, and I_{fi} the oscillator strength given by (8.7). The values of I_{fi} were assumed to be equal to those of I_{if}. The lifetime of an emission can be calculated from the reciprocal of the *A* coefficient.

Table 8.6 shows the energy position and lifetime of the Mn^{2+} emission in YPO_4: Mn, Zr, determined by experiment and calculation. The Mn^{2+} emission is assigned to the electron transitions from the $^6A_1$ ground states to the $^4A_2$ lowest excited states, as mentioned above. The spin-orbit splitting among $^6A_1$ sublevels is less than 3.30×10^{-5} eV. Since the highest-energy Γ_7 sublevel of $^6A_1$ can be sufficiently populated by thermal excitation from the lowest-energy Γ_6 sublevel, the values of g_i and g_f were put to 4 and 6, respectively. The energy position of the Mn^{2+} emission was determined by an average of the energy difference between $^4A_2$ and $^6A_1$ states. The effective refractive index was set to be 1.636 (Jellison et al. 2000).

In Table 8.6, the energy position of the Mn^{2+} emission is very close to the mean energy of the $^4A_2 \rightarrow {}^6A_1$ transition, obtained from the calculated results of the cluster (b). The reason for very little difference between experiment and calculation would be due to either the overestimation of the electron–electron repulsion energy

Table 8.6 Energy position and lifetime of a Mn^{2+} emission in YPO_4: Mn, Zr. The results of calculation were obtained from the cluster (b)

	Calculation	Experiment[a]
Energy position (eV)	2.77	2.60
Lifetime (ms)	5.44	13.0

[a]Kitaura et al. (2007)

or the underestimation of the crystal field splitting. Perhaps, it may be due to Stokes shift induced by photoexcited state of a Mn^{2+} ion. On the other hand, the lifetime was determined to be 5.44 ms using the calculated results of the cluster (b). This value is close to the lifetime (= 13 ms) determined from the optical experiment at room temperature (Kitaura et al. 2007). Therefore, the theoretical calculation based on the DVME method has a potential for the prediction of both energy position and lifetime of TM emissions in various phosphor materials.

8.5 Summary

The advantage of relativistic DV-Xα and DVME methods for investigations of the electronic structures and optical properties of TM impurities in solids was demonstrated in this chapter. The electronic structure analysis of a Mn^{2+} ion in YPO_4: Mn, Zr provided us valuable information on the interaction of Mn-3*d* electrons with ligands in this phosphor. From the evaluation of orbital populations and orbital deformation parameters, it turned out that the 3*d* wave functions of Mn^{2+} ions is more spread out when they are introduced in the YPO_4 lattice. The spatial expansion of the Mn-3*d* wave function was explained by the effect of covalency in Mn-O bonds.

The relativistic DVME method was very useful for better understanding of optical properties of YPO_4: Mn, Zr. The most noticeable feature was that it can analyze very weak absorption bands due to spin-forbidden transitions of Mn^{2+} ions. The comparison of experimental and theoretical absorption spectra gave us a significant evidence to clarify the whole picture of an enhancement in the Mn^{2+} emission of YPO_4: Mn, Zr. From the evidence, it was proposed that the introduction of a Mn^{2+} ion into the Y^{3+} site is promoted by the addition of Zr^{4+} ions. The enhancement in the Mn^{2+} emission of YPO_4: Mn, Zr was reasonably explained according to this basic idea.

Theoretical prediction of luminescence properties of Mn^{2+} ions in YPO_4: Mn, Zr was also attempted as the first trial using the results of the relativistic DVME method. As for the energy position and lifetime of the Mn^{2+} emission, good agreement was found between experiment and calculation. It is therefore expected that the theoretical calculation utilizing the DVME method can be successfully applied to the material design of new functional phosphors.

Acknowledgements One of the authors (M. K.) acknowledges the financial support by a Grant-in-Aid for Scientific Research (C) (No. 20613013) from Japan Society for the Promotion of Science. Partial Grants was also provided by the Nippon Sheet Glass Foundation for Material Science and Engineering, and the Society for Discrete variational *X*α. A part of this work was supported by the Joint Studies Program of the Institute for Molecular Science.

References

Adachi H, Tsukada M, Satoko C (1978) Discrete variational Xα cluster calculation. I. Application to metal clusters. J Phys Soc Jpn 45:875–883

Boulanger D, Parrot R, Cherfi Z (2004) Cluster model for radiative transition probabilities of d^5 ions in tetrahedral symmetry: case of Mn^{2+} ions in the common cation series ZnS, ZnSe, and ZnTe. Phys Rev B 70:075209(1–12)

Costa GKB, Pedro SS, Carvalho ICS, Sosman LP (2009) Preparation, structure analysis and photoluminescence properties of $MgGa_2O_4$:Mn^{2+}. Opt Mater 31:1620–1627

Dondi M, Matteucci F, Cruciani G (2006) Zirconium titanate ceramic pigment: Crystal structure, optical spectroscopy and technological properties. J Solid State Chem 179:233–246

Fazzio A, Caldas MJ, Zunger A (1984) Many-electron multiplet effects in the 3d impurities in heteropolar semiconductors. Phys Rev B 30:3430–3455

Fowler WB, Dexter DL (1962) Relation between absorption and emission probabilities in luminescence centers in ionic solids. Phys Rev 128:2154–2165

Jellison GE Jr, Boatner LA, Chen C (2000) Spectroscopic refractive indices of metalorthophosphates with the zircon-type structure. Opt Mater 15:103–109

Kaneyoshi M, Nakazawa E (2005) Luminescence of YPO_4:Zr and YPO_4:Zr, Mn under vacuum ultraviolet excitation. J Electrochem Soc 152:H80–H83

Kitaura M, Nakajima Y, Kaneyoshi M, Nakagawa H (2007) Effect of Zr^{4+} addition on photoluminescence properties of YPO_4:Mn^{2+}. Jpn J Appl Phys 46:6691–6695

Lakshman SVJ, Reddy BJ (1973) Optical absorption spectrum of Mn^{2+} in rhodonite. Physica 66:601–610

Mehra A (1971) Comparative ligand field studies of manganese(II) spectra. J Phys Chem 75:435–437

Milligan WO, Mullica DF, Beall GW, Boatner LA (1982) Structural investigations of YPO_4, $ScPO_4$, and $LuPO_4$. Inorg Chim Acta 60:39–43

Mohler RL, White WB (1994) Influence of structural order on luminescence of oxide spinel: manganese activated spinels. Mater Res Bull 29:1109–1116

Ogasawara K, Iwata T, Koyama Y, Ishii T, Tanaka I, Adachi H (2001) Relativistic cluster calculation of ligand-field multiplet effects on cation $L_{2,3}$ x-ray-absorption edges of $SrTiO_3$, NiO, and CaF_2. Phys Rev B 64:115413(1–5)

Parrot R, Boulanger D (2001) Influence of the molecular spin-orbit interaction on the orbital triplet levels $^4T_1$ and $^4T_2$ of Mn^{2+} in ZnS. Physica B 308–310:928–931

Shannon RD (1976) Revised effective ionic radii and systematic studies of interatomic distances in halides and chalcogenides. Acta Cryst A 32:751–767

Slater JC (1974) Quantum theory of molecules and solids, vol 4. McGraw-Hill, New York

Smith BA, Zhang JZ, Joly A, Liu J (2000) Luminescence decay kinetics of Mn2+-doped ZnS nanoclusters grown in reverse micelles. Phys Rev B 62:2021–2028

Su F, Ma B, Ding K, Li G, Wang S, Chen W, Joly AG, McCready DE (2006) Luminescence temperature and pressure studies of Zn_2SiO_4 phosphors doped with Mn^{2+} and Eu^{3+} ions. J Lumin 116:117–126

Tamatani M (2007) Phosphor handbook. In: Yen WM, Shionoya S, Yamamoto H (eds) Principal phosphor materials and their optical properties, 3rd edn. CRC Press, Boca Raton

Tanabe Y, Sugano S (1954a) On the absorption spectra of complex ions. I. J Phys Soc Jpn 9:753–766

Tanabe Y, Sugano S (1954b) On the absorption spectra of complex ions. II. J Phys Soc Jpn 9:766–779

Watanabe S, Kamimura H (1989) First-principles calculations of multiplet structures of transition metal deep impurities in II-VI and III-V semiconductors. Mater Sci Eng B 3:313–324

Watanabe S, Ogasawara K (2008) Experimental and first-principles analysis of 4f-5d absorption spectra for Ce^{3+} in $LiYF_4$ considering lattice relaxation. J Phys Soc Jpn 77:084702(1–7)

Watanabe S, Ogasawara K, Yoshino M, Nagasaki T (2010) Experimental and first-principles analysis of f^n-$f^{n-1}d^1$ absorption spectra and multiplet energy levels of Pr^{3+} and Nd^{3+}, and U^{3+} in $LiYF_4$. Phys Rev B 81:125128(1–18)

Chapter 9
Applications of DV-Xα Method for New Material Design in Dye-Sensitized Solar Cell

Dong-yoon Lee and Yang-Soo Kim

9.1 Introduction

Dye-sensitized solar cell (DSSC) is a sort of electrochemical device using nano-porous oxide electrode, dye, electrolyte, and nano catalyst, which is different from other solar cells with p-n junction of semiconductor (Grätzel 2000; Arakawa 2001). Recently, the efficiency and performance of conventional DSSC have been enhanced to above 11 % by adopting newly developed nano materials (Chiba et al. 2006). These advances are mostly related to surfaces of nano-sized electrode, and interfaces between electrode and electrolyte. However, unfortunately it is very difficult to understand the operating mechanisms of chemical and physical reactions included in these surfaces and interfaces, which are responsible for key phenomena, because useful tools for direct in-situ observation of nano structures have not been found until now. Sometimes, this difficulty brings about the hardness of new material design for further development. In this point, the electronic structure calculations based on the first principles method can give powerful guide without experiments by providing useful data including electron energy levels, density of state, charge transfer, binding conditions between atoms, etc.

The electronic structure for bulk state can be easily calculated by most of calculation methods with different algorism. However, there are not many programs to show the comparatively reasonable calculations for surface state and interface structure, because of the severe distortion of electron density and bonding state in calculation area. The first principle calculation methods based on density

D.-y. Lee (✉)
Nano-hybrid Energy Materials Research Center of Korea Electrotechnology Research Institute, Changwon City 641-120, Kyeongnam, Republic of Korea
e-mail: dylee@keri.re.kr

Y.-S. Kim
Korea Basic Science Institute, Suncheon 540-742, Republic of Korea
e-mail: kimyangsoo@kbsi.re.kr

T. Ishii et al. (eds.), *The DV-Xα Molecular-Orbital Calculation Method*,
DOI 10.1007/978-3-319-11185-8_9

functional theory (DFT), which uses one electron approximation in electron density field, have some merits for the calculation of distorted structure. The discrete variational X_α (DV-Xα) method, which uses Hartre-Fock-Slater approximation and full potential molecular orbital, can be used as good tools for the calculations of electronic structures in interfaces and surfaces of DSSC, providing bonding state, electronic energy levels and charge transfer (Ellis et al. 1976; Adachi et al. 1978). Main concerns of this work on DSSC are concentrated on the energy level structure near the Fermi level in TiO_2 surface, charge transfer in interface between iodide and carbon-based catalyst, and design of optical wavelength conversion materials with ZnO.

A nano-porous TiO_2 electrode is an oxide semiconductor and plays a role of path to transport carrier electrons made in dye molecules, which are adsorbed on the surface of the TiO_2 electrode and generate exited electrons by photo-sensitization. The surface state of TiO_2 controls the density and value of surface energy levels. The open voltage and the short circuit current of solar cell are deeply related to the surface levels. The most important factor to affect on the surface level of TiO_2 is the surface doping of foreign elements, especially transient metals. The change of energy levels by transient metals and elements of IV family was calculated and discussed in this work.

Graphene and carbon nano tube (CNT) can be used as catalysts of DSSC, replacing Pt, which is deposited as nano particles on counter electrode against working electrode of TiO_2 (Cha et al. 2010; Seo et al. 2010). The theoretical estimation on catalytic effect of materials is difficult because of absence of direct evaluation criterion. It can be only estimated by indirect data such as band gap. The DV-Xα method is useful to evaluate this effect, because it provides information on direct charge transfer and reliable electron density distribution. In this work the electron charge transfer from graphene or CNT to iodide in interface of electrode/ electrolyte was calculated and discussed.

Optical wavelength conversion technology has been tried to convert rays, which have wavelength out of acceptable range in solar cell, into visible rays. We call as down-convert when ultraviolet rays change to visible rays and as up-convert in the case of infrared to visible. The down-converting can occur by passing ultraviolet rays through transparent oxide doped with special element. In our case we tried to use ZnO thin film as base material and Y as doping element. To investigate the possibility of ZnO:Y material as proper convertor, DV-Xα method was used for detailed calculation of energy level and bond structure near Fermi level.

9.2 Calculations

The first-principles molecular orbital (MO) calculations by the DV-Xα method (Ellis et al. 1976; Adachi et al. 1978) using the program code SCAT (Ellis et al. 1976) were performed in order to obtain the electronic structure, chemical bonding state and the degree of charge transfer on surfaces or interfaces of material

Table 9.1 The fractional coordinates of independent atomic sites, the valence state and the Wyckoff letter in the structure of TiO_2

Materials	Space group	Atom	Valence state	x	y	z
TiO_2 (rutile)	P42/mnm	Ti	4	0	0	0
		O	−2	0.3049	0.3049	0
Graphene	P63/mmc	C1	–	0	0	1/4
		C2	–	1/3	2/3	1/4
ZnO	P63mc	Zn	2	1/3	2/3	0
		O	−2	1/3	2/3	0.3829

systems in DSSC. Calculations are focused on the energy level structure of TiO_2 surface doped with transition metals and IV family, charge transfer from graphene or CNT counter electrode to iodide, and the dopant level of ZnO. Numerical atomic orbitals (NAO) were employed as the basis functions. They were generated and optimized by solving the radial part of the Schrödinger equation for a given environment at each iteration of the self-consistent calculation. All the elements of Hamiltonian and the overlap matrices were calculated numerically on the basis of the discrete variational integration scheme. The detailed computational procedure is described in the literature (Adachi et al. 1978) and the space group and the lattice constant of materials used in the present study is given in Table 9.1. Coordinates of CNT were generated with 1.41–1.42 Å of C-C bond length by the personally coded program. The type of CNT was single wall CNT (6,0) through all calculation because it provide enough curvature different from graphene.

All calculations in DV-Xα method should be executed with cluster models and full potential of all orbital included in the cluster model. The proper building of models is very important in this method, because calculated results changes with models and, sometimes, shows unreasonable values regardless of the convergence of calculations. Therefore, various models in every material were tried to obtain acceptable data in the view of crystal chemistry and energy level structure. In the case of TiO_2 and ZnO the Madelung field was applied over enough large volume to obtain environment similar to bulk and to eliminate the edge effect, which brings distorted charge distribution and energy levels. Hydrogen was attached to each carbon atom in the edge position of models in graphene and CNT to eliminate edge effect.

9.3 Effect of Surface Doping on Chemical Bonding State and Energy Level of TiO_2

The application of the first principles method to DSSC involves two purposes; material design for enhancement of performance, such as increase of efficiency, open voltage, short circuit current and fill factor; understanding of mechanisms acting between components, especially in interfaces between corresponding

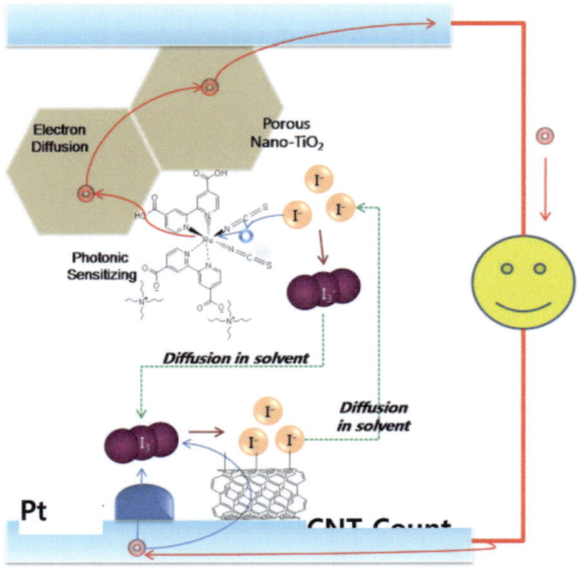

Fig. 9.1 The schematic plot of dye-sensitized solar cell, showing materials and interfaces

materials. Materials and interfaces can be identified in Fig. 9.1, showing the schematic plot of DSSC. DSSC composed from materials such as TiO_2, dye, electrolyte, counter electrode and transparent conducting oxide (TCO). And it has many interfaces such as TCO/TiO_2, TiO_2/dye, dye/electrolyte, TiO_2/electrolyte, electrolyte/counter electrode and counter electrode/TCO.

The main properties of porous nano-TiO_2 in DSSC are decided by surface states or surface energy levels, and surface area. TiO_2, which is a sort of insulator in pure state, has the properties of semiconductor by the introduction of surface energy levels (Lee et al. 2004). Surface levels affect on electric conductivity of oxide semiconductor and the open voltage of DSSC. The open voltage of DSSC is decided by the difference between conduction band edge (CBD) of TiO_2 and redox level of iodide/triiodide couple in electrolyte at counter electrode. The density of surface levels increases with surface area. The high density of surface levels increases electric conductivity of TiO_2, but decreases the open voltage because it modifies CBD to lower level. Therefore, the surface area of nano-TiO_2 should be controlled to moderate value. Surface levels change with surface doping of foreign atoms. It is well known that existence of foreign atom, especially transient metals such as Fe and Ni, decreases the open voltage. Here the effect of surface doping by transient metals and VI family elements will be shown.

The cluster models used calculations by DV-Xα method were the $(Ti_{15}O_{56})^{52-}$ model for the calculations of the bulk state and the $(Ti_{11}O_{34})^{24-}$-O^a model, as shown in Fig. 9.2. The O^a is the surface oxygen, which lies on the surface Ti ion in Fig. 9.2b. The effect of transition metal doping on the surface state was investigated using $(MTi_{11}O_{34})^{24-}$-O^a model (M = foreign metal ion). In this model, a foreign metal ion replaces the Ti ion in center position of the (110) surface plane. The atomic orbitals used in this work are 1*s*-4*p*, 1*s*-4*p* and 1*s*-2*p* for Ti, M and O, respectively.

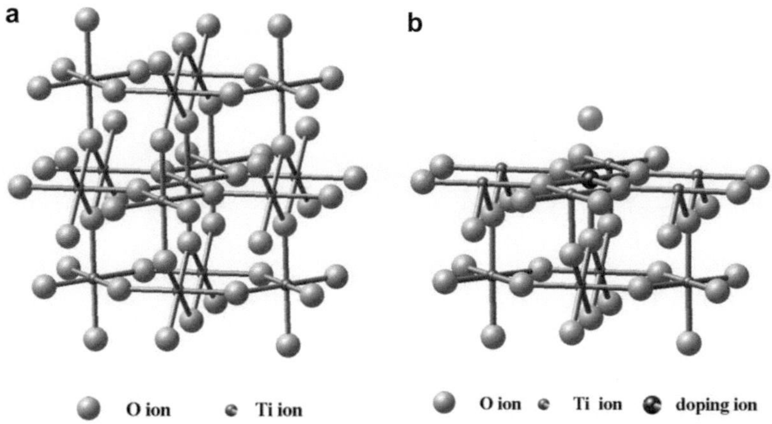

Fig. 9.2 Cluster models used for the calculation of electronic structure of rutile-TiO_2; (**a**) $(Ti_{15}O_{56})^{52-}$ cluster for the calculation of bulk state and (**b**) $(MTi_{10}O_{34})^{24-}$-O^a for the surface calculation. The O^a atom means the adsorbed surface oxygen loosely bonded to TiO_2 surface. A foreign metal ion replacing Ti ion lies below the surface oxygen

The calculation using small cluster models should be carried out with special care in association with the non-neutrality of electrostatic charge and the imperfect bonding of atoms positioned near the edge of a cluster. To overcome this problem, the Madelung field should be applied over enough extended volume in order for atoms in edge to have an electrostatic environment similar to that of atoms in bulk state. Figure 9.3 shows the atomic structure of (110) (1 × 1) surface, which is stable surface found in sapphire (Ramamoorthy et al. 1994; Kasowcki and Tait 1979). In Fig. 9.3a, Ti atoms have two sorts of site with different surrounding; a site having six folded bonds with oxygen atoms and a site having five folded bonds with one dangling bond in the site of surface oxygen. O^a in figure is not an oxygen atom in TiO_2 structure, but from foreign source. This oxygen or other anions can be strongly adsorbed on the fivefolded Ti atom because of dangling bond. This adsorbed atoms or atomic groups are related to the source of donar levels that is responsible for properties of semiconductor in nano-TiO_2.

In this calculation, the Madelung field having the size of 6 × 6 × 6 unit cells was applied on bulk TiO_2 models. Clusters were cut from real crystals and embedded in a Madelung potential that had been generated by point charges located at the external atomic sites. For surface calculations, the Madelung potential was limited to the (110) surface plane, as shown in Fig. 9.3b. Therefore, the Madelung field was not applied out of the surface plane, and then O^a atom, which lies on outer space of the (110) plane, does not undergo the influence of the Madelung field. In the DV-Xα method, this restriction of the Madelung field in the surface differentiates a surface calculation from a bulk calculation.

Calculations using octahedron TiO_6 model were executed to understand the orbital mixing of Ti and O and explain generation of surface energy level acting as a donar level. Figure 9.4 shows the calculated energy level diagrams of TiO_6 for

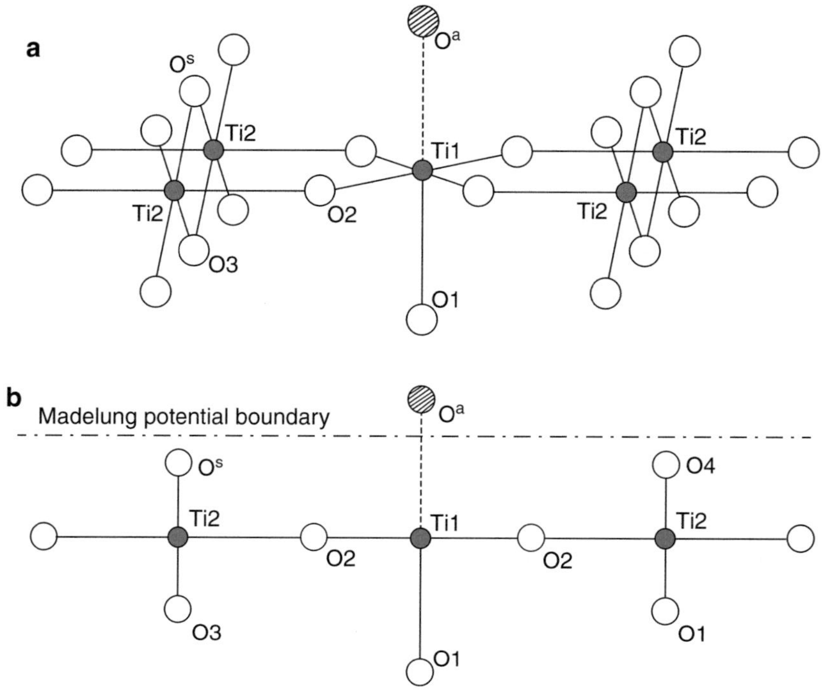

Fig. 9.3 The schematic arrangement of atoms in (110) (1 × 1) surface of rutile TiO_2 is shown in (**a**). The Madelung potential field was applied within surface boundary, as shown in (**b**), except surface oxygen atom, which could be considered as foreign atom

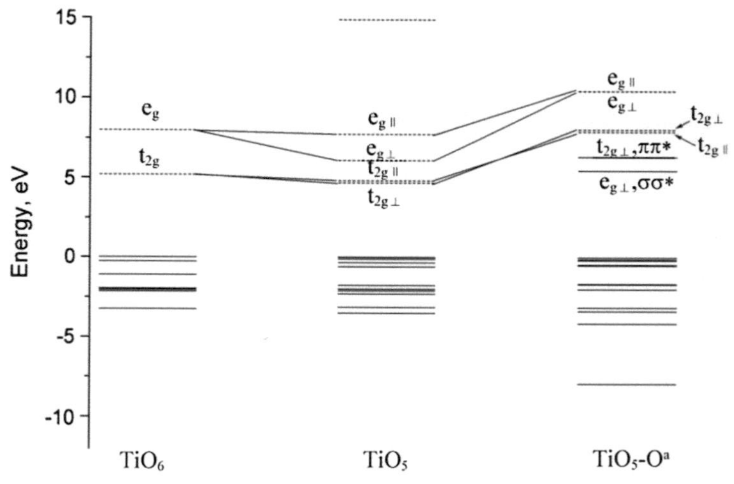

Fig. 9.4 The calculated energy level diagram of TiO_6 for bulk, TiO_5 for surface and TiO_5-O^a for surface with adsorbed oxygen. ⊥and || mean the perpendicular and parallel direction to the (110) surface

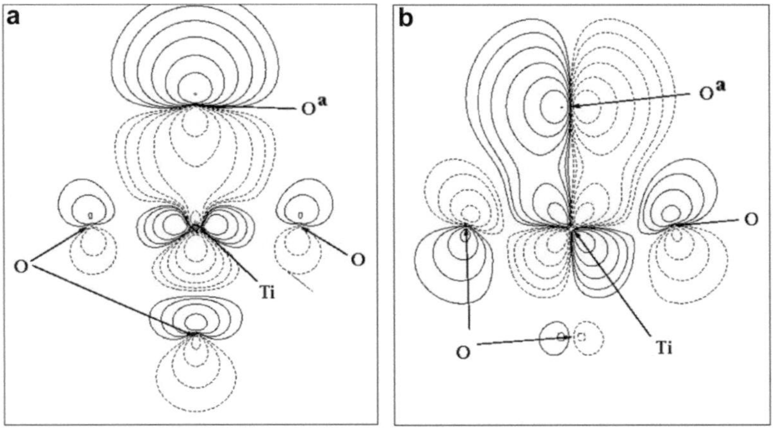

Fig. 9.5 Wave functions of $e_{g\perp}$(**a**) and $t_{2g\perp}$ (**b**). σ bonding and σ* anti-bonding appear in $e_{g\perp}$ (**a**) and π bonding and π* anti-bonding in $t_{2g\perp}$ (**b**)

bulk, TiO_5 for surface and TiO_5-O^a for surface with foreign adsorbed oxygen. The 3d level is composed of t_{2g} of trifle degeneracy and e_g of double degeneracy. In metallic oxides with octahedron structure, it appears t_{2g} level at lower position of conduction band and e_g level above t_{2g} level, as shown in Fig. 9.4. If one oxygen atom is removed from surface as TiO_5 model, t_{2g} and e_g were separated to $t_{2g\perp}$ and $e_{g\perp}$ perpendicular to surface and $t_{2g||}$ and $e_{g||}$ parallel to surface. If O^a is adsorbed on this surface of TiO_5, $e_{g\perp}$ and t_{2g} are formed as surface levels. These surface levels are mixed levels of Ti 3d and O 2p and have bonding and anti-bonding through axis of O^a and Ti atom. In wave functions shown in Fig. 9.5, $e_{g\perp}$ has strong σ bonding state between Ti 3d and Oa 2p and σ* anti-bonding state between Ti 3d and O 2p of internal oxygen. In the case $t_{2g\perp}$, π bonding and π* anti-bonding are found between Ti 3d and O 2p. From these results, it can be understood that strong covalent bonding is formed between Ti and adsorbed oxygen and this strong interaction makes the surface donar levels below the conduction band.

In DSSC, the effect of surface doping in TiO_2 nano electrode has been investigated and discussed for long time since the beginning of DSSC (Ko et al. 2005; Wijayarathna et al. 2008). The $(MTi_{10}O_{34})^{24-}$-O^a model for this calculation was built by replacing Ti atom on surface to a metallic atom (M) in surface model of $(Ti_{11}O_{34})^{24-}$-O^a. The calculated energy level diagrams with doping of transient metal are shown in Fig. 9.6. In this figure, atomic number of dopant element increases from left to right. All dopant levels are found between the conduction band and the valence band of TiO_2. These levels decrease with atomic number of doping element and, finally, approach to near the valence band in Fe^{3+}, Co^{3+} and Ni^{2+}. Lowering of dopant levels lowers the surface level playing a role of donar level and increases the density of surface state because the dopant ion interacts with surface oxygen that is a source of surface level. This lowering of surface levels means the drop of open voltage in DSSC, decided by the difference between surface

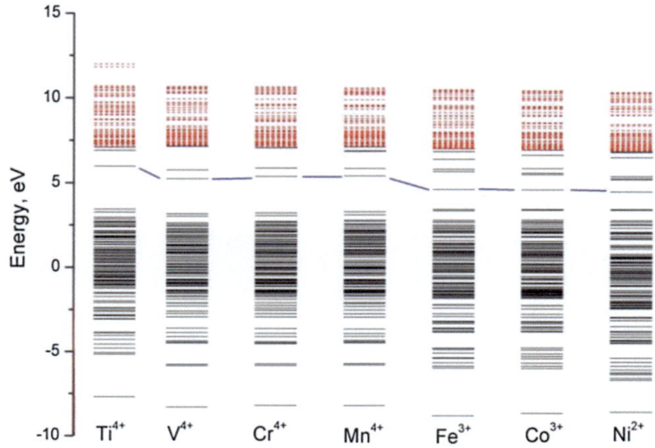

Fig. 9.6 Energy level diagrams calculated with $(MTi_{10}O_{34})^{24-}$-O^a (M = transient metal) model shows the dopant levels between conduction band and valence band of TiO_2. The dopant level decreases with the atomic number of doping element and brings about the lowering of open voltage in DSSC. *Dotted line* and *solid line* mean electron unoccupied level and electron occupied level, respectively

level and iodide redox level. Actually, it is well known in DSSC industry that a little contamination of TiO_2 by transient metal will bring bad performance of DSSC.

Recently, there were reports that using of silica-modified TiO_2 could enhance the efficiency of DSSC (Park et al. 2010; Hoshikawa et al. 2006). It is expected that doping of IV family elements involving Si shows different effect on surface level structure from transient metals because they have the properties of semi-metal instead of pure metallic properties. The density of state (DOS) calculated using $(MTi_{14}O_{56})^{52-}$(M = Si, Ge, Sn) model for bulk doping and $(MTi_{10}O_{34})^{24-}$-O^a model for surface doping are shown in Fig. 9.7.

The doping of IV family elements doesn't change the band gap of bulk TiO_2, although details of orbital mixing change. In pure bulk TiO_2 the lower levels are composed from orbitals mixed with Ti 3d and O 2p. By bulk doping, outer s orbital of IV family elements such as Si 3s, Ge 4s and Sn 5s is mixed to the lower part of conduction band, as shown in Fig. 9.7. These mixed states don't change the band gap between conduction band edge and valence band edge at least in our model. However, surface doping severely changes mixing interaction near the Fermi level. The contribution of Si 3s, Ge 4s and Sn 5s to DOS near conduction band edge decreases to negligible extent compared to other partial DOSs. However, p orbital component such as Si 3p, Ge 4p and Sn 4p interact with O 2p and make dopant levels at the center of forbidden band between the conduction band edge and the valence band edge. That is, it is expected that the surface doping of IV family element also lowers the surface levels toward the valence band edge similar to the case of transient metal, although the position of surface levels, DOS and mixing action of orbitals are different each other. These results deny the possibility of increasing the band gap of TiO_2 by doping of IV family and the increase of open

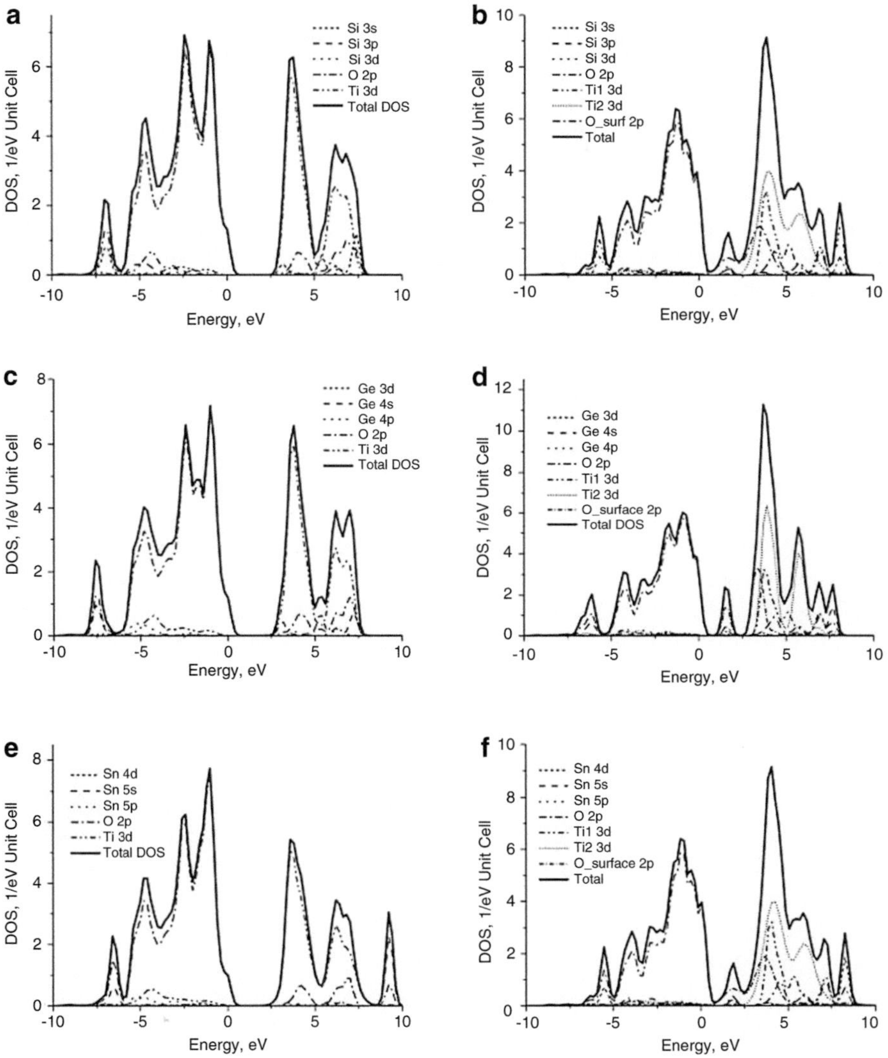

Fig. 9.7 The calculated density of state using $(MTi_{14}O_{56})^{52-}$ (M = Si, Ge, Sn) model for bulk doping and $(MTi_{10}O_{34})^{24-}$-O^{a} model for surface doping. (**a**), (**c**) and (**e**) are DOS of bulk states and (**b**), (**d**) and (**f**) are DOS of surface doped states

voltage, proposed in experimental articles (Park et al. 2010; Hoshikawa et al. 2006). Therefore, it is more reasonable that enhanced efficiency appeared in experiments should be attributed to the responsibility of other experimental factors such as sintering effect or change of Schottky barrier.

9.4 Theoretical Evaluation on Catalytic Effect of CNT and Graphene with Iodide by Charge Transfer

The counter electrode of DSSC is a sort of catalyst, which transfer electron to electrolyte by reducing triiodide/iodide couple. Nano-sized Pt is usually used as a counter electrode. However, new catalysts replacing Pt have been developed because Pt is expensive and need high temperature coating process for manufacturing. CNT and graphene is good candidates for this purpose. The superior catalytic properties of CNT and graphene have been recently reported in the field of DSSC and fuel cell (Cha et al. 2010; Seo et al. 2010; Koo et al. 2006; Strelkoa et al. 2000; Maldonado et al. 2006; Shao et al. 2008; Msezane et al. 2010). However, it is not clearly elucidated how CNT and graphene show catalytic effects. It is said that defects of CNT are the source of electrochemical activation in association with Raman spectrum investigation showing high defect peak in good catalytic materials. The evaluation on catalytic effect of CNT and graphene has been theoretically tried using the variation of band gap. In this work DV-Xα method, which has merits for calculations of chemical bonding and charge distribution because of using full potential, is used to calculate the direct charge transfer through interfaces of CNT/iodide and graphene/iodide. To estimate catalytic ability of materials with the charge transfer is more persuasive than with the difference of band gap because the charge transfer gives direct criterion for catalytic activity in DSSC.

To understand catalytic reaction, first of all, it is necessary to know iodide/triiodide redox couple in DSSC. Iodide plays a role to transfer electron from counter electrode to dye molecule adsorbed on nano-TiO_2. The reduction rate and diffusion rate of iodide/triiodide couple at the counter electrode surface are, sometimes, a controlling factor in entire series of reactions due to its slowest process compared to other reactions. Therefore, in this work calculations are focused on the charge transport phenomena between triiodides and the count electrode of CNT and graphite. The conventional charge transport mechanism of iodide system is known as diffusion-adsorption-desorption type. However, recently it is reported by experimental measuring that the real charge transport rate is higher than the values expected by conventional transport mechanism. Moreover, solid electrolyte having very low ion diffusion rate also shows high charge transport rate. Recently, new direct transfer mechanism model without adsorption/desorption was suggested by Grotthuss and in this model electrons transfer by direct hopping from iodide to iodide. Direct observation by experiment is impossible in this theme, and only theoretical calculations can give answers. DV-Xα calculations were used for the affirmation of Grotthuss mechanism and the evaluation on catalytic effect of CNT and graphene by this new transport model.

Figure 9.8 shows calculated the difference charge density map and the net charges of six iodides serially arranged from the surface of single-wall carbon nanotube (SWCNT). The difference charge density map is drawn by calculating charge difference between charge density of free atoms and calculated charge density. Net charge (netc) is the total integrated value of charges involved in each ion.

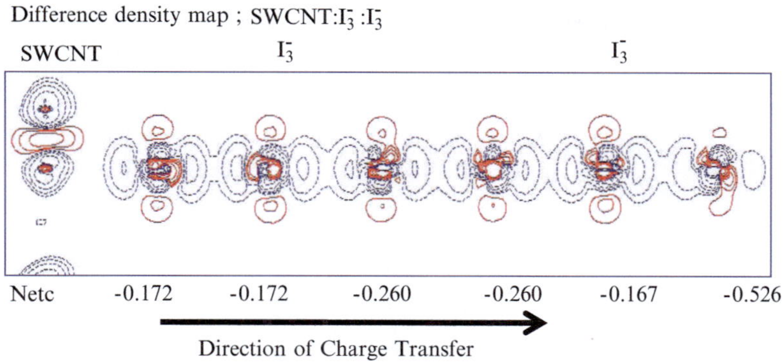

Fig. 9.8 Net charge and difference charge density map of atoms in series of iodide on CNT electrode

Negative value means that corresponding ion receives electrons from other ions. Triiodide has the form of linearly arranged three iodine atoms apart from 3 Å each other and has total charge of one electron negative. In this figure, netc of negative charges increases with distance from counter electrode surface to the end of iodide serial and the far most iodide from surface has highest negative net charge. From this result it is possible to consider that charges can transfer through iodide links from counter electrode to iodide, and then from iodide near surface to outer iodide. The iodides, which lie far from electrode surface, can obtain extra negative charges without any process of adsorption on the electrode surface and diffusion through electrolyte. It is expected that this fact can be accepted as the theoretical verification of Grotthuss type charge transport mechanism.

Although triiodides adsorbed on counter electrode as Fig. 9.8, actual redox reaction proceeds like following:

$$I_3^- \rightarrow 2I^0 + I^- \rightarrow 2I^0 + I^- + 2e^- \rightarrow 3I^-$$

In this process it is considerable that the ion directly adsorbed onto surface of electrode is only I^0. Therefore, in this work calculations were executed with models of CNT/I^0 and graphene/I^0 in various circumstances.

Figure 9.9 shows the models of graphene used for calculations. All models have hydrogen ending in all edge atoms to eliminate the distortion of charge density in edge. Models having different sizes such as $C_{37}H_{15}$, $C_{73}H_{21}$, $C_{12}1H_{27}$ and $C_{150}H_{32}$, were tested for calculations to decide optimal size of model and obtain energy level structure of pure graphene. The calculated energy level diagrams are shown in Fig. 9.10. The band gap between the highest occupied molecular orbital (HOMO) and the lowest unoccupied molecular orbital (LUMO) decreases from 1.73 to 0.53 eV with number of carbon atoms involved in models. Although it is expected that the decrease of the band gap with increasing model size will continue and the exactness of calculation will also increase with model size, it was not possible to make final conclusion because of calculation limit related to computer program.

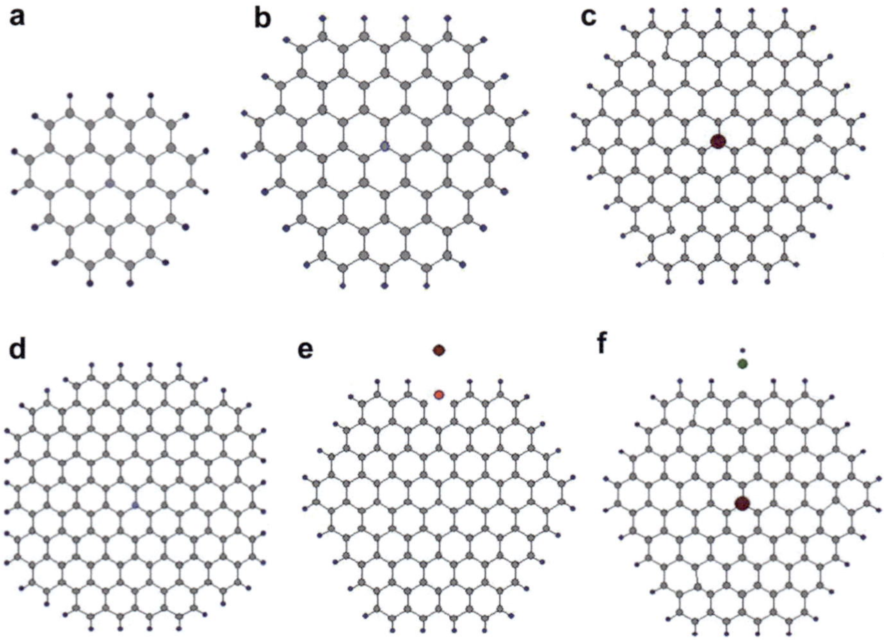

Fig. 9.9 The models for calculations of the graphene and graphene/I^0. (**a**) $C_{37}H_{15}$, (**b**) $C_{73}H_{21}$, (**c**) $C_{121}H_{27}$-I^0, (**d**) C_{150}H32, (**e**) $C_{120}H_{27}$-O-I^0, (**f**) $C_{121}H_{27}$-$(OH)^-$-I^0. Large circle in the center is iodide, middle size is carbon and smallest sized circle is hydrogen

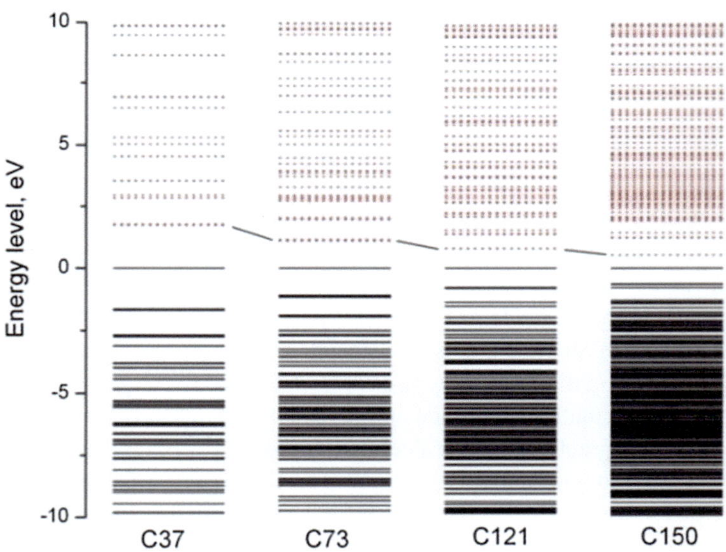

Fig. 9.10 The energy level diagram of graphene calculated with models of various sizes. All models have hydrogen ending in all edge carbon atoms

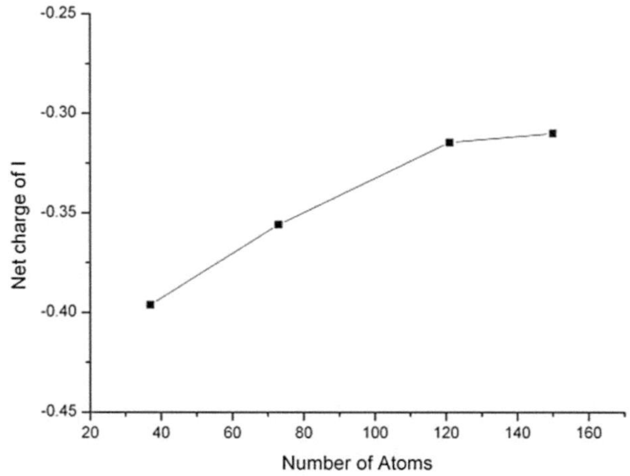

Fig. 9.11 Net charge transfer from graphene to iodide with number of carbon atoms involved in model. Netc is saturated above 120 carbon atoms

Charge transfer from graphene to I^0, which appears as net charge (netc) of iodide, can be estimated as a measure of catalytic effect. It can be accepted as more reasonable evaluation than the lowering of band gap used in other calculation methods. The change of netc with model size is shown in Fig. 9.11. In this calculations, distance between carbon and iodide was 3.5 Å, giving highest net charge transfer. It can be known that netc of iodide is saturated in models composed from above 121 carbon atoms. From this result the model with 121 carbon atoms was used as optimum sized model through all calculation for graphene, considering the limit of program.

Similar calculations were executed for selecting optimum sized model of CNT. The type of CNT used in this work was SWCNT (6,0) having small curvature enough to distinguish from graphene having infinite curvature. Figure 9.12 shows the selected optimum model of $C_{120}H_{12}$, which contains nearly same number of carbon atoms. The optimum distance between carbon and I^0 was decided as 3.1 Å, which was shorter than that of graphene.

Net charge transfer appears as netc of iodide in this calculation. The netc of iodide in the interface of graphene/I^0 or CNT/I^0 varies with distance between carbon and iodide, as shown in Fig. 9.13. Curve of netc has negative maximum with distance. The charge transfer in negative maximum from carbon to iodide in CNT/I is bigger than that of graphene. From this result it is expected that CNT having small curvature has more catalytic effect than graphene. However, this expectation should not be directly accepted for the superiority of CNT to graphene, because it comes from calculations with models, which have no defect and impurities. The final conclusion about the ability of catalyst should be decided after calculations with models having defect and impurities because real materials have enough density of such imperfect structures. Examples of such calculation were tried in this work using models shown in Fig. 9.9e and f and Fig. 9.12c. The netc changes by the existence of defects and adsorbed foreign atoms or groups. Some results were shown in Table 9.2, showing comparison between the pure model

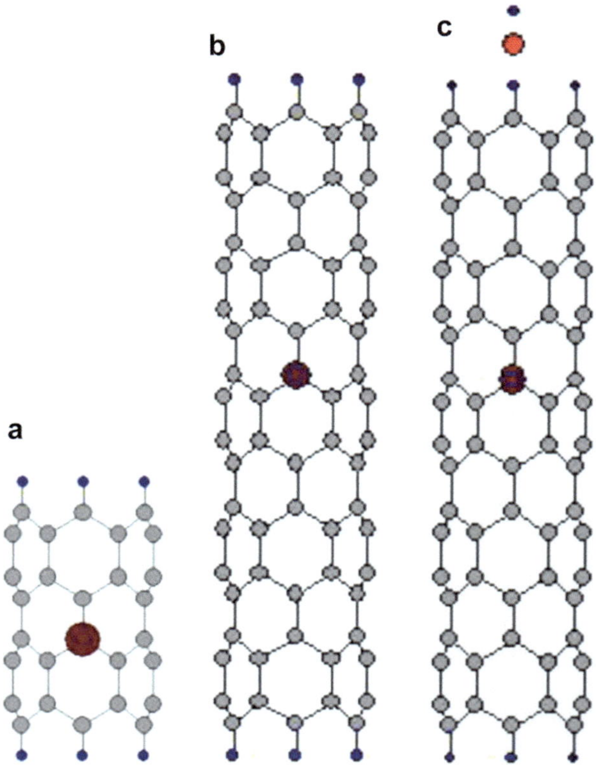

Fig. 9.12 Models used for the calculation of CNT. (**a**) $C_{48}H_{12}$-I^0, (**b**) $C_{120}H_{12}$-I^0 and (**c**) $C_{120}H_{12}$-OH-I^0 (for adsorption of hydroxyl group) Large circle in the center is iodide, middle size is carbon and smallest sized circle is hydrogen

without a defect or a dopant and models with OH^- adsorption. In this table, adsorption of hydroxyl group at the end of graphene or CNT changes the netc of iodide by noticeable amount, compared to that of pure state without defects. In the both case of CNT and graphene, the net charges transferred to iodide decrease with the existence of OH^- group. The catalytic effect of CNT is also superior to that of graphene with reduced values similar to pure models. From this it is expected that treatments introducing hydroxyl group to CNT and graphene will be harmful for advancing catalytic effect of these materials. Like this example, sometimes, the degree of charge transfer severely changes with defects such as vacancy and disordering, impurities such as oxygen doping and metallic substitution, or adsorbed ion groups such as carboxyl group and hydroxyl group. It is expected that it is possible to find and design good catalytic carbon-based materials by the DV-Xα method using proper models in various situations. However, results obtained in this study need to be modified and corrected by further calculations using more accurate models because of disregarding solvent molecules, which solvate iodide and electrode surface.

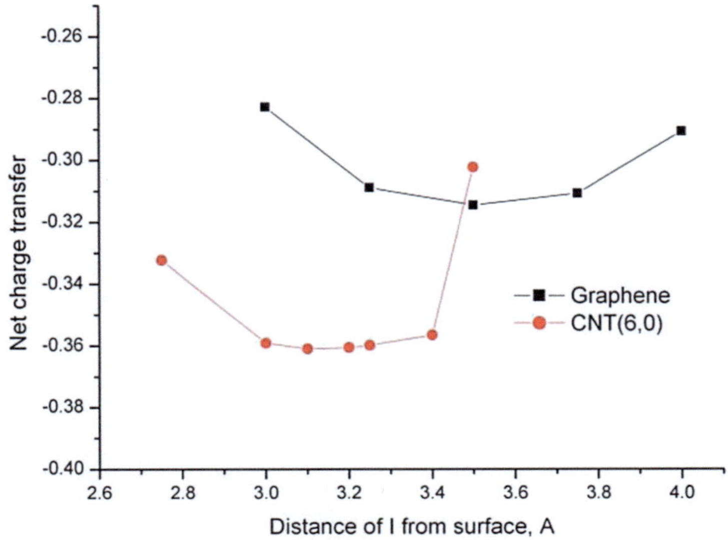

Fig. 9.13 The change of netc of iodide in interface of grapheme/I and CNT/I. CNT has more negative charge than graphene. This means CNT can be more catalytic than graphene without defect or impurities

Table 9.2 Comparison of calculated netc between models with no defects and models with OH^- group

	Model	No defect	Adsorption of OH^-
Graphene	$C_{121}H_{27}\text{-}I^0$	−0.315	–
	$C_{121}H_{27}\text{-}(OH)^-\text{-}I^0$	–	−0.258
SWCNT (6,0)	$C_{120}H_{12}\text{-}I^0$	−0.361	–
	$C_{120}H_{12}\text{-}OH^-\text{-}I^0$	–	−0.311

9.5 Design of Optical Down-Convertor Based on ZnO

ZnO is a sort of oxide semiconductor, having the wide band gap of 3.0–3.5 eV, and widely used as transparent conducting oxide of solar cell and other optical purposes. ZnO has not enough conductivity in natural state for device applications, but can be turned to comparatively high conducting n-type semiconductor by doping of Al, Ga or In (Lee et al. 2005). Optically ZnO can be used as an absorbent of ultraviolet ray because of absorbing the ray of 325 nm wavelength. If absorbing ultraviolet ray can be converted to visible ray, DSSC can use this converted ray for photovoltaic generation. It is possible to change the absorbing or evolving wavelength of ZnO by doping. In this work it was tried to convert 325 nm ultraviolet ray to ray in visible area by doping of Y.

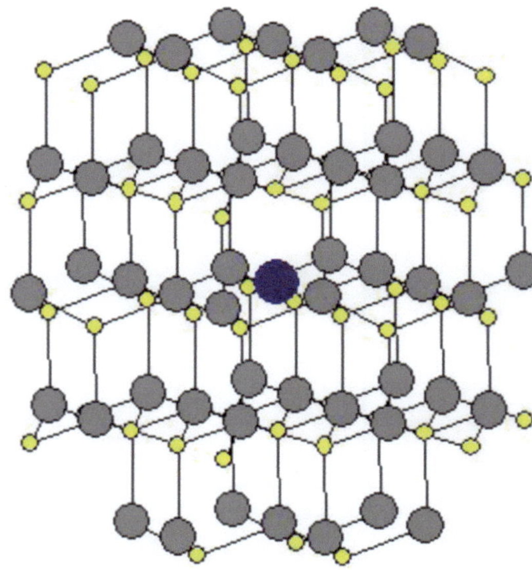

Fig. 9.14 The model for calculations of ZnO doped with Y, $[YZn_{50}O_{53}]^{4-}$. Y appears as largest ball in the center of the model. Zn has middle size and O is smallest balls

The model used for calculations of DV-Xα method is $(YZn_{50}O_{53})^{4-}$, as shown in Fig. 9.14. Y appears in center of model. The Madelung field over extended region of $3 \times 3 \times 3$ unit cells was applied to embed the model.

The energy level diagram, the density of state and overlap population of pure ZnO without Y doping were calculated, as shown in Fig. 9.15. The calculated band gap of pure ZnO appears as 3.29 eV, which is similar to expected value and guarantees reliability of the used model and the calculation method. Orbital related to electric conductivity and donar level is Zn 4s with anti-bonding component. When III family elements such as Al, Ga and In is doped, s orbital in the outer shell of doping elements replaces Zn 4s and is mixed with O 2p orbital. This mixing orbital decreases the band gap and increases electrical conductivity (Lee et al. 2005). In the case of Y doping Y 5s is mixed into the site of Zn 4s and makes mixed levels of Y 5s – O 2p, as shown in Fig. 9.16. This new formed level is occupied with electrons and can play a role of trapping level for optical transition. The energy difference between Y-doping level and HOMO was calculated as 2.95 eV. When rays having photon energy of above 3.29 eV enter into ZnO, electrons in HOMO level are excited to LUMO level. This excited electron lowers its energy to Y-doped level, and then recombines to a hole in HOMO level. By this recombination, visible ray corresponding to 2.95 eV evolves. By this process ultraviolet ray converts to visible ray in ZnO:Y.

This calculation result was experimentally testified. Thin film of ZnO doped with Y was manufactured by the spin coating. Down-converting phenomena were checked by measuring fluorescent spectrum by monochromator. In this experiment it was observed that incident wavelength of 295 nm was converted to 422 nm wavelength, which was very close to 415 nm wavelength of 2.95 eV (Kim et al. 2006).

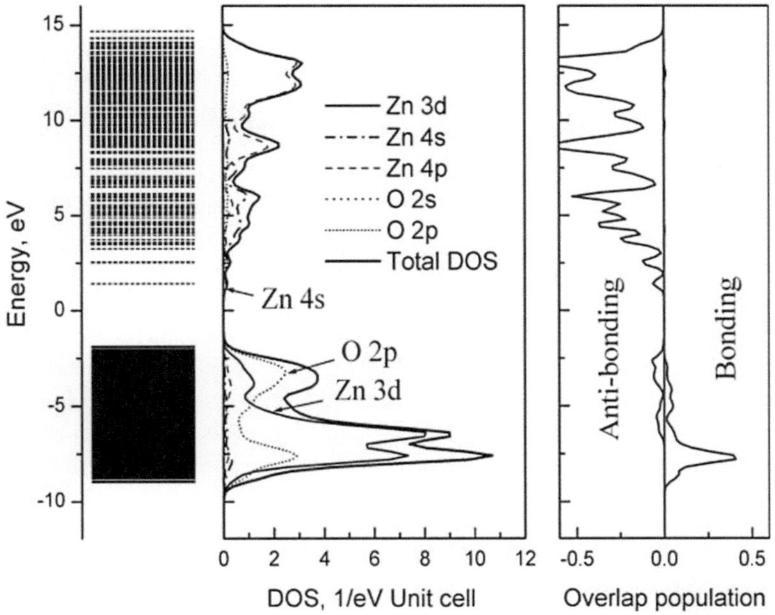

Fig. 9.15 The energy level diagram, the density of state and the overlap population of pure ZnO without doping. *Dotted line* in Energy level diagram means unoccupied level and *solid line* means occupied level. ZnO has not donar level in pure state

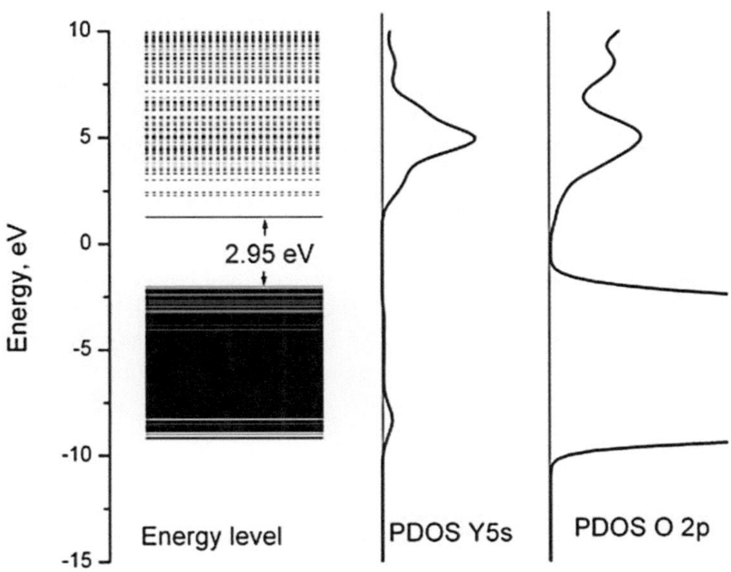

Fig. 9.16 The energy level diagram and the partial density of state of pure ZnO doped with Y. *Dotted line* in Energy level diagram means unoccupied level and *solid line* means occupied level. Y doping makes the electron occupied donar level at LUMO position of ZnO

Conclusions

The first principles calculations using DV-Xα method were executed to understand the surface electronic structure and estimate the charge transfer through interfaces for the purpose of designing new materials for DSSC.

The electronic structure of TiO_2, which is related to open voltage of DSSC, was calculated with the $(Ti_{15}O_{56})^{52-}$ model for the calculations of the bulk state and the $(Ti_{11}O_{34})^{24-}$-O^a model under the Madelung field having the size of 6 × 6 × 6 unit cells. By using small models, it can be identified that strong covalent bonding is formed between Ti and adsorbed oxygen and this strong interaction makes the surface donar levels below the conduction band. The effect of surface doping in TiO_2 nano electrode has been investigated using the $(MTi_{10}O_{34})^{24-}$-O^a model. The calculated energy level diagrams with doping of transient metal showed that all dopant levels were found between the conduction band and the valence band of TiO_2 and the energy of theses levels decreased with atomic number of doping element toward the valence band. Fe^{3+}, Co^{3+} and Ni^{2+} doping brought doping level to nearly valence band. It is expected that this lowering of dopant levels decreases open voltage in DSSC by decreasing surface levels. The doping of IV family elements, of which effect was in debate, also brought similar results to the case of transient metal doping, although the doping levels were found in center position of band gap.

The catalytic effect of CNT and graphene as counter electrode of DSSC was estimated by the degree of charge transfer between electrode and iodide. The optimum model was selected as $C_{121}H_{27}$ by calculations with models with different sizes. The calculated optimum distance between graphene and iodide was 3.5 Å giving the highest net charge transfer. In the case of SWCNT (6,0), the calculated optimum model was $C_{120}H_{12}$ and the optimum distance of iodide from CNT was decided as 3.1 Å, which was shorter than that of graphene due to its small curvature. The calculated charge transfer from carbon to iodide in CNT/I is bigger than that of graphene in negative maximum. From this result it is expected that CNT having small curvature has more catalytic effect than graphene. Defects and adsorbed ionic groups can greatly change the catalytic effect. The netc changes with the existence of defects and adsorbed foreign atoms or groups. It was calculated that adsorption of hydroxyl group decreased charge transfer and brought a little harmful effect as catalyst.

The calculation on ZnO doped with yttrium was executed to develop new down-converting material for DSSC. The model used for calculations of DV-Xα method was $(YZn_{50}O_{53})^{4-}$ with the Madelung field over extended region of 3 × 3 × 3 unit cells. Doping of Y formed an electron occupied dopant level in the position of 2.95 eV from LUMO, which is lower than HOMO level of pure ZnO. It was expected that the existence of this level

(continued)

would convert rays above 3.29 eV to visible ray of 2.95 eV, which is corresponding to 415 nm wavelength. This expectation was testified by an experiment, which showed that ultraviolet rays were converted to 422 nm visible rays similar to expectation by the calculation.

Acknowledgement This work was supported by the "New & Renewable Energy Core Technology Program" of the Korea Institute of Energy Technology Evaluation and Planning (KETEP) grant funded by the Ministry of Trade, Industry & Energy, Republic of Korea (No. 2011 T100100678).

References

Adachi H, Tsukada M, Satoko C (1978) Discrete variational Xα cluster calculations. I. Application to metal clusters. J Phys Soc Jpn 45:875

Arakawa H (2001) Recent advances in research and development for dye-sensitized solar cell. Siemusi Publisher, Tokyo, p 54

Cha SI, Koo BK, Seo SH, Lee DY (2010) Pt-free transparent counter electrodes for dye-sensitized solar cells prepared from carbon nanotube micro-balls. J Mater Chem 20:659–662

Chiba Y, Islam A, Watanabe Y, Komiya R, Koide N, Han L (2006) Dye-sensitized solar cells with conversion efficiency of 11.1 %, Japanese. J Appl Phys 45(25):L638–L640

Ellis DE, Adachi H, Averill FW (1976) Molecular cluster theory for chemisorption of first row atoms on nickel (100) surfaces. Surf Sci 58:497

Grätzel M (2000) Perspectives for dye-sensitized nanocrystalline solar cells. Prog Photovolt Rev Appl 8:171

Hoshikawa T, Ikebe T, Yamada M, Kikuchi R, Eguchi K (2006) Preparation of silica-modified TiO_2 and application to dye-sensitized solar cells. J Photochem Photobiol A Chem 184:78–85

Kasowcki RV, Tait RH (1979) Theoretical electronic properties of TiO_2 (rutile) (001) and (110) surfaces. Phys Rev B 20(12):5168–5177

Kim HJ, Lee DY, Song JS (2006) Characterization of yttrium doped zinc oxide thin films fabricated by spin-coating method. J Korean Elec Electron Mater Eng 19(5):457–463

Ko KH, Lee YC, Jung YJ (2005) Enhanced efficiency of dye-sensitized TiO_2 solar cells (DSSC) by doping of metal ions. J Colloid nd Interf Sci 283:482–487

Koo BK, Lee DY, Kim HJ, Lee WJ, Song JS, Kim HJ (2006) Seasoning effect of dye-sensitized solar cells with different counter electrodes. J Electroceram 17:77–80

Lee DY, Lee WJ, Song JS, Koh JH, Kim YS (2004) Electronic surface state of TiO_2 electrode doped with transition metals, studied with cluster model and DV-Xα method. Comput Mater Sci 30:383–388

Lee DY, Lee WJ, Min BK, Kim IS, Song JS, Kim YS (2005) Electronic state of ZnO doped with elements of IIIB family, calculated by density functional theory. J Korean Elect Electron Mater Eng 18(7):589–593

Maldonado S, Morin S, Stevenson KJ (2006) Structure, composition, and chemical reactivity of carbon nanotubes by selective nitrogen doping. Carbon 44:1429–1437

Msezane A, Felfli Z, Sokolovski D (2010) Novel mechanism for nanoscale catalysis. J Phys B At Mol Opt Phys 43:201001–201006

Park KH, Gu HB, Jin EM, Dhayal M (2010) Using hybrid silica-conjugated TiO_2 nanostructures to enhance the efficiency of dye-sensitized solar cells. Electrochim Acta 55:5499–5505

Ramamoorthy M, King-Smith RD, Vanderbilt D (1994) Defects on TiO_2 (110) surfaces. Phys Rev B 49(11):7709–7715
Seo SH, Kim SY, Koo BK, Cha SI, Lee DY (2010) Influence of electrolyte composition on the photovoltaic performance and stability of dye-sensitized solar cells with multiwalled carbon nanotube catalysts. Langmuir 26(12):10341–10346
Shao Y, Sui J, Yin G, Gao Y (2008) Nitrogen-doped carbon nanostructures and their composites as catalytic materials for proton exchange membrane fuel cell. Appl Catal B Environ 79:89–99
Strelkoa VV, Kutsa VS, Thrower PA (2000) On the mechanism of possible influence of heteroatoms of nitrogen, boron and phosphorus in a carbon matrix on the catalytic activity of carbons in electron transfer reactions. Carbon 38:1499–1524
Wijayarathna TRCK, Aponsu GMLP, Ariyasinghe YPYP, Premalal EVA, Kumara GKR, Tennakone K (2008) A high efficiency indoline-sensitized solar cell based on a nanocrystalline TiO_2 surface doped with copper. Nanotechnology 19:485703–485707

Chapter 10
Microscopic Approach to Water by Using the DV-Xα Method, and Some Innovative Applications

Sunao Sugihara, Takashi Igarashi, Chikashi Suzuki, and Kenji Hatanaka

10.1 Introduction

Although water is for living organisms, the basic sciences of water is not completely understood. Water-molecules are considered to be bound together by hydrogen bonds, what then happens when these hydrogen bonds are broken? We will discuss the effects of breaking of hydrogen bonds in water and the ways that we can utilize the resulting energy of water with broken hydrogen bonds.

First, let us examine the scientific background. The crystal structure, density, and structure of water at various pressures have been discussed by Mishima and Stanley (1998) and by Elington and Dehencedetti (2001), who reported that water becomes more disordered when compressed, and that directional attractions (hydrogen bonds) combine with short-range repulsions to determine the relative orientation of neighboring molecules. Novoa et al. (1997) showed that $H_3O_2^-$and H_3O^+ are involved in a proton-transfer mechanism with a free-energy of 0.23×10^{-23} to 0.48×10^{-23} kJ in the THz region.

There have been many studies on water from the structural point of view and several studies on functional waters have been reported in biological fields. For example, the function of intraprotein water in the membrane protein bacteriorhodspin has been

S. Sugihara (✉)
Kanagawa University, Rokkaku bashi, Kanagawa ku 221-8686, Yokohama, Japan

Sugihara Institute of Science and Technology, Kamariya nishi, Kanazawa ku 236-0046, Yokohama, Japan
e-mail: natsuyama41@gmail.com

T. Igarashi • C. Suzuki
Nuclear Science and Engineering Directorate, Japan Atomic Energy Agency, Ibaraki-ken 319-1195, Japan

K. Hatanaka
MCM Co. Ltd., Uehonmachi, Tennoji Ku 543-0001, Osaka, Japan

T. Ishii et al. (eds.), *The DV-Xα Molecular-Orbital Calculation Method*,
DOI 10.1007/978-3-319-11185-8_10

studied (Garczarek and Gerwert 2006), and the impeding mechanisms of the H_3O^+ ion in relation to water in aquaporins in cell membranes has been examined (Yasui and Mastushita 2010).

However, these previous structural studies do not deal with interactions with other substances, as occur in the case of MICA (***Mi***nimal ***Ca***talyst) water and MICA energy (see Sect. 10.3). One interesting recent study was conducted by Lin et al. (2008) who examined the terahertz vibration–rotation–tunneling (VRT) spectroscopy of the deuterated water tetramer. The made a combined analysis of the vibrational bands at 4.1 and 2.0 THz, and they discussed their results for the dynamics of the water tetramer in relation to those of the water trimer, and they also attempted to quantify the many-body forces in the condensed phases of water. Furthermore, this group reported a 'first principles' potential energy surface for liquid water deduced from the VRT spectroscopy of water clusters (Goldman et al. 2005).

There are many reports on applications of X-ray absorption spectroscopy (XAS) in studies on the structure of liquid water and ice. For example, Leetmaa et al. (2010) used an approach that involved the development of a calculation method. Studies by soft X-ray spectroscopy in conjunction with extended X-ray absorption fine structure (EXAFS) analysis and molecular-dynamics simulations (Tokushima et al. 2008) showed the existence of mixed structures of water and ice as a result of the formation of tetrahedral crystal and strongly distorted hydrogen-bonded species; however, the discussion was limited to structures arising from hydrogen bond networks. Furthermore, Leetmaa et al. (2010) performed a theoretical approximation to XAS of liquid water and ice, and they discussed the absolute energy scale in vibrational modes by comparison of the results of periodic calculations with those of experiments. Some interesting results were introduced by their consideration of the full-core with an excited electron in the lowest state.

Fourier-transform IR (FTIR) spectroscopic studies relate to terahertz analysis in the mid-IR region. Millo et al. (2005) have shown that hydrogen-broken bonds generate at least two kinds of water species: molecules with broken O–H bonds and molecules with broken lone-pair electrons. A spectral analysis of water in the mid-IR region (360–4,000 cm^{-1}, equivalent to 11–120 THz) showed that the enthalpy difference for the hydrogen bond obtained through OH-stretching analysis corresponds with the value calculated from bending data (Freda et al. 2005). Several recent studies have been conducted by using instrumentation with nonlinear optics. THz waves were generated by stimulated polariton scattering through phonon–polariton interaction in a lithium niobate ($LiNbO_3$) crystal as a result of the action of the laser (Kawase et al. 1996). Parametric oscillation (0.3–2.5 THz) has also been generated by using a $LiNbO_3$ crystal (Kojima et al. 2003).

In 1932, a theoretical study of a quantum mechanics treatment of the water molecules was reported (Coolidge 1932) in which mathematical development of electron reactions involving three centers is described. The binding energy was calculated to be 3.5 V, although the actual energy is about 10 V; this discrepancy was not resolved. It is difficult to obtain the binding energy of water by the first-principles method or in a mathematical manner even when the oxygen atom is introduced, although taking the ionic state into account reduces the discrepancy arising from the omission of any wave function relating to ionic characteristics. In relation to the energy of the hydrogen bond, Scott and Vanderkooi (2010) discussed the distance potential energy surface of the water dimer. The theoretical calculations showed O–O separation in the range 0.25–0.40 nm have a smaller effect on the system energy and charge than do changes in the hydrogen-bond angles of the water dimer. Adachi (1993) studied the electronic structures of the single water molecule and pentamer in terms of charge transfer and contour maps of the wave function.

Another interesting macroscopic phenomenon relating to water was reported by Fuchs et al. (2007, 2008), who examined the dynamics of floating water under high dc voltages of up to 25 kV, where a bridge is formed in a similar manner to that present in electrohydrodynamic sprays or jets, although the mechanisms have not been clearly elucidated Woisetschläger et al. (2010).

As well as water, it is interesting to examine the dinitrogen molecule (N_2) in relation to activated water. Discussions of "active nitrogen" go back to 1911 when Shrutt described a chemically active modification of nitrogen in afterglow gas that he referred to as "active nitrogen" (for details, see Wright and Winkler 1968). This is now recognized as an emission of the first positive band system, $N_2(B\,^3\Pi_g) \rightarrow N_2(A\,^3\Sigma_u{}^+)$, of the nitrogen spectrum, and other band systems also contribute to this emission, together with some shorter-lived nitrogen afterglows that corresponds to other forms of "active nitrogen". Active nitrogen is generally produced by an electric discharge through nitrogen, and many methods have been reviewed (Wright and Winkler 1968). However, these forms of active nitrogen have a high energy (up to 9.764 eV), which is referred to as the dissociation energy of the ground state of nitrogen, with a corresponding uncertainty in the heat of formation of the $N(^4S)$ atom. Our "active nitrogen" has an energy of less than 1.5 eV and will be further discussed in the Results and the Discussion sections.

In the ground state, the electronic structure of the nitrogen atom is $1s^22s^22p^3$, with three $2p$ electrons distributed among the p_x, p_y, and p_z orbitals with various spins. A forbidden absorption-band system of N_2 in the vacuum-ultraviolet region has been discussed by Tanaka et al. (1964) in relation to the basis state at 8.3 eV. Nakamatsu et al. (1993) examined X-ray absorption near-edge structure of N_2 by using the DV-Xα molecular orbital method to show the shape of the resonance for the N_2 molecule and to explain electron scattering by the attractive potential of the molecule. A molecule correlation diagram for nitrogen has been calculated by means of self-consisting field molecular orbital (SCF MO) calculations using 20-term basis set of Slate-type functions by Mulliken (1972b); as result, the equivalent distance was calculated to be 1.06 Å. Note that Mullikan studied the atomic orbitals of dissociating N_2 states at high energy levels, such as $3\sigma_u$ (several eV).

Here, we will discuss hydrogen bonding in minimal catalyst water, and the effects caused by water with broken hydrogen bonds at the microscopic level, as well as at the macroscopic level in daily life. Minimal catalyst water (MICA) as named by Hatanaka (1990–1993), is obtained by modifying ordinary water by a special process that involves high-pressure treatment. MICA water can be produced from tap water or any other type of water by high-pressure treatment (0.5–3 MPa). During this process, hydrogen bonds between the water molecules gradually break, so that MICA contains free electrons that are considered to be in a plasma-like state in the water molecule (Sugihara 2009). Materials activated by MICA energy can behave as catalysts by transferring the energy associated with the broken hydrogen bonds.

We are interested in the physical and chemical behaviors of electrons in water after breaking of hydrogen bonds, which we refer to as activation of water; even an understanding of biological processes requires a knowledge of the electronic structures of the water molecule. This is why we refer to nitrogen, which can be energized by activated water. Our basic hypothesis is that activated water emits electromagnetic waves in the region from the infrared through to the terahertz region, although we have not yet been able to identify these directly. Because electrons must play an important role after breaking of hydrogen bonds in MICA, we must study details of the electronic behavior of the water molecule. For this purpose, we carried out the first-principles calculation to identify the electronic structure of the water molecule. We also attempted to deduce the behaviors of electrons and electromagnetic waves in water and in MICA-energized substances by these methods. We used sophisticated techniques such as nuclear magnetic resonance (NMR) spectroscopy and equipment such as the superconducting quantum interference device (SQUID) to examine the behavior of the electrons and nuclei. We also used terahertz and Fourier-transform infrared (FTIR) spectroscopy to examine the electromagnetic waves, and we performed isotopic analyses of ^{1}H and ^{2}H in nuclei of water. Analyses of macroscopic properties of water, such as its dielectric properties, can show how water is changed by activation with terahertz radiation.

To calculate the energy levels in the electronic structures of two water molecules, we used the discrete variational Xα potential (DV-Xα) method Adachi (1993). MICA water can activate or provide energy to other substances such as solids, liquids, or gases through electromagnetic waves, and it can transport this energy to other substances through the functions of its electrons. One important type of activation is that of nitrogen gas in the air or of nitrogen atoms in organic substances by means of MICA energy. We calculated the energy levels of N_2 molecule, and we used the results to elucidate the mechanism for the activation of N_2 by MICA energy; however this mechanism has not been fully clarified.

We propose a concept involving electromagnetic waves that is new to the chemistry and physics of water, and we describe several devices that we have developed as applications of the basic concept of MICA water and its energy. We also propose that polyethylene film processed by MICA energy is useful for keeping foods fresh, and the mechanism by which it operates involves activation of N_2 in the air. Furthermore, the ranges of uses of substances activated by MICA energy can be expanded to a variety of fields in a number of industries.

10.2 Calculations and Processes

10.2.1 Materials

MICA water was produced by passing tap water (or any other type of water) at 80–90 *L*/m through basalt or magnetic with natural magnetism at a pressure of 0.5–3.0 MPa to cause a burst detonation at the surface of the water. Furthermore, here is an example to show how another substance can be energized by water.; a raw materials such as pellets of acrylic- styrene resin were energized by immersion in water processed as described above, then the pellets were injection molded to form the pot, and any kinds of materials and/or substances can be available to treat for the energy from the processed water, for instances, such as solid; metal, ceramics, glass, organic substance as well as liquid.

10.2.2 Computational Methods

The DV-Xα method was used to calculate the electronic structures of two water molecules. The basis functions were O*1*s-*3*s and H*1*s-*2*s and the number of sampling points was 30,000. The symmetry was C_{2v}, and one analysis was performed for a change of overlap population (BOP) by expansion of the O-H bond, and another was performed for a change in BOP by excitation of electrons after breaking of the hydrogen bond. The function of the number of excited electrons was also analyzed. The energy level was determined by applying structural optimization. The total energy was also calculated by total-energy spin DV-Xα (TESDA), as developed by Nakagawa (2002, 2003). Furthermore, we also calculated the energy-level structures of the N_2 molecule and as a function of the N–N distance for the case when one electron is excited. The value of the well potential α was 0.7 and the number of sample points was 20,000; the basic functions were *1s–4p*.

10.2.3 Spectroscopic Analysis and Measurements

10.2.3.1 Terahertz Spectroscopy and FTIR

The terahertz spectrum of water was measured at 1–20 THz by using a crystal of MgO: $LiNbO_3$. We also measured dielectric constant of water and some other substances by means of terahertz time-domain spectroscopy (TDS). Furthermore, we used the lowest-energy regions of FTIR spectroscopy (spectra recorded on an FT/IR-6100 spectrometer JASCO, Tokyo) for terahertz spectrum analysis in the region below about 600 cm^{-1} (20 THz).

To study the hydrogen nucleus, we used NMR spectroscopy (JNM-ECA 600 MHz: JEOL Ltd., Tokyo) on water, oil (^{1}H), and several organic nitrogen compounds (^{15}N). We also performed an isotopic analysis of MICA water and we compared the results with those for other samples of commercial waters. The activated substances were studied by FTIR spectroscopy in the terahertz region, and in the case of water and plastics by terahertz spectroscopy.

In studies of the effects, MICA activation on car exhaust gases, a sample of copper-plated plastic (CPP) and a fluorescent lamp were activated by energy-transfer from MICA water, and measurements were performed with a F16P5 gas analyzer (Ferrer, Wallingford, CA). Furthermore, one person recorded data on fuel consumption during 3 years when he drove a Toyota car designed for use with hi-octane gasoline but actually fueled with regular gasoline. The deodorization equipment, comprising an activated ceramic honeycomb, was introduced in a room containing the experimental equipment.

10.2.3.2 NMR Spectroscopy and SQUID

NMR spectroscopy (JNM-ECA 600 MHz; JEOL Ltd., Tokyo) was used to record the shape of the spectrum and the energy shift. The external magnetic field applied was 14.1 Tesla (10^4 Gauss).

The resonance frequency (of the order of kHz) in ^{1}H, ^{13}C and ^{15}N were observed on each sample. For example, the ^{1}H nucleus can absorb at 600 MHz and ^{15}N absorbs at 60 MHz in accordance with the following equation; $2\pi\nu_N = \gamma_N H_0$ (ν_N: inherent resonance frequency of nucleus, γ_N: gyromagnetic ratio. H_0: external magnetic field).

The SQUID (Superconducting Quantum Interference Device) can detect a small magnetic fields of the order of 10^{-8} to 10^{-9} times the strength of the geomagnetic field (10^{-4} T) at the temperature of liquid helium. Because water itself could not be measured at these temperature, a glass was activated with the processed water (activated water) and nonactivated glass was used as a control for the magnetization analysis. The equipment was operated at the external magnetism up to 50 kOe (5 T), and the magnetization (emu) is illustrated along with the field in the form of the H–M curve, which shows the magnetic hysteresis.

10.3 Results and Discussions

10.3.1 Basic Aspects of the Electrical Properties of Water

In terms of the electrical resistivity of water, it is known that theoretical maximum value is approximately $1.8 \times 10^5\,\Omega \cdot m$ at 25 °C, and important feature of water is its polar nature. Generally chemical properties of water are discussed in terms of ions such as $H_3O_2^-$ and H_3O^+, as shown in Fig. 10.1. We can also measure the dielectric properties of water. Because oxygen has a higher electronegativity than hydrogen,

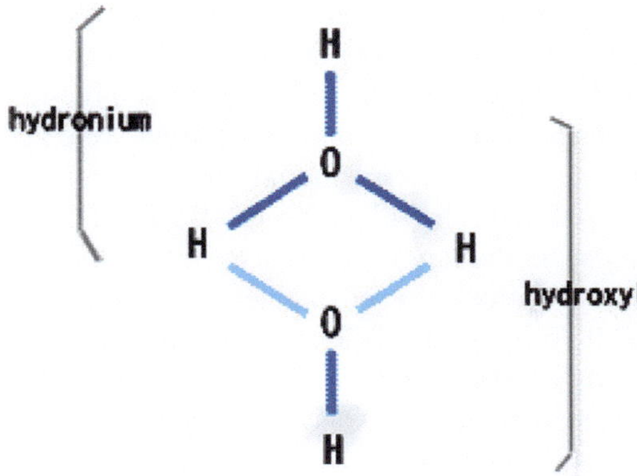

Fig. 10.1 General view of a bonded water dimer. Water is generally considered to constitute large molecule with many hydrogen bonds

the region of the water molecule where the oxygen atom is situated carries a partial negative charge. This results in the formation of a dipole, and as a result, the formation of a hydrogen bond.

The hydrogen bond is a relatively week bond compared with the covalent bonds that exist within the water molecule itself. The bond strength is 0.05–0.3 eV, depending on the temperature, pressure, magnetic field, and environment, and is characterized by the local dielectric constant, as discussed below. This type of bond occurs both in inorganic molecules, such as water, and in organic molecules, such as DNA. Many of the characteristic properties of water, such as its triple point, compressibility, electrolysis properties, solvent power, cleaning, power, and the ability to form emulsions, arise from the presence of hydrogen bonds. Many researchers have performed structural examinations of water [for example, Scott and Vanderkooi (2010)]. The XAS spectroscopy of water and ice was examined by Cavelleri et al. (2002) who produced some interesting results on electron transfer in terms of contour plots of typical excited orbitals in the dimer-D molecular orbital.

We now propose that effects of broken hydrogen bonds in water should be discussed precisely. The concept of "electromagnetic waves" in relation to MICA water involves the region from the near-infrared (10^{14} Hz) through the terahertz region (10^{12} Hz); in the latter case the wavelength is of the order of micrometers. Electromagnetic waves in the terahertz region have wavelengths that are between those of light and those of radio waves. This energy can resonate with various substances; for instance, the signal from the hydrogen bond of the carboxyl radical (COOH) has a wavelength of 3×10^{-3} m, that from the rotation of the water molecules has a wavelength of 2×10^{-4} m, and that associated with the angle between the atoms in the amino radical NH_2• has a wavelength of 3×10^{-6} m. These energies of radicals and molecules can therefore resonant with MICA energy in the region from the near-infrared to the terahertz regions.

As discussed earlier, we can cause hydrogen bonds between molecules to break by applying special processes, and we are interested in the physical and chemical behaviors of electrons after breakage of the hydrogen bonds (which we call "activation" of water).

Our hypothesis is that activated water (MICA energy), as described above, emits electromagnetic waves in the region from the infrared through terahertz, although we have not been able to observe this directly. The electrons freed by breaking of hydrogen bonds must play an important role in this process, and the energy must be transferred to other materials by radiation and/or conduction. As a basic idea, we postulate that a coding of “0” and “1”, in a similar manner to information, might be the key to the transfer of “information” and that the electrons possess physical properties other than “some form of energy”. We therefore performed NMR, SQUID, and FTIR analyses by using an isotopic approach with ^{1}H and ^{2}H for electrons and nuclei interacting with electromagnetic waves.

Our main interest was to know which electrons produced the observed effects in terms of quantum chemistry and electromagnetic properties and how they did so.

10.3.1.1 Computational Results

In considering the breaking the hydrogen bonds, we performed computational calculations to examine the excitation of electrons in a pair of water molecule (the water dimer). Figure 10.1 illustrates a general view of bonded water in a computational analysis in which two O–H groups are regarded as forming a hydrogen bond. In Fig. 10.2, the hydrogen bonds are shown by the dotted lines. The left-hand figure shows a plot of the bond-overlap population (BOP) versus the change in distance between the molecules Δx (Å) in the ground state. Even after breaking hydrogen bonds, the lowest energy in BOP can be calculated, leading to be stable state, meaning that the smaller the BOP, the more stable the molecule is.

The right-hand figure shows a plot of the BOP versus the number of excited electrons in hydrogen bonds. The degree of covalency of the hydrogen bonds is less

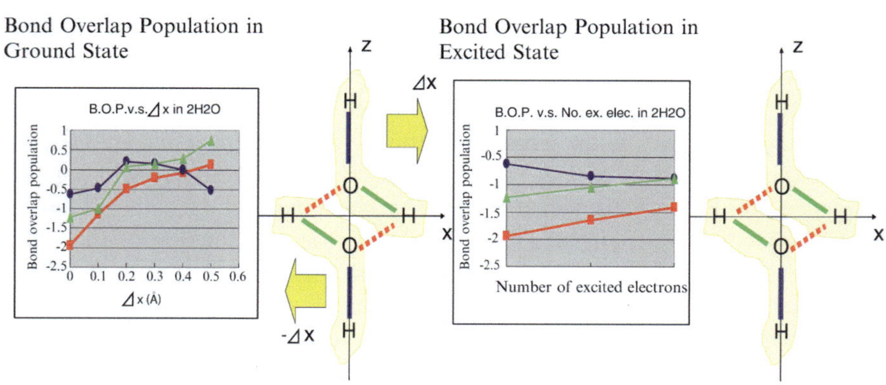

Fig. 10.2 *Left*: Bond overlap population (BOP) versus the change in distance along the x-axis of the ground state. *Right*: BOP vs. the number of electrons in the excited state for two molecules of water with two broken hydrogen bonds. The BOP represents (*vertical axis*) represents the degree of covalency of the bond between atoms. *Red dotted lines* (*solid in graph*) indicate hydrogen bonding, and *green* and *blue* ones indicate another O-H bonds

than that of other O–H bonds. Furthermore, it is interesting to note that, on increasing the number of excited electrons, the Coulombic repulsion becomes smaller, and the molecular structure of water results in a lower energy state.

Adachi (1993) examined electronic structures of the water pentamer [$(H_2O)_5$] and single water molecule, H_2O by using DV-Xα method, and he found that the levels of molecular orbitals and the contour map of H_2O molecules predict that charge transfer occurs from the hydrogen atom to the oxygen atom according to the wave functions; The orbital that overlaps functions with the φ_+^{1s} orbital of hydrogen is the $2p_z$, of O2p atomic orbital, and the bonding molecular orbital φ_{a1} (a_1; irreducible representation of the symmetry of the orbital in group theory) is formed by this interaction; the φ_{a1}^* anti-bonding orbital is also formed. Furthermore, the orbital that interacts with φ_-^{1s} is the $O2p_x$ orbital, and the $O2p_y$ does not interact with nonbonding H_2 molecular orbital according to Adachi's explanation. Addition to them, Fig. 10.3 shows the molecular orbital levels (1), the $1b_2$:*yz* plane and

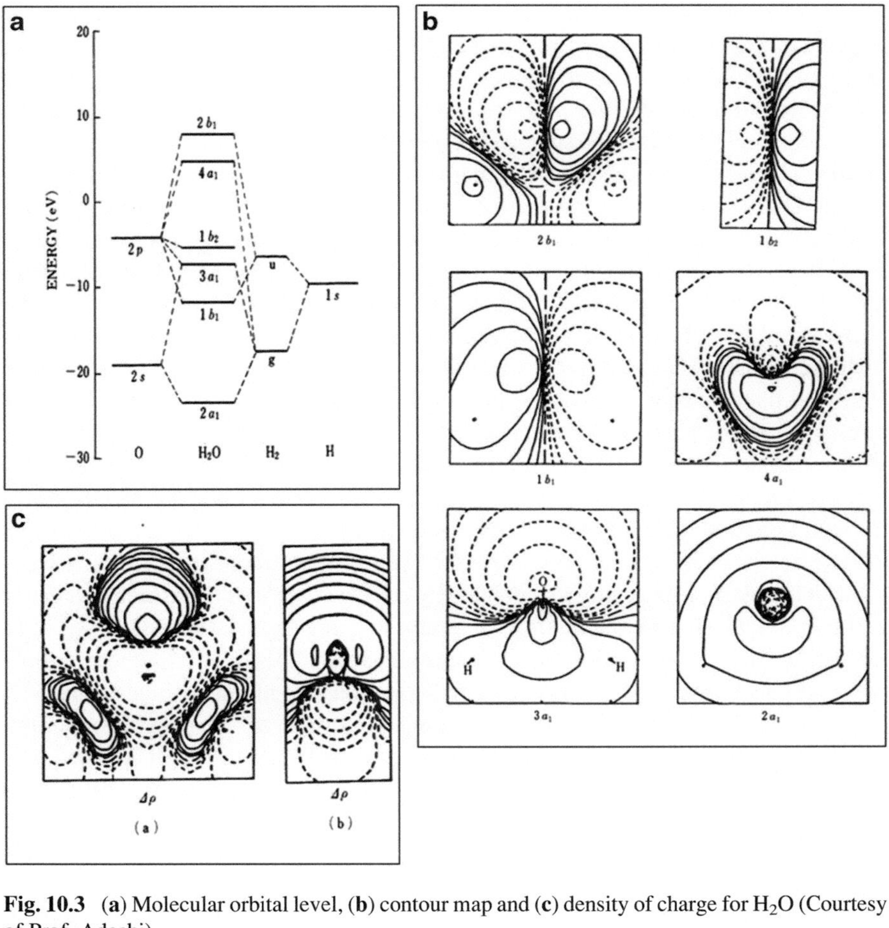

Fig. 10.3 (**a**) Molecular orbital level, (**b**) contour map and (**c**) density of charge for H_2O (Courtesy of Prof. Adachi)

others: *xz* planes in contour maps (2), and charge density (3) for H_2O. The differences in charge increases the distance between O and H, and in the opposite space of O and H as illustrated in Fig. 10.3a, b, respectively. The (a) forms a covalent bond and (b) forms due to $1b_2$ electron and $3a_1$ electron in the energy levels as shown in Fig. 10.3(1) in which electrons can be regarded to function with proton as an activated water after braking hydrogen bonding. These models of a water molecule are basically understandable, and the overlapping of hydrogen atom with the $O2p_x$ orbital may give rise to the observed bond angle of 104.5° in the water molecule rather than one of 90°.

10.3.1.2 Dielectric Properties

Two molecules of water have been shown to form hydrogen bonds easily. As illustrated in Fig. 10.1, the hydrogen bonds are broken in the dotted line in two positions. From chemical point of view, H_3O^+ (the hydronium ion) is acidic and is capable of sterilization and removal of active enzymes, whereas $H_3O_2^-$ (in which a hydroxy ion radical is bonded with a molecule of H_2O) possesses cleaning abilities due to its interface activities. In addition, MICA water possesses energy as the result of processing by pressure and minerals. The degree of covalency of the hydrogen bonds is less than that in other O–H bonds. Furthermore, it is interesting to note that with an increasing number of excited electrons, the Coulombic repulsion becomes smaller and the molecular structure of water results in a lower energy state. According to Muro-Oka (1990), charges on oxygen in an O–H hydrogen bond are attracted by hydrogen. From the results of these calculations, we propose the existence of free electrons in a plasma-like state in MICA water containing broken hydrogen bonds. The altered molecular structure of water results in a lower energy state and the electrons still have spin (magnitude of $1\hbar/2$, $\hbar = h/2\,\pi$ (h = Planck's constant). These plasma-like electrons (collective electrons) contain energy that can be transferred as radiation, and when MICA water approaches the surface of a solid or liquid, the atoms of the substance absorb energy from the MICA water through radiation or, if contact occurs by conduction.

The nature of this radiation is discussed below. The frequency of MICA electrons in the plasma-like state is calculated by analogy to that of a plasmon in a metal surface. The plasma frequency in our system is given by the expression $\omega_p^2 \approx 4\pi\, n\, e^2/(m^*\, \varepsilon_\infty)$, and found to be equal to approximately 1 THz (ε_∞ is 4; other symbols are used conventionally: $m^* = m_0$ of a free electron, n = number of electrons).

The electromagnetic energy is the ranges from near infrared through the terahertz region. Essentially, MICA energy is soft, as we will show later. The perturbation of electromagnetic waves of the MICA-electron may initiate the transition. The energy quantum for transition from a level m to a level n can be formulated as follows:

$$\hbar\,\omega_n = E_m - E_n \tag{10.1}$$

Therefore, a transition to a higher level accompanies the absorption of a quantum. Thus, a substance can be energized by MICA electron through radiation (electromagnetic waves) or by conduction.

Recently, on the basis of FTIR spectroscopic studies in the mid-IR region, Millo et al. (2005) showed that breakage of hydrogen bonds generates at least two kinds of water species: molecules with broken O–H bonds and molecules with broken lone-pair electrons. Let us consider a substance to which energy has been transferred from MICA electrons. The real part of the relative permittivity ε_r will therefore change. The dielectric loss (tan δ) is introduced, and the loss factor (ε_r') is given as follows:

$$\varepsilon_r' = \varepsilon_r \tan\delta \tag{10.2}$$

Furthermore, ε_r' and ε_r are often described in terms of a complex number:

$$\varepsilon_r^* = \varepsilon_r - i\,\varepsilon_r' \tag{10.3}$$

where ε_r^* is the complex relative permittivity. The real part ε_r is the relative permittivity (dielectric constant Re) and the second (imaginary) term of the complex value (dielectric constant Im) corresponds to the loss factor. The dielectric parameters of normal mineral water and MICA water have been measured by means of time-domain spectroscopy (TDS).

The fact that there is a decrease in ε_r and /or ε_r^* means that the loss factor and/or the relative permittivity decreases; as a result, the energy of MICA electron will be effectively transmitted to other substances. The results are shown in Fig. 10.4: the values of ε_r (Re) for normal water and MICA water at 1 THz are 4.5 and 4.2, respectively.

Furthermore, the values of ε_r^* (Im) are 2.7 and 2.4 for normal water and MICA, respectively. These facts show that because of its lower dielectric constant ε_r', MICA water is electrically more conductive than conventional mineral water and the lower value of ε_r^* also shows a smaller loss factor or lower energy loss from the medium leading to a greater phase velocity for electrons in MICA than for electrons in normal water. $\boldsymbol{\nu} = c/\sqrt{\boldsymbol{\varepsilon}}$ (c: velocity of light, ε: dielectric constant). The energy of an electron is therefore increased by about 5–6 % as the result of the MICA processes. This has been confirmed by means of Raman spectroscopy of MICA-treated plastic; this showed that the energy was shifted to values about 5 % higher than those in untreated substances. The intensity of the peak was also higher in MICA-treated plastic. The difference in energy between the spectrum of the control sample and that of the activated one was calculated to be about 0.6 meV (0.15 THz), although MICA energy is essentially weak. We measured the frequency dispersion of the reflected electromagnetic wave in the terahertz region as produced by MICA electrons.

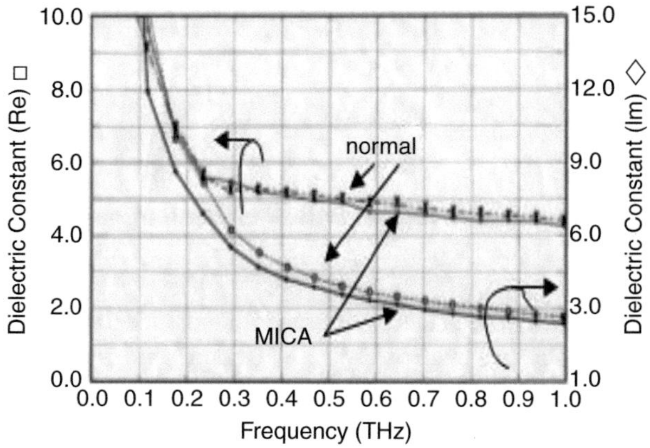

Fig. 10.4 Complex dielectric constants of MICA water and normal mineral water; the upper two curves show the dielectric constant (Re: real part) on the left-hand axis, and lower two curves show the dielectric constant (Im: imaginary part) on the right-hand axis. Re (□) and Im (◇) of MICA water, and Re (×) and Im (○) of normal mineral water. The vertical axis on the left show dielectric constant (Re = real part) at 0.0–10.0, and dielectric constant (Im = imaginary part) at 0.0–5.0 in numeral. The axis of abscissas shows the frequency in numeral of 0.0–1.0

10.3.1.3 NMR Analysis of Water

In NMR analysis, the nuclear spins of ^{1}H in H_2O (^{15}N in nitrogen compounds) change depending on the external magnetic field to present a certain energy difference between the up and the down spin. This can resonate with externally applied radiowave energy, depending on the energy difference (nuclear magnetic resonance). The NMR spectra are shown in Fig. 10.5. The half-width (HW) curves indicate the width of spectrum of the water molecule, showing that postulate that MICA water is more active than the control. Furthermore, we can measure a relaxation time of T_2 (spin–lattice relaxation time), and the half-width value is obtained from the following equation:

$$T_2 = 1/(\pi \cdot \Delta\nu), \qquad \Delta\nu \ = half - width,\ \ Hz \tag{10.4}$$

The larger the value of T_2, the narrower is the width of the spectrum (Δv). The value of T_2 for the control and MICA samples were 0.016 and 0.035 s, respectively. We have therefore shown that molecules of MICA water move more freely due the presence of smaller clusters as a result of breaking of hydrogen bonds.

The narrower half-width is, the lager T_2 is obtained, and the larger relaxation time means longer distance between collisions resulting in a smaller molecule when you take the same volume. For example, the values of T_2 for control water and MICA water are 0.016 and 0.035, respectively. When one regards that water will become smaller size of molecule, it is easier to understand many a macroscopic phenomena.

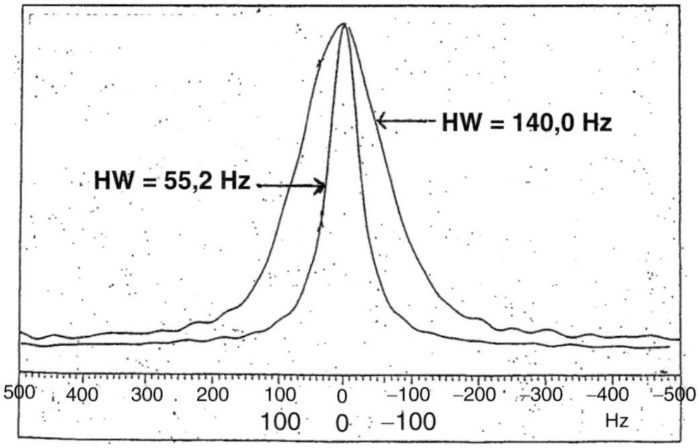

Fig. 10.5 NMR spectra of MICA water and control water showing half-widths ($\Delta\nu$). The half-width (HW) curves indicate the width of spectrum of the water molecule, showing that postulate that MICA water is more active than the control. Furthermore, we can measure a relaxation time of T_2 (spin–lattice relaxation time), and the half-width value is obtained from the following equations: $T_2 = 1/(\pi \cdot \Delta\nu)$, $\Delta\nu = half - width,\ Hz$

10.3.1.4 Isotopic Analysis of Hydrogen

The ratio between ^{2}H and ^{1}H given by the equation;

$$\delta\mathrm{H} = \left({}^2\mathrm{H}/{}^1\mathrm{H}\right)_{\text{sample}}/\left\{\left({}^2\mathrm{H}/{}^1\mathrm{H}\right)_{\text{standard}} - 1\right\} > \times 1000 \tag{10.5}$$

δ is defined by this formula as the ratio of deuterium and hydrogen in the sample compared with that in the standard (i.e., the control). The amount of deuterium in natural water is only 0.015 % and the extra neutron present in ^{2}H cannot disappear after the process of activation of water by MICA. One electron is freed after breaking of the hydrogen bond from a water molecule, thereby increasing the negative value of δH.

$$\begin{aligned} \delta H &= \left\{\left({}^2H/{}^1H\right)_{sample}/\left({}^2H/{}^1H\right)_{standard} - 1\right\} - 1 \\ &= \{(2n+e)/n\ /(2n+e)/(n+e) - 1\} < 0 \end{aligned} \tag{10.6}$$

The value of δH for activated water (MICA water) has a value of −65.21, whereas that for Tokyo tap water is −58.71, as shown in Table 10.1. Water from mountain sources shows higher values of δH.

10.3.1.5 THz and FTIR Spectrum of Water

Terahertz wave (THz) is usually defined to be electromagnetic wave of 30 μm to 1 mm, thus it exists between radio wave to infrared radiation and it is transparent to plastic, paper, gum, woods, tooth, born and dry foods.

Table 10.1 Isotope analysis of water which shows a relatively high value of $\delta^{18}O$ and δ^2H (the definitions are described in Sect. 10.3.1.4 of the text) (After private communication by Institute of Isotope Analysis in Yokohama)

Name of water	$\delta^{18}O$	δ^2H	Place
MICA water	−10.79	−65.21	Wakayama (Japan)
Tap water	−8.58	−58.72	Tokyo (Japan)
Glacier cap	−19.85	−141.69	Alaska (USA)
Crystal geyser	−15.32	−109.71	Wiede (USA)
Whistler	−17.00	−120.02	Canada
Hachiman daira	−11.75	−72.3	Iwate (Japan)
Fukusui	−9.56	−60.58	Tenshin (China)
Premium soda	−9.46	−49.99	Yamazaki (Japan)
Nagara river	−9.17	−51.60	Gifu (Japan)
Hiden tenryo water	−8.74	−54.52	Ooita (Japan)

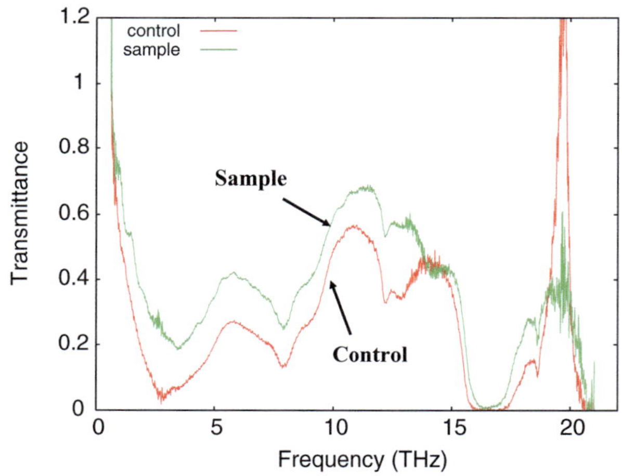

Fig. 10.6 THz transmittance of MICA water and untreated water (Courtesy Dr. Ohno in Riken at Sendai (2007))

THz wave are generally absorbed by water, but MICA water is slightly more transparent than control water, because of its slightly higher energy compared with the control, as shown in Fig. 10.6; here, the vertical axis shows the transmittance and the horizontal axis shows the terahertz frequency.

Terahertz waves are generally absorbed by water, but MICA water is slightly more transparent than control so that MICA (sample) seem to transmit straight.

One application of terahertz waves is in medical treatment (especially of organs at shallower depths in the body) and they can take the place of X-rays in inspecting the body. On the other hand, polyethylene is generally transparent to terahertz waves as previously noted, whereas the transmitted peaks become lower in the energized polyethylene; in other words, the energized film absorbs terahertz waves (Fig. 10.7).

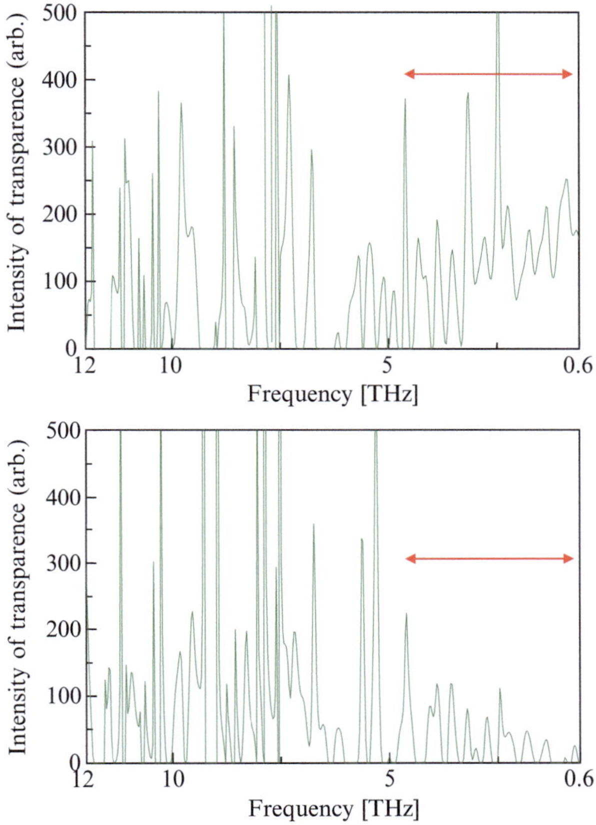

Fig. 10.7 FTIR spectra in the THz region of the untreated polyethylene film (thickness 0.02 mm) (upper spectrum) and MICA-activated polyethylene film (lower spectrum). Activated film absorbs more as compared with control film

In the figure for FTIR spectroscopy, the vertical axis represents the arbitrary intensity of transmittance and the horizontal axis indicates the frequency in the terahertz regions. The polyethylene was energized when it was in pellet form by exposure to MICA water, and it was then formed into a film by injection molding. Thus the film can be energized by the energy transmission which will be discussed in an application in Sect. 10.3.3.3.

10.3.1.6 Magnetic Properties of Glass Energized by MICA Energy

Very low levels of magnetism can be detected by SQUID. Water itself cannot analyzed and a captioned specimen should be activated by the MICA water since the SQUID can only work under liquid He condition. A common glass was activated by MICA process and the treated sample was compared with an untreated control. The atoms in the glass were effected at a very low level in high external magnetic fields of up to 50 kOe (5 T). The electrons in the glass activated by MICA process will increase the momentum (J) resulting in larger magnetic moment (μ_B) according to the equation; $\mu_B = \gamma J$, where γ means gyromagnetic ration.

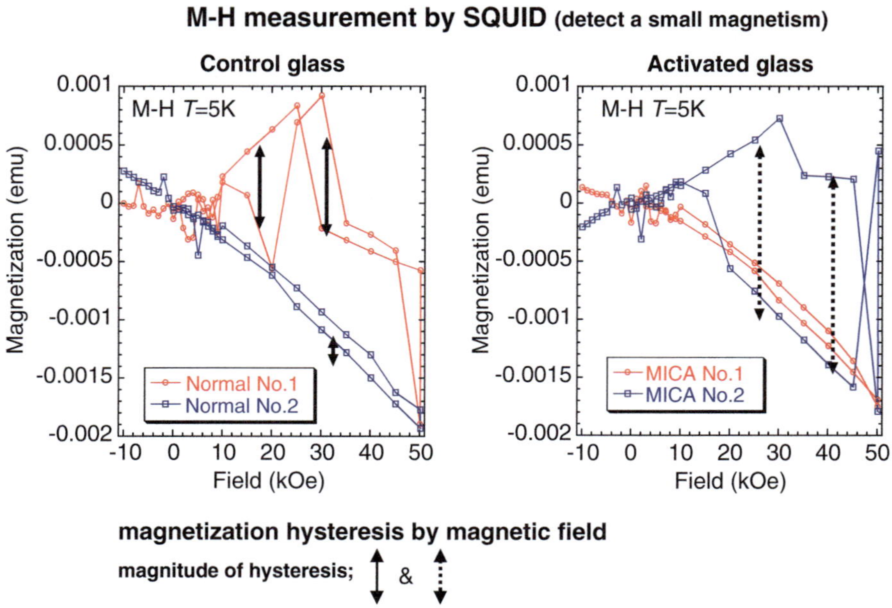

Fig. 10.8 Hysteresis in magnetization of MICA-treated glass and control glass at 5 K and magnetic field is up to 5 T (50,000 G). Glass possesses usually diamagnetism showing magnetic susceptibility of ~ -10^{-5} emu. Electron's spin is affected by magnetic field and residual magnetization appears when the field goes down, consequently magnetic moment due to spin increases to resulting in paramagnetism (+ ~ 10^{-6} emu)

The results are shown in Fig. 10.8 in which indicate a hysteresis between increasing magnetic field and decreasing it. The borosilicate glass was subjected to MICA treatment and the treated sample was compared with a control sample. The untreated glass was normally diamagnetic. On the first treatment, it appeared to be magnetized, but no hysteresis appeared on the second occasion. However, this phenomenon was opposite to the result for MICA glass, where the glass showed no magnetization on the first exposure to a magnetic field and a large magnetization on the second exposure. This is an interesting result. We believe that MICA glass has a certain energy that cancels low magnetization in the first measurement, because the electron spinning with a certain energy in the MICA glass resonate with the external magnetic field so that, subsequently, a larger magnetization appears in the second measurement. Furthermore, magnetic moment due to spin increases to resulting in paramagnetism (+~10^{-6} emu) when the magnetic field goes down.

10.3.2 Activation of Nitrogen

10.3.2.1 Computational Results for Dinitrogen (N_2)

Previously, Mulliken (1972a) have reported calculations on the ground state of the nitrogen molecule by SCF MO computation using a 20-term basis set of Slater-type functions in which the internuclear distance R from 0.01 au to the equilibrium distance R_e (=2.0134 au) in 17 stages distances. Although the scenario is not still well understood, we can see new aspects of MICA water as the catalyst through calculations on N_2 activation. By using the total energy spin DV-Xα program, we calculated the total energies of dinitrogen in its activated and nonactivated states by considering the spin of the molecule as a function of the length of the N–N bond. These calculations showed that, for certain bond lengths, activated nitrogen can exist in stable forms that lose energy slowly. Activated nitrogen should emit radiation below the frequency of infrared rays, and is stabilized on lowering of the energy, as the gap changes from 3 to 1.5 eV, which corresponds to that of infrared rays.

The results of our calculations are shown in Fig. 10.9, in which the total energy (eV) is plotted against the N–N distance (Å). When one electron is excited, the total energy level and the N–N distance change. The energy level of the conduction (unoccupied) level becomes lower by an amount that depends on the change in the N–N distance, and the potential of the free electron changes as a result. The value a in N(a↓)–N(a↑) describes a continuous change in spin from 0 to 3. We identified three metastable states at various N–N distances. The energy of the nonmagnetic

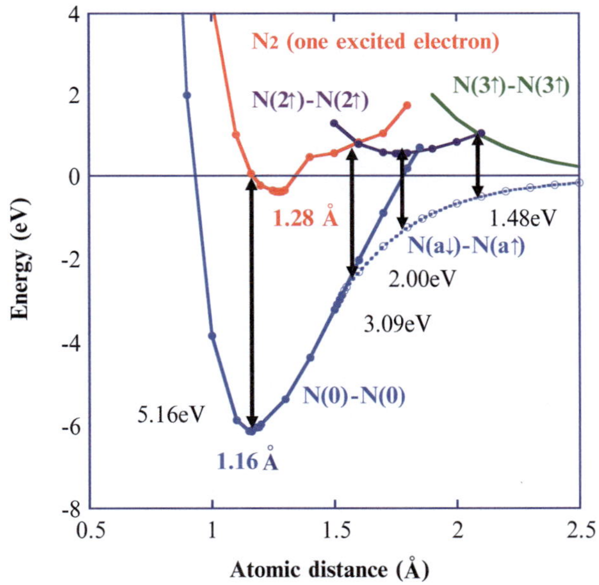

Fig. 10.9 Plots of the total energy versus atomic distance showing, how the emitted energy changes with the N-N distance (The *arrows* indicate the difference in energy from the ground state)
Interatomic distance 1.28 Å: 5.16 eV[N(0)-N(0) and N_2^* (one excited electron)]
Interatomic distance 1.60 Å: 3.09 eV [N(a↓)-N(a↑) and N(2↑)-N(2↑)]
N_2 curve (*in red*) crosses with N(2↑)-N(2↑) *purple color*
Interatomic distance 1.75 Å: 2.00 eV [N(a↓)-N(a↑) and N(2↑)-N(2↑)]
Minimum point of N(2↑)-N(2↑)
Interatomic distance 2.10 Å: 1.48 eV [N(a↓)-N(a↑) and N(3↑)-N(3↑)]
N(2↑)-N(2↑) crosses with N(3↑)-N(3↑) in *green color*

ground state N(0)–N(0) is a minimum at a bond distance of 1.16 Å, which is 7 % more than the experimental value (1.0998 Å) (Kagakubenran 1993), and the difference in the energy between

N(0)–N(0) and N_2^* (with one excited electron) is 5.16 eV. At a larger distance, the curve for one-electron-excited N_2 crosses that of the N(2↑)–N(2↑) at 1.60 Å, where the difference of the ground state is 3.09 eV. The state N(2↑)–N(2↑) has a minimum energy at a bond distance of 1.75 Å as a result of lowering of the energy; the difference between the ground state of N(a↓)–N(a↑) and the minimum state of N (2↑)–N(2↑) is then 2.0 eV. Finally, the curve for N(2↑)–N(2↑) shows a larger bond distance and crosses the curve N(3↑)–N(3↑) at 2.1 Å, where the difference in the energy is 1.48 eV. The N(3↑)–N(3↑) state can exist around 2.5 Å, emitting a frequency corresponding to the terahertz region.

The X-ray absorption near-edge structure (XANES) spectrum of N_2 has been reported (Nakamatsu et al. 1993) and, a strong peak appears at an energy level above that of the vacuum level around the absorption edge in the XANES. It has been suggested by Nakamatsu et al. (1993) that the excited electron is scattered by the potential energy and that a quasi-stable energy level exists. Therefore, as our calculation also shows, the excited electron may participate in a certain meta-stable state.

The absorption edge seems to appear at around 5 eV in the N ***K*** X-ray absorption spectrum of N_2 (Bianconi et al. 1978). According to our calculations, when one electron is excited, the lowest energy level at a bond distance of 1.28 Å corresponds to the difference of the energy of 5.16 eV from the ground state. Because the difference of the energy roughly describes the absorption edge, this result coincides with our calculation of the dependence of the energy level on the N–N distance. Furthermore, the theories considered are postulated for a dissociation energy of ground-state molecular nitrogen of 9.76 eV, at which Wright and Winkler (1968) state that the corresponding high association energy for ground-state nitrogen atoms makes it unnecessary to postulate any recombination of electronically excited atoms, and the primary process may be the recombination of $N(^4S)$ atoms along a $^5\Sigma_g^+$ potential energy curve. There is an interesting description by Pauling and Wilson (1950) who calculated that the bonding energy of N_2 is 110.8 kcal/mole (4.8 eV) from the figures for the bonding energy of N–N [38.4 kcal/mol (1.7 eV)] and the energy for the process $2\ N \rightarrow N_2$ [226.0 kcal/mol, (9.8 eV)].

The changes in the energy levels of the highest occupied molecular orbital (HOMO) and the lowest unoccupied molecular orbital (LUMO) with changes in the N–N distance are plotted in Fig. 10.10. As the N–N distance increases and the potential of the free electron changes, the energy of the conduction level (LUMO) increases and that of the HOMO decreases, approaching that of the LUMO.

Figure 10.11 shows the N_2 molecular contour map (L) and the formation of the N_2 molecule orbital (R), as calculated by Adachi (1993). In terms of energy levels, the symmetrical orbital is designated g (gerade) and u (ungerade) for unsymmetrical orbital. The orbital, σ_g is a bonding orbital of a σ bond and σ_u is an anti-bonding orbital (these have the opposite meanings for π-bonds). When there is a small

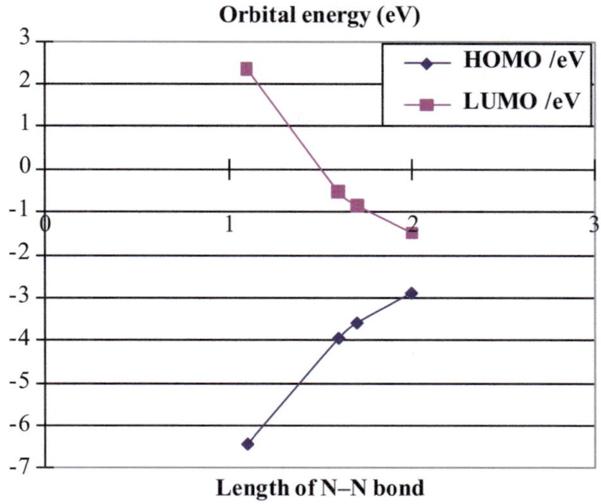

Fig. 10.10 Plots of the energy of the HOMO and LUMO versus the N-N distance (Å)

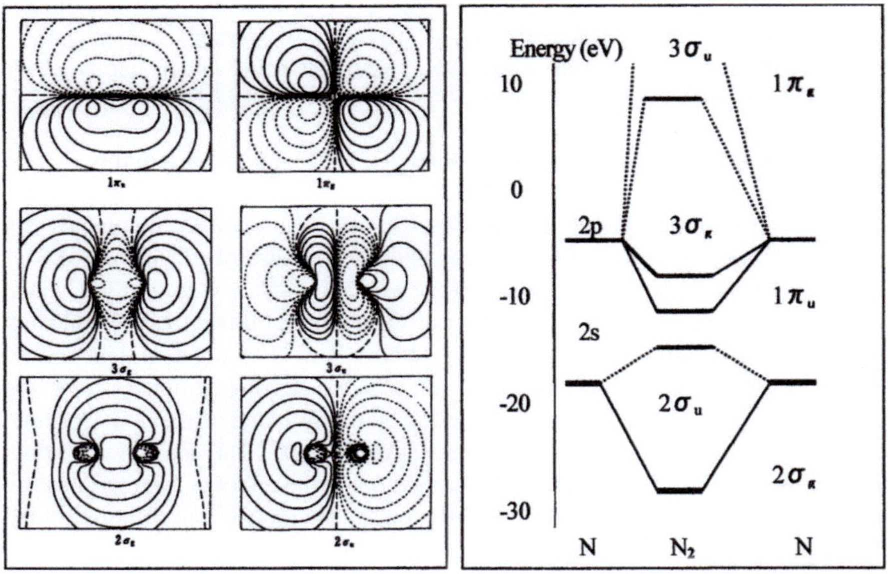

Fig. 10.11 N_2 molecule contour map (L) and formation of N_2 molecule orbital (R) as discussed in Sect. 10.3.2.1 of Computational results for dinitrogen (Courtesy of Prof. Adachi)

difference in energy level between 2s and 2p, the two orbitals hybridize resulting in an interaction between orbitals of the same symmetry. What the $2\sigma_g$ orbital is a bonding orbital, the $2\sigma_u$ orbital is an anti-bonding orbital between N 2s in the N_2 molecule, as can be seen in the contour maps on the left-hand side of Fig. 10.11.

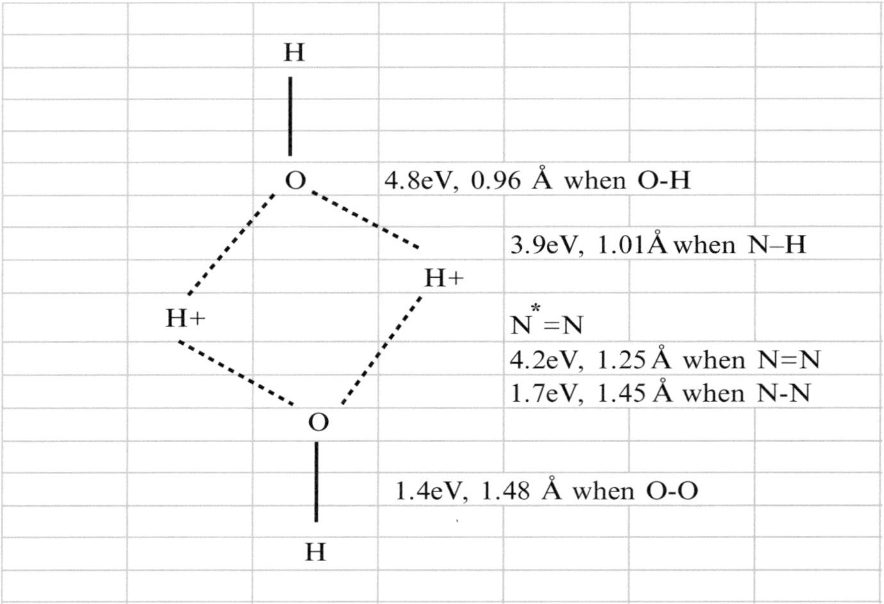

Fig. 10.12 Model for the attack by –N* = N on an O–H$^+$ bond (*dotted line*: hydrogen bond) The values of energy of the covalent bond and the bond lengths are taken from Emsley (1998). The energy of the covalent bond is indicated by converting from kJ/mol into eV

10.3.2.2 Possibility of Activation of Nitrogen in Organic Compounds

Before describing the functions of activated nitrogen in organic compounds, we will look at the interaction between it and hydrogen bond-broken water molecules (Fig. 10.12). The figure shows a model for attack by –N*=N on an O–H$^+$ bond (dotted line: hydrogen bond). The values of the energies of the covalent bond and the bond lengths are taken from Emsley (1998). The N–H bond is short with a relatively weak energy, and this must assist in transferring protons in cells, as discussed below. Wernet et al. (2004) examined the structure of the first coordination shell in liquid water by XAS Auger spectroscopy and X-ray Raman scattering, showing that the spectra analyses were carried out to the effect of NH_3 termination after saturating the dangling O-H by an NH_3 molecule and lone pair electrons of the N oriented toward the free O-H.

Three nitrogen-containing organic compounds were studied by ^{15}N and ^{1}H NMR spectroscopy: aniline ($PhNH_2$), benzonitrile (PhCN), and glycine ($H_2NCH_2CO_2H$). The compounds were processed by MICA energy, and we compared the results to those for control compounds. We observed that ^{15}N chemical shifts approach the standard value more closely in the MICA-treated samples, as shown in Table 10.2, although the differences are relatively small, which means that nitrogen in the MICA sample closes to the standard leading to resonate aptly. In relation to ^{1}H (Table 10.3), the shift is less than that of nitrogen, but the electron affinity of the CN group of benzonitrile becomes weaker, resulting in generation of the electron

Table 10.2 Chemical shift of ^{15}N in organic compounds on NMR spectroscopy
They are aniline ($PhNH_2$), benzonitril (PhCN) and glycine ($H_2NCH_2CO_2H$) processed by MICA energy. ^{15}N chemical shifts approach the standard value more closely in the MICA-treated samples, which means that nitrogen in the MICA sample closes to the standard leading to resonate aptly

	Chemical shift (δ, ppm) of ^{15}N			
			Difference from standard	
Substance	Control	MICA processed	Control	MICA
Benzonitrile (PhCN)	257.882	257.846	8.882	8.846
Aniline ($PhNH_2$)	98.146	98.366	150.854	150.634
Glycine ($H_2NCH_2CO_2H$)	3.435	3.466	1.285	1.254

Table 10.3 Chemical shift of ^{1}H in organic compounds on NMR spectroscopy
Aniline ($PhNH_2$), benzonitril (PhCN) and glycine ($H_2NCH_2CO_2H$) were processed by MICA energy, and compared to the energy for control compounds. The electron affinity of the CN group of benzonitrile becomes weaker, resulting in generation of the electron density of hydrogen in the benzene ring hereafter it can likely be resonated. On the other hand, the = C-N group in aniline has a lower electrons from benzene ring which means less resonance. resonance of amino group with the benzene ring in aniline due to a near strength of covalent bond, and $2p^2$ of lone pair can move on πorbital

	Chemical shift (δ, ppm) of ^{1}H			
			Difference from standard	
Substances (amino acid)	Control	MICA processed	Control	MICA
Benzonitrile (PhCN)	7.628~7.429	7.607~7.412	5.698~5.499	5.677~5.482
Aniline ($PhNH_2$)	4.109	4.149	2.179	2.219
Glycine ($H_2NCH_2CO_2H$)	–	–	–	–

density of hydrogen in the benzene ring hereafter it can likely be resonated. On the other hand, the = C-N group in aniline has a lower electrons from benzene ring which means less resonance. On the other hand, resonance of amino group with the benzene ring in aniline due to a near strength of covalent bond, and $2p^2$ of lone pair can move on π orbital. For better understanding these, we illustrate relation between transferring of electrons and strength of covalent bond, by considering electron affinity, too, as shown in Fig. 10.13. Furthermore, Fig. 10.14 shows that the NMR peaks of MICA-treated glycine were sharper than those of the control, meaning that the molecule (amino radical) can have larger momentum being active by MICA energy as discussed in later; a mode of vibration known as inversion, which plays the role of one of active nitrogen.

If we look at cell in organs, these contain properties of the aquaporin family (AQP) which are membrane-protein (Kozono et al. 2002), that consist of a series of amino acids, such as aspartic acid, praline and alanine. Their size is approximately a few Å at the narrowest point (the pore) and they are several Angstroms long. An interesting property of aquaporins is their permeability to a water and glycerol without any ion permeability. Furthermore, water molecules can be transport through the pores of aquaporin at a rate of 3×10^9 molecules per second. AQP1

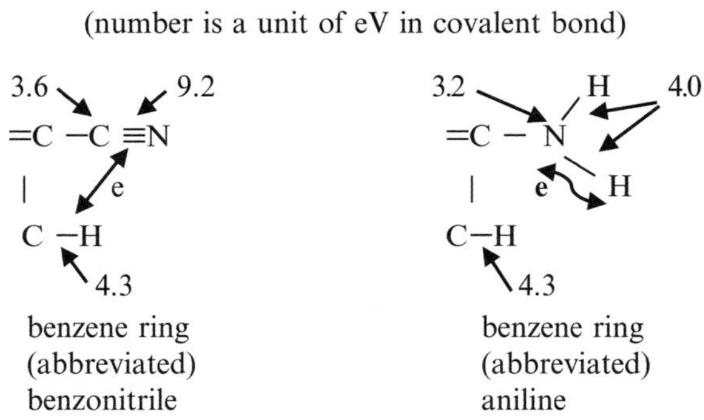

Fig. 10.13 Electron charge transfer in organic compounds containing nitrogen
Electron affinity; N (3.04) > C (2.55) > H (2.20), and electron can transfer toward nitrogen. Electron density must be large around group-C≡N, and electron movement in hydrogen between benzene ring and-C≡N, whereas this movement is less in aniline

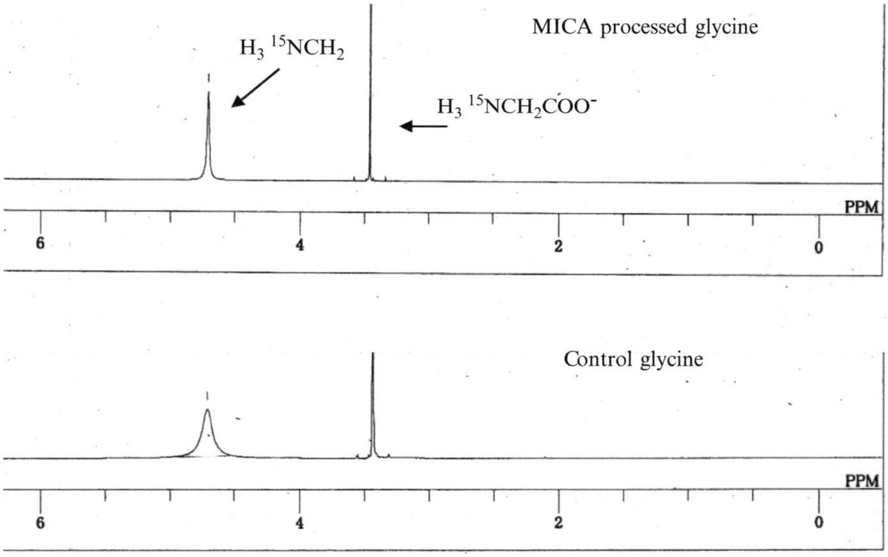

Fig. 10.14 NMR for glycine
The NMR peaks of MICA-treated glycine were sharper than those the control. That means the amino radical can have longer momentum being active by MICA energy as discussed in Sect. 10.3.2.2 (Possibility of activation of nitrogen in organic compounds) in detail

exists as a tetramer, each unit containing its own aqueous pore, and the selective mechanism for permeation by water through the channel pores is reported to be proton blockage, although the mechanisms of this permeability has not yet been fully clarified (Yasui 2004; Yasui and Mastushita 2010).

The nitrogen in the amino acid might play an important role in gating of water in the pore, where N–H (bond length of 1.01 Å and energy 3.9 eV, as described previously) could interact with water with broken hydrogen bonds, although the angles of –N–H or $–NH_2$ are also important for reaction or permeation. The bond angles vary depending on the groups attached to the nitrogen atom, and "three-covalent nitrogen" denotes a pyramidal R_3N compound of the kind NRR′R″, which has a mode of vibration known as inversion (Cotton and Wilkinson 1977). A nitrogen atom oscillating through the plane of the three R groups can turn inside out, and the lone pair is involved in hydrogen bond to other molecules. For the nitrogen atom to cross from one side of the plane to the other, the molecule passes through a potential-energy barrier to inversion of 6 kcal/mol (0.26 eV); this has a similar strength to the hydrogen bond, and its frequency is only 0.024 THz.

In relation to the proton-transfer mechanism, Novoa et al. (1997) reported $H_3O_2^-$ (hydroxyl ion) and H_3O^+ (hydronium) were involved with free energy values of 0.23×10^{-23} to 0.48×10^{-23} kJ in the terahertz region (~5 THz).

10.3.3 Other Activated Substances with Various Applications in Industry

Water is necessary to everybody on the earth and is a basic material for daily life. In addition to this, activated water (MICA water and its energy) has many functions and affects a variety of substances. As described above, we have studied the physics, chemistry, and biology of water as basic science. We will now discuss the properties of materials activated by the MICA process, their functions, and relating companies which make a business, during a test and in consideration for a few years in Table 10.4.

One can recognize that MICA technology can be applicable to any substances and industries. The origin is just tap water, and anything are not added although the pressurized water passes through the columns containing basalts, magnetite, etc. so that the mineral can be taken into. On the other hand, we can understand why there are many applications if we regard the physical side of this water. We believe that the molecules of the processed water become smaller resulting in distinctly active functions and large surface area. The basic idea relating to this water is a cycle or series of reduction-oxidation, and energy transmittance of activated water to another substance or material.

There are so many genres as mentioned above; here we will present some of them, (1) wrapping film for maintain foods fresh, (2) honeycomb ceramics for clean air in a room or reducing odor, (3) cupper plated plastic for reducing automobile exhaust gases and (4) oil for anti-oxidation. The water itself is better permeability into a skin of body since the molecule is small so that the water can be easily absorbed into any cell in a body, although here we do not discuss the effect to a body functions through energized cloths. And we do not present the item of four relating to emulsion fuel with water, although this it also due to a small molecule of water.

Table 10.4 Activation by MICA water and its application with results activated materials and effects
Actual results by a various field of company containing in test, and some of them have been industrialized

		Makers or groups
Materials	Functions	Manufacturing or in course of test
1. Plastic form	Keeping fresh vegetables and meats	Two makers and analysis company
Wrapping film		
2. Honeycomb	Removing odor	Two ventures and steel company
Ceramics	Reducing volatile organic compound	
SUS		
Cloths		
3. Aluminum	Elimination of car exhaust gases	Venture and car company (in USA and Sweden)
Cupper-plated plastic		
4. Water and oil	Emulsion fuel	Venture (in Korea)
5. Oil (salad oil and engine oil)	Anti-oxidation	Restaurant chain store
		Tempura restaurant
6. Paint and cement	Protection of stain	Tent company
	Improvement of wetting property	
7. Cloth and toiletry	Improvement of metabolism	Cosmetic company

10.3.3.1 Activation of Cement, Aluminum, and Ferrite

The substances to which MICA-energy was imparted were aluminum metal, Portland cement, and an Fe_3O_4–based magnet, and these were analyzed by terahertz spectroscopy, as shown in Fig. 10.15. Both treated and untreated samples of aluminum metal showed almost the same results. Terahertz waves were reflected more by MICA-treated cement than by conventional cement. The magnet consisted of an alloy and Fe_3O_4, and was therefore partially ceramic in nature, so its reflectance was intermediate between that of cement (which is 100 % ceramic) and that of aluminum (which is 100 % metal). Terahertz radiation cannot usually penetrate metals or water. In general, terahertz electromagnetic waves are reflected by the electrons in a metal that has a plasma frequency. Terahertz surface plasmon (TSP) propagation on a metal sheet has been reported by Jeon and Grischkowsky (2006) in discussing TSP pulse propagation dispersion.

10.3.3.2 Ceramics and Metal for Car Exhaust Gases

Catalytic converters in automobile exhaust gas systems function only after combustion of gasoline in the engine. Pt, Pd, and Rh are all used in controlling emissions of nitrogen oxides (NO_x) in Japan (Ueda 1997), but few reports of data obtained on real automobiles are available. Recently, there has been a trend to install

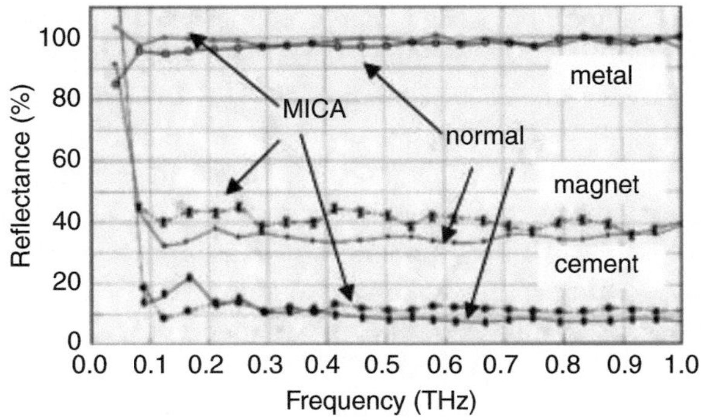

Fig. 10.15 THz frequency dependences of the reflectance
Normal and activated samples of a metallic Al plate, Portland cement, and a magnet; Al: normal (□) and MICA (○); Magnet: normal (☆) and MICA (×); and Poland cement: normal (□) and MICA (※). Vertical axis shows reflectance in numeral of 0–100 (%), and abscissas shows frequency (THz) in numeral of 0.0–1.0

Table 10.5 Results of testing a CPP device in an automobile
One model of an automobile (Nissan Cube, UA-BZ11) fitted with the device and another identical model lacking the device were subjected to idling tests for 1 h in a vinyl tent and the resulting emissions were analyzed. The detail are discussed in Sect. 10.3.3.2. CPP indicates cupper-plated plastic; with activated and without activated

	CO_2 (%)	O_2 (%)	CO (%)	HC (ppm)	NO_x (ppm)
Normal	4.9	14.9	0.20	51.9	22.8
CPP	4.5	15.0	0.14	52.0	21.8
Activated CPP	3.6	16.8	0.08	19.2	13.6
Change (%)	−26.5	12.8	−6.0	−63.0	−40.3

catalytic converters closer to the engine, which ensures that catalyst functions immediately when the engine starts. The Toyota and Daihatsu group has developed new materials, known as super-intelligent catalysts, comprising perovskite oxides, such as $LaFe_{0.57}Co_{0.38}Pd_{0.05}O_3$ (Nishihata et al. 2002). Although these super-intelligent catalysts (as they are known) represent the state of the art for petrol engines, and the Pd is regenerated, the catalysts still use precious metals, and perovskite oxides are complicated materials. Tanabe et al. (2002) reported that a catalyst containing Pt supported on a thermally stable oxide (zeolite and CeO_2–ZrO_2) on which hydrocarbons become adsorbed and migrate to the Pt surface, and the zeolite relates to the deodorization described in Sect. 10.3.3.4 (Deodorization with activated ceramics), not only reduction of exhaust gases but also deodorization.

Below, we describe the applications of MICA in photochemical removal of pollutants from automobile exhausts (Table 10.5) reported in Sugihara and Hatanaka (2009) and in deodorization of air. Our pollution-reduction system for automobile exhaust gases has two distinct characteristics: the first is that it does not require any particular catalyst, and the second is that it is positioned ahead of the engine (on or inside the air filter), resulting in a change in the properties of the air taken in by the engine. We found that a copper-plated plastic (CPP) device (5×5 cm) activated by MICA-energy reduced emissions of CO_2 and NO_x from an automobile, showing drastic reductions of CO_2, hydrocarbon and NO_x resulting in gasoline saving indeed. In a separate experiment, we also found that an MICA-activated fluorescent lamp deodorized organic solvents in an automobile paint plant and in a perfume company. As we mentioned in Sect. 10.3.2.1, our active nitrogen has an energy of less than 1.5 eV, as shown in Figs. 10.9 and 10.10, which is not necessarily a dissociation energy as discussed above (see Sect. 10.3.2.2). We will henceforth use the designation N_2^* for active nitrogen.

Nitrogen (N_2^*) in the air activated by MICA devices is introduced through the intake into the engine where various hydrocarbons (HCs) in the gasoline are burned by air. The bulk of typical gasoline consists of hydrocarbons between 5 and 12 carbon atoms per molecule, for example:

$$C_8H_{18} \rightarrow \quad C_nH_m, \quad CO_2, \quad CO, \quad H_2O, \quad NOx$$

In our system, N_2^* may form higher hydrocarbon compounds,$<h$ HC $>N_2^*$ instead of NO_x, and these compounds reacts with CO_2, NO_x, etc., and are then is burned to form lower HCs denoted to be $<l$ HC$>$;

(i) $l\ HC + h\ HC + O_2 + N_2 \rightarrow (h\ HC)N_2^* + l\ HC + CO + CO_2 + NO_x + H_2O$
(ii) $(hHC)N_2^* + O_2 \rightarrow (C \text{ or } l\ HC) + N_2 + H_2O$
(iii) $(hHC)N_2^* + CO \rightarrow (C \text{ or } l\ HC) + N_2 + O_2 + H_2O$
(iv) $(hHC)N_2^* + CO_2 + NO_x \rightarrow (C \text{ or } l\ HC) + N_2 + O_2 + H_2O.$

As we can understand from these chemical formulas, N_2 and H_2O are formed resulting in better combustion. In other words, both reduction and oxidation occurred in the engine space. This is a much simpler way of reducing emissions of CO_2 and NO_x than, for example, the method proposed by Nishihata et al. (2002).

One model of an automobile (Nissan Cube, UA-BZ11) fitted with the device and another identical model lacking the device were subjected to idling tests for 1 h in a vinyl tent and the resulting emissions were analyzed as shown in Table 10.5 (measurement instrument; FERRER F16P 5 Gas Analyzer; GXT International, NY, USA). Furthermore, the humidity inside the tent was 65 % compared with 50 % in the tent of the automobile without the MICA device. Those results suggest the gasoline was burned nearly completely in the engine of the automobile fitted with the device. One end of a vinyl hose was connected to the exhaust pipe of an idling automobile, the opposite end was placed in water in a pail, and the water was analyzed by the standard method JIS-K0102 45.2: the total nitrogen contents of the water were 4.1 and 7.4 mg/L for the automobile with the CPP device and the automobile without the device, respectively.

We have also measured the levels of volatile organic compounds (VOC) in automobile exhaust following the installation of CPP devices activated by the MICA process, as shown in Fig. 10.16. The CPP device clearly reduced VOC levels after 4 h of driving, probably by the same mechanism as described above.

Furthermore, we demonstrated another car resulting in reduction of CO_2; 30.0 %, CO; 62.1 %, HC; 71.0 % and NO_x; 95.4 %, and oxygen increased by

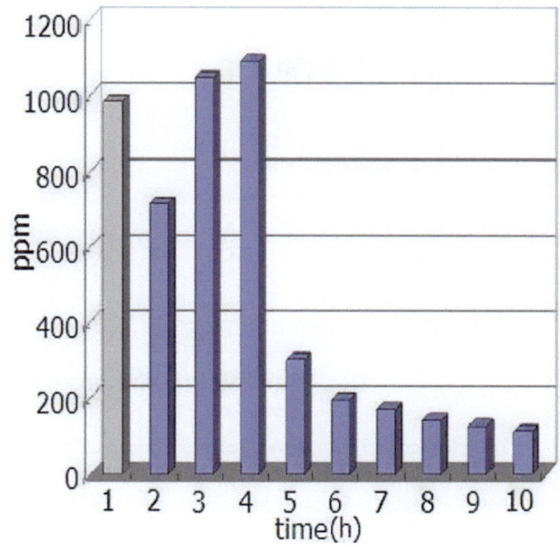

Fig. 10.16 Measurement of VOC emissions in automobile exhaust fumes before and after installing a Cu-plating plastic devices activated by the MICA process
In the first 2 h after installation of the device, adhered soot or microparticles in the engine are eliminated by the MICA-treated air engine room, and are eliminated. A steady state exists subsequently

13.6 % as well as in Table 10.5. These results of reduction rate naturally depend upon the kind of car, distance covered by travelling, etc.

As the result of the near complete combustion of gasoline that can be achieved with a MICA device, the distance driven per liter can be increased. This was tested by driving several types of automobiles in a city. For an 11-year old, 2-L-capacity, four-wheel-drive automobile, the average fuel consumption was 9.7 km/L after 13,000 km. This represents a 20 % improvement compared with a similar vehicle without the device. A 16-L diesel-engine tractor showed a 3 % improvement and less variability ($\sigma_n = 0.02$ km/L) in fuel consumption, with a minimum of 2.43 km/L and a maximum of 2.48 km/L. For comparison, the truck without the device showed $\sigma_n = 0.16$ km/L with a range of 2.06–2.7 km/L. We believe that better and more-stable fuel consumption is important for the transport business.

10.3.3.3 Activated Polyethylene for Keeping Foods Fresh

A several kinds of polymer films have been examined in the terahertz region (up to 2 THz) by Nagai and Fukasawa (2004) and Nagai (2005). In this region, the transmittance of polyethylene (PE) is more than 80 %, and is almost independent of the frequency unlike poly(vinyl chloride) (PVC), polyamide-6 (PA-6) or poly (methyl methacrylate) (PMMA), for which the absorption increases with increasing frequency. Plastics are normally transparent to terahertz waves, as discussed above. We examined the absorption behaviors of MICA-treated and nontreated polyethylene films in the terahertz region (Fig. 10.7). The nontreated polyethylene (upper graph) was more transparent at frequencies below 5 THz than was the MICA-treated film (lower graph). We postulate that the MICA film can absorb terahertz energy by means of a resonance.

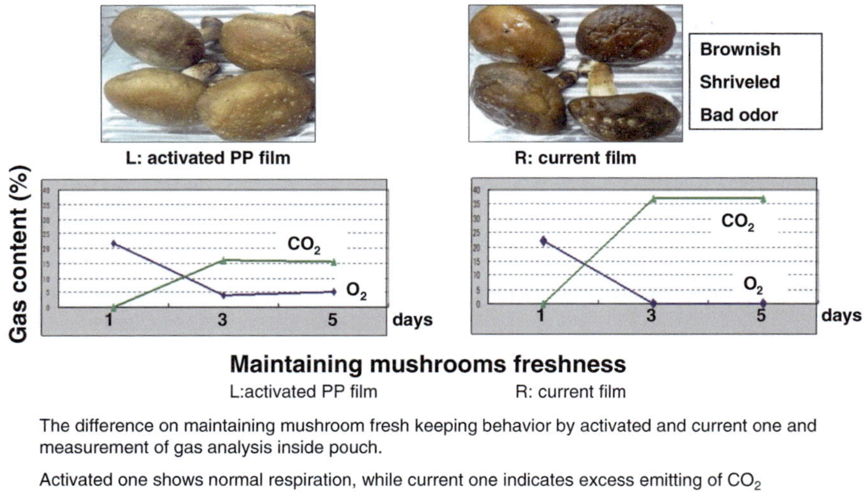

Fig. 10.17 Differences in maintenance of freshness stored in a MICA-activated polyethylene (PE) film and a control PE film, and analyses of oxygen in the PE pouch. The activated film shows normal respiration whereas control contained CO_2 only and no oxygen(O_2)

We found that the air inside film is changed by irradiation with terahertz waves and that this causes the formation of an ambient antioxidant that keeps vegetable, meat and other foods fresh. Figure 10.17 shows photographs of mushrooms kept in control and MICA–treated samples of polyethylene film for 5 days, and analyses of the gases inside the film. The mushrooms in the MICA-treated film could respire by taking in oxygen as well as carbon dioxide, whereas the mushrooms stored in the control film could only take in carbon dioxide, and they became spoiled with a bad odor. This is because the film treated with MICA energy can protect the mushrooms from attack by oxygen and nitrogen. Several users have also reported that MICA-treated polyethylene film is very effective in keeping meat fresh and juicy with no color changes. We also experienced the beef to maintain fresh resulting in no change of color and less drip which was reported in Sugihara et al. (2011), and the reasons regarding with freshness of beef is also due to energized air (nitrogen in particular) in the active film.

10.3.3.4 Deodorization with Activated Ceramics

Platinum catalysts supported on ceramics such as honeycomb-type zeolite are widely used in deodorization of exhaust gases at high temperatures, such as 600 K. We have developed ceramics (zeolite) energized by MICA energy that do not need rare-earth catalysts or platinum catalysts and high temperatures. Terahertz and/or far-infrared electromagnetic waves from an activated honeycomb resonate with –COOH and CH_3 groups, etc. MICA-energized ceramics operate at room temperature without a catalyst and show 80–90 % reduction of xylene, toluene, and formaldehyde, for instances, 416 ppm at the entrance and 65 ppm at exit. Deodorization equipment on

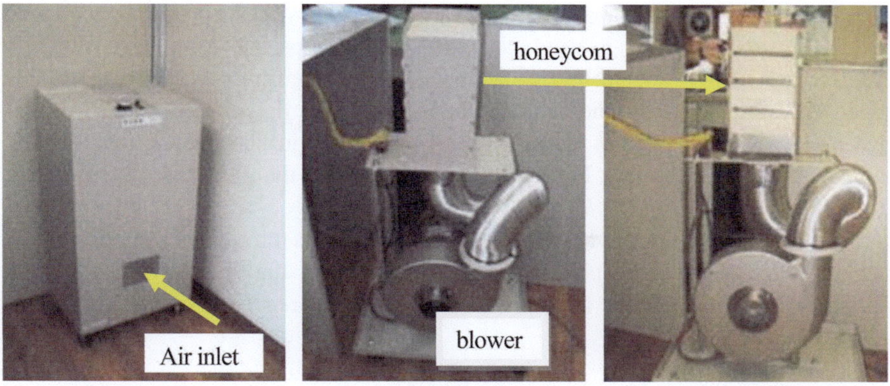

Simple outside | inside(blower) | four activated honeycomb

Fig. 10.18 Equipment for Volatile Organic Compound (after Amari Co. Ltd catalogue 2010) Only activated honeycomb and no platinum catalyst and perform under room temperature. Usually, Pt-coated honeycomb absorb odor and dissociate by the Pt catalyst under high temperature like 600 K. THz and/or far-infrared electromagnetic wave from the activated honeycomb resonate with –COOH and CH_3, etc., in the volatile organic compound

this principle has been manufactured for removal of toluene, xylene and formaldehyde in industry and in household, as shown in Fig. 10.18.

Furthermore, we demonstrated 40 activated fluorescent lamps (100 W) to reduce VOC at the car-painting factory where they used the paints with solvent of xylene and toluene; 420 and 214 ppm of VOC before and after changing the lamps, respectively (0.49 % reduction). Let us discuss the mechanisms for reduction of VOC; toluene, 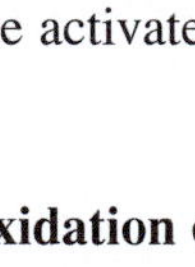, and ortho-xylene 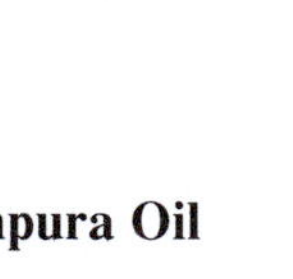 where the bonding strength of CH_3 radicals close to one of infrared and thus methyl radicals are apt to be broken rather than benzene ring due to resonate with radiation (infrared to terahertz) from the activated lamps.

10.3.3.5 Anti-oxidation of Salad Oil and Tempura Oil

Oxidized salad oil, tempura oil, and the like can be revitalized by exposure to MICA energy provided from a MICA-treated container or by mixing of used oil with MICA-treated oil. Salad oil consists of 45 % linoleic acid [diunsaturated; (9*Z*,12*Z*)-Me$(CH_2)_4$CH=CHCH=CH$(CH_2)_7CO_2H$], oleic acid [monounsaturated; (*Z*)-$CH_3(CH_2)_7$CH=CH$(CH_2)_7CO_2H$], and palmitic acid [fully saturated; Me $(CH_2)_{14}CO_2H$]. Unsaturated bonds can be changed to saturated ones by exposure to MICA energy, leading to a straighter structure and transparency to terahertz waves. The main constituent of tempura oil is linoleic acid.

Usually, the number of double bonds in the unsaturated fatty acid decreases on oxidation when the oil is used, and aldehyde groups are formed resulting in degradation. NMR spectroscopy can be used to determine the ratio of double bonds to single bonds. The ratio in the original oil, determined by carbon-13-NMR, was 15.4 %, whereas the value was 14.4 % in oil just before disposal (usually after five cycles of usage). Furthermore, by ^{1}H-NMR, the ratio was 10.0 % in the five-times used oil and 9.2 % in the original.

The another means of evaluating degradation of oil is through its anisidine value, where aldehyde groups produced by oxidation of the oil are reacted with 4-methoxyaniline in acetic acid and thus the chemical reaction occurs as follows:

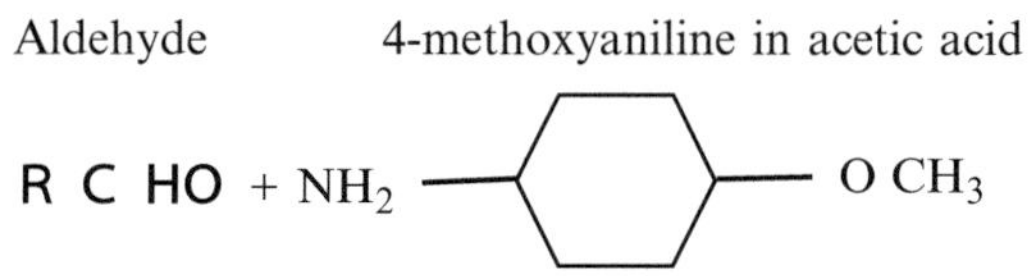

RCHO is generated by oxidation of oil. By this reaction, absorbance coefficient is changed to result in anisidine value, which presents a lower value after MICA processing of oxidized oil. and the absorbance coefficient of the product is measured. Mixing used oil with 1 % MICA-treated oil decreased the anisidine value by 8 % (8 h after heating). Furthermore, the value became to 5.5 after MICA treatment for 16 h.

Therefore, activated nitrogen in air may play a role in preventing oxidation of oil or in protecting oil from approach by oxygen, and also smaller water molecule and reducing circumference of the water may facilitate for anti-oxidation as often mentioned.

Conclusion

We have studied the electronic structures of water with broken-hydrogen bonds by using the DV-Xα method. According to calculation results, the proton and electron even after breaking hydrogen bond can exist stably in a water molecule, and we propose that the plasma-like state in proton/electrons pair emits electromagnetic waves in the region from the near infrared through the terahertz region. Water with broken-hydrogen bonds (MICA water) activates other substances by transfer of energy. These basic phenomena were indirectly detected by NMR, FTIR, THz measurement, isotope analysis and SQUID for magnetization. However, we have still researched water to approach it's kernel. In application, we postulate that devices treated with MICA energy can even activate dinitrogen in air and in organic compounds. By using MICA-treated devices, we have achieved remarkable reductions in CO_2 levels and elimination of NO_x as a result of improved combustion of gasoline in automobile. We also succeeded in deodorization and reduction of VOCs; this process has been applied in various industries. Furthermore, we

(continued)

obtained good results in terms of keeping foods fresh through the use of MICA-activated polyethylene film, which suppresses oxidation. The underlying concept of the process is related to the presence in electromagnetic energy supplied by MICA water /energy. As a future work, we will report the results that we have approached the kernel of water.

Acknowledgements We express our gratitude to Professor H. Adachi for use of the DV-Xα program and to Dr. K. Nakagawa for use of the TESDA program and also to Dr. J. Yasui of Canon for many discussions. For the sophisticated NMR analysis we thank Dr. T. Takayama and Mr. T Igarashi at Kanagawa University, and we express our gratitude to Professor M. Yasui of Keio University for discussions regarding hydrogen bonds in the membrane protein aquaporin. We thank Mr. A. Hanawa of the Institute of Isotope Analysis Co. Ltd. for performing the isotopic measurements. We also express our gratitude to Professor M. Nakao and Mr. H. Ishitani of Kyushu University of Technology for performing SQUID measurements, and to Dr S. Ohno of Tohoku University for performing terahertz measurements.

References

Adachi H (1993) Introduction of quantum materials chemistry: approach from DVXα method. Sankyo Shuppan, Tokyo, pp 54–61, 70–74, 161–166

Bianconi A, Petersen H, Brown FC, Bachrach RZ (1978) *K*-shell photoabsorption spectra of N_2 and N_2O using synchrotron radiation. Phys Rev A At Mol Opt Phys 17:1907–1911

Cavelleri M, Ogasawara H, Pettersson LGM, Nilsson A (2002) The interpretation of X-ray absorption spectra of water and ice. Chem Phys Lett 364:363–370

Coolidge AS (1932) A quantum mechanics treatment of the water molecule. Phys Rev 42:189–209

Cotton FA, Wilkinson GR (1977) Advanced inorganic chemistry. Interscience, New York, pp 323–326

Elington JR, Dehencedetti PG (2001) Relationship between structural order and the anomalies of liquid water. Nature 409:318–321

Emsley J (1998) The elements, 3rd edn. Clarendon Press, Oxford, pp. 52, 142, 148

Freda M, Pilluso A, Santucci A, Sassi P (2005) Transmittance Fourier-transform infrared spectra of liquid water in the whole mid-infrared region: temperature dependence and structural analysis. Appl Spectrosc 59:1155–1159

Fuchs EC, Woisetschläger J, Gatterer K, Maier E, Pecnik R, Holler G, Eisenkölbl H (2007) The floating water bridge. J Phys D Appl Phys 40:6112–6114

Fuchs EC, Gatterer K, Holler G, Woisetschläger J (2008) Dynamics of the floating water bridge. J Phys D App Phys 40:185502–185506

Garczarek F, Gerwert K (2006) Functional waters in intraprotein proton transfer monitored by FTIR difference spectroscopy. Nature 439:109–112

Goldman BN, Leforestier C, Saykally RJ (2005) A "first principles" potential energy surface for liquid water from VTR spectroscopy of water clusters. Phil Trans R Soc A 363:493–508

Hatanaka K (1991) EU 0421563; (1990) JP 1786552; (1993) US 5034138; (1991)

Jeon T-I, Grischkowsky D (2006) THz Zenneck surface wave (THz surface plasmon) propagation on a metal sheet. Appl Phys Lett 88:061113–3

Kagakubenran (1993) Edited by Nihonkagakkai

Kawase K, Sato M, Taniuchi T, Ito H (1996) Coherent tunable THz -wave generation from $LiNbO_3$ with monolithic grating coupler. Appl Phys Lett 68:2483–2485

Kojima S, Tsumura N, Takeda WM, Nishizawa S (2003) Far-infrared phonon–polariton dispersion probed by terahertz time-domain spectroscopy. Phys Rev B Condens Matter Mater Phys 67:035102–035105
Kozono D, Yasui M, King LS, Agre P (2002) Aquaporin water channels: atomic structure and molecular dynamics meet clinical medicine. J Clin Invest 109:1395–1399
Leetmaa M, Ljungberg MP, Lyubartsev A, Nilsson A, Pettersson LGM (2010) Theoretical approximation to X-ray absorption spectroscopy of liquid water and ice. J Electron Spectrosc Relat Phenom 177:135–157
Lin W, Han JX, Takahashi LK, Harker HA, Keutsch FN, Saykally RJ (2008) Terahertz vibration-rotation-tunneling spectroscopy of the water tetramer-d_8: combined analysis of vibrational bands at 4.1 and 2.0 THz. J Chem Phys 128:094302–094311
Millo A, Raichlin Y, Katzir A (2005) Mid-infrared fiber-optic attenuated total reflection spectroscopy of the solid–liquid phase transition of water. Appl Spectrosc 59:460–466
Mishima OR, Stanley HE (1998) The relationship between liquid, supercooled and glassy water. Nature 392:164–168
Mulliken RS (1972a) Rydberg and valence-shell character as functions internuclear distance in some excited states of CH, NH, H_2, and N_2. Chem Phys Lett 14:141–144
Mulliken RS (1972b) The nitrogen molecule correlation diagram. Chem Phys Lett 14:137–140
Muro-Oka's (1990) Graduate dissertation. Hyogo University of Teacher Education
Nagai N (2005) Analysis of the industrial materials by THz spectroscopy. Rev Laser Eng 33:848–854 (In Japanese)
Nagai N, Fukasawa R (2004) Abnormal dispersion of polymer films in the THz frequency region. Chem Phys Lett 388:479–482
Nakagawa K (2002) Total energy calculation of molecules using DV-Xα molecular orbitals. Bull Soc Discrete Variational Xα 15:121–125
Nakagawa K (2003) Total energy calculation of molecules using spin-DV-Xα molecular orbitals. Bull Soc Discrete Variational Xα 16:93–97
Nakamatsu H, Mukoyama T, Adachi H (1993) DVXα molecular orbital method for X-ray absorption near edge structure: application to N_2 and CrO_4^{2-}. Jpn J Phys 32:23–25
Nishihata Y, Mizuki J, Akao T, Tanaka H, Uenishi H, Kimura M, Okamoto T, Hamada N (2002) Self-regeneration of a Pd-perovskite catalyst for automotive emissions control. Nature 418:164–166
Novoa JJ, Mota F, Perez del Valle C, Plana M (1997) Structure of the first solvation shell of the hydroxide anion: a model study using $OH–(H_2O)_n$ ($n = 4, 5, 6, 7, 11, 17$) clusters. J Phys Chem A 101:7842–7848
Pauling L, Wilson EB (1950) Introduction to quantum mechanics with application to chemistry (translated into Japanese, Hakusuisha, Tokyo, 1950)
Scott JN, Vanderkooi JM (2010) A new hydrogen bond angle/distance potential energy surface of the quantum water dimer. Water 2:14–28
Sugihara S (2009) Analysis of water using DV-Xα method and innovative applications. Bull Soc Discrete Variational Xα 22:284–291
Sugihara S, Hatanaka K (2009) Photochemical removal of pollutants from air or automobile exhaust by minimal catalyst water. Water 1:92–98
Sugihara S, Suzuki C, Hatanak K (2011) The mechanisms of activation of substances by minimal catalyst water and application in keeping foods fresh. Water 3:87–94
Tanabe T, Hatanaka T, Tuji M, Shinjo R (2002) NO_x selective catalytic reduction over Pt supported catalyst promoted by zeolite and CeO_2–ZrO_2. R & D Rev Toyota CRDL 37(3):25–30
Tanaka Y, Ogawa M, Jursa AS (1964) Forbidden absorption-band system of N_2 in the vacuum-ultraviolet region. J Chem Phys 40:3690–3700
Tokushima T, Harada Y, Takahashi O, Senda Y, Ohashi H, Pettersson LGM, Nilsson A, Shin S (2008) High resolution X-ray emission spectroscopy of liquid water: the observation of two structural motifs. Chem Phys Lett 460:387–400

Ueda K (ed) (1997) Handbook of energy and resources. Japan Society of Energy and Resources, Tokyo, pp 1229–1231 (in Japanese)

Wernet P, Nordlund D, Bergmann U, Caverlleri M, Odelius M, Ogasawara H, Näslund LÅ, Hirsch TK, Ojmäe L, Glatzel P, Pettersson LGM, Nilsson A (2004) The structure of the first coordination shell in liquid water. Science 304:995–999

Woisetschläger J, Gatterer K, Fuchs EC (2010) Experiments in a floating water bridge. Exp Fluids 48:121–131

Wright AN, Winkler CA (1968) Active nitrogen. Academic Press, New York, p 30 and pp 140–152

Yasui M (2004) Molecular mechanisms and drug development in aquaporin water channel diseases: structure and function of aquaporins. J Pharmacol Sci 96:260–263

Yasui M, Mastushita K (2010) Recent topics in aquaporins: molecular mechanism of proton repulsion. Kidney 33:12–15 (In Japanese)

Chapter 11
Electronic Structure and Chemical Bonding of $Li_{1.1}Nb_{0.9}O_{2-y}$ as a Negative Electrode Material for Lithium Secondary Batteries

Yang-Soo Kim, Chan-Seo You, Ri-Zhu Yin, Changkeun Back, Dong-yoon Lee, Masataka Mizuno, and Soon-Ki Jeong

11.1 Introduction

During the past 20 years, much effort has been undertaken to obtain lithium secondary batteries with higher energy densities, and great progress has been made (Sony lithium ion battery performance summary 1994; Murphy and Cason-Smith 1990; Anani et al. 1987; Choi et al. 2004; Idota et al. 1997). However, there is still an increasing demand for further lightweight and small-size rechargeable batteries for portable electronic devices. In addition, recent environmental concerns about global warming and the need to reduce CO_2 emissions have provided us with a strong driving force for developing large scale lithium secondary batteries with high energy density for use in electric vehicles. Energy density is the ratio of a

Y.-S. Kim • C.-S. You
Korea Basic Science Institute, Suncheon 540-742, Republic of Korea

R.-Z. Yin
R&D center, Samsung SDI Co. Ltd., 508, Sungsung-dong, Cheonan-si 330-300, Chungnam-do, Republic of Korea

C. Back
ESS Division, Central Research Center, Hanwha Chemical Co. Ltd., Yusung 305-804, Republic of Korea

D.-y. Lee
Nano-hybrid Energy Materials Research Center of Korea Electrotechnology Research Institute, Changwon City 641-120, Kyeongnam, Republic of Korea

M. Mizuno
Center for Atomic and Molecular Technologies, Graduate School of Engineering, Osaka University, 2-1 Yamada-oka, Suita 565-0871, Osaka, Japan

S.-K. Jeong (✉)
Department of Chemical Engineering, Soonchunhyang University, Asan 336-745, Chungnam, Republic of Korea
e-mail: hamin611@sch.ac.kr

T. Ishii et al. (eds.), *The DV-Xα Molecular-Orbital Calculation Method*,
DOI 10.1007/978-3-319-11185-8_11

battery's energy delivery capability to its weight or volume, measured in watt-hours per kilogram or watt-hours per liter. It is widely considered the most important metric of battery technology. It determines the size and weight of the battery system. Higher energy density translates into smaller, lighter battery systems or applications with longer run-times or ranges. One of the important requirements of negative electrode materials is that they should have a lithium secondary battery with a large energy density; also, the redox potential at the negative electrode should be as low as possible (Denis et al. 1997). Li-ion rechargeable batteries are presently made with Li transition metal oxide as the positive electrode material and carbonaceous material as the negative electrode material (Sony lithium ion battery performance summary 1994). Li-ion rechargeable batteries can offer a higher energy density, higher average voltage, and higher power density than existing rechargeable batteries. However recent portable devices are becoming smaller, slimmer, lighter, and more multi-functional and customers are demanding longer operating times. The energy density of currently available batteries is insufficient to meet these need. To overcome this situation, a number of researchers have recently focused on two activities: (i) improving the existing positive materials like Co-, Ni-, or Mn-based oxides, or searching for new ones, or (ii) searching for alternatives to carbonaceous negative electrode materials, which alternatives should have large capacity and strong reducing power. The second approach constitutes the issue of this paper. We report a novel V-based material that can be used as a negative potential material (Yin et al. 2008). A unique feature of the vanadate materials upon lithiation is their amorphization which is lower than 1 V; this low level leads to a different shape of the voltage profile between the first and second lithiation processes. For practical cell applications, in order to achieve a different voltage profile between the first and second cycle, it is necessary to accurately control the cell balance for safety and reliability.

In the present work, we report another new oxide, $Li_{1.1}Nb_{0.9}O_{2-y}$, as a novel active material for the negative electrode of lithium secondary batteries. To begin with, it was predicted by first principles calculations that the redox reactions of lithium ion on $Li_{1.1}Nb_{0.9}O_{2-y}$ material would occur at low potentials. Then, this was confirmed experimentally by electrochemical measurements. The initial and fully charged $Li_{1.1}V_{0.9}O_{2-y}$ compounds show R-3m and P-3m1 hexagonal structures, respectively, in the XRD analysis. However this is not the case for $Li_{1.1}Nb_{0.9}O_{2-y}$.

Despite extensive studies, material scientists are still unsatisfied because very basic questions remain unanswered: Why are there phase-transitions during the charge and discharge of $Li_{1.1}V_{0.9}O_{2-y}$? Why is this not the case for $Li_{1.1}Nb_{0.9}O_{2-y}$? What are material parameters most necessary to obtain good capacity and performance? In the present study, we aim to understand the differences among these isomorphic compounds from the viewpoint of the changes in chemical bonding associated with Li-intercalation.

Two kinds of first-principles calculations were used. Molecular orbital (MO) calculations by the DV-Xα method using the program code SCAT were performed in order to obtain the electronic structure and chemical bonding information of the lithium niobium oxide. The ab initio total-energy and molecular dynamics program Vienna ab-initio simulation package (VASP) was used to calculate the open-circuit voltage (OCV) and the general gradient approximation (GGA) exchange. The first-principles calculations were carried out to clarify the remarkably low reaction potential.

11.2 Experimental

The sample formulated as $Li_{1.1}Nb_{0.9}O_{2-y}$, was prepared by heating a mixture of powders of $Li_{1.1}Nb_{0.3}O_{1.3}$ and NbO at a molar ratio of 6:3.6 in an alumina tube at 1,000 °C in H_2 environment for 60 h. $Li_{1.1}Nb_{0.3}O_{1.3}$ was obtained by heating a mixture of Li_2CO_3 and Nb_2O_5 at a molar ratio of 3.3:0.9 in an alumina tube at 800 °C in air for 48 h. These reactions are represented by the following Eqs. (11.1) and (11.2).

$$3.3\,(Li_2CO_3) + 0.9\,(Nb_2O_5) = 6\,(Li_{1.1}Nb_{0.3}O_{1.3}) + 3.3\,(CO_2) \quad (11.1)$$

$$6(Li_{1.1}Nb_{0.3}O_{1.3}) + 3.6\,(NbO) = 6\big(Li_{1.1}Nb_{0.9}O_{2-y}\big) \quad (11.2)$$

The structure of $Li_{1.1}Nb_{0.9}O_{2-y}$ was characterized by X-ray diffraction (XRD) using a Panalytical MPD (Cu Ka radiation). The data was collected in the range of 5–100° under step scanning mode at 40 kV/30 mA, 0.021 deg/step, and 10 s/step exposure time. Rietveld refinement of the collected XRD data was performed using the Reflex (Accelrys Inc. Ver. 5.0) Rietveld (DBWS) code.

Composite electrodes were used for the charge/discharge tests. The test electrode was prepared by coating a mixture of the $Li_{1.1}Nb_{0.9}O_{2-y}$ powder (80 wt%) and a carbon black (10 wt%) and a poly (vinylidene difluoride) binder (10 wt%) on copper foil. The charge and discharge tests were carried using conventional three-electrode cells at a constant current of 18.1 mA g^{-1}. The electrolyte solutions consisted of 1 mol dm^{-3} (M) $LiClO_4$ dissolved in 1:1 (by volume) mixtures of ethylene carbonate (EC) and diethyl carbonate (DEC) (EC + DEC) (Kishida Chemical Co., Lithium Battery Grade). The counter and reference electrodes were lithium foil. CV was performed between 3.0 and 30.0 V at a sweep rate of 0.5 mV s^{-1}. All electrochemical measurements were carried out in an argon-filled glovebox (Miwa, MDB-1B + MM3-P60S) with a dew point <-60 °C. All potentials are referred to as volts vs. Li/Li^+.

11.3 Computational Procedure

The electronic structures and chemical bonding of lithium niobium oxide were calculated by the DV-Xα method (Averill and Ellis 1973; Ellis et al. 1976) using the program code (Adachi et al. 1978). The population analysis was made according to

Mulliken (Mulliken 1955). All calculations in the DV-Xα method had to be executed with cluster models and the full potentials of all orbitals were included in the cluster model. The proper building of models is very important in this method, because calculated results change with models and, sometimes, show unreasonable values regardless of the convergence of calculations. Therefore, various models for every material were tried in order to obtain acceptable data in the view of crystal chemistry and energy level structure. In the case of $Li_{1.1}Nb_{0.9}O_{2-y}$, the Madelung field was applied over a sufficiently large volume to obtain an environment similar to that of the bulk state and to eliminate the edge effect, which brings distorted charge distribution and energy levels.

The ab initio total energy and molecular dynamics program VASP (Kresse and Hafner 1993; Kresse and Furthmuller 1996a, b), based on the density functional theory pseudo-potential plane wave method and the GGA exchange and correlation functional, was used to obtain the OCV. First-principles calculation on lithium transition metal oxides of positive electrodes for lithium-ion batteries have been performed by many researchers (Ceder et al. 1997; Mishra and Ceder 1999; Reed and Ceder 2002; Koyama et al. 2003).

In general the open-circuit voltage $E_{OCV}(x)$ depends on the differences in the Li chemical potential between the anode Li metal and the cathode. In this paper, we have investigated the possibility of using metal oxide powders for anodic electrode as active materials and have determined their electrochemical performances through the calculation. In this paper, first-principles calculation for the voltages was carried out as follows:

$$OCV = \frac{-\Delta G}{zF} = \frac{-\left(E^{Li}_{cathod} - E^{Li}_{anode}\right)}{zF} \tag{11.3}$$

Here F is the Faraday constant and z is the ion transported charge in the electrolyte. In the lithium ion battery (LIB) system, Li^+ is the ionic conductor, which means that $z = 1$ for the LIB system.

11.4 Results and Discussions

In a previous work, we reported that average potentials for the reversible lithium insertion within $Li_{1.1}V_{0.9}O_2$, determined theoretically and experimentally were 0.93 and 0.3 V, respectively (Yin et al. 2008; Linden 1995). These low potentials mean that there is a new possibility of active material as a negative electrode is present in $Li_{1.1}M_{0.9}O_2$ (M = transition metal) compounds. In this view, further research has been conducted to consider various $Li_{1.1}M_{0.9}O_2$ compounds.

Table 11.1 gives the average potential for lithium insertion in $Li_{1.1}M_{0.9}O_2$ obtained by first principle calculations; in these calculations, $Li_{2.1}M_{0.9}O_2$ was regarded as the charged state. The detailed calculation process has been described elsewhere (Yin et al. 2008; Kim et al. 1998; Koyama et al. 1999a, b).

Table 11.1 Average potential for lithium insertion into various $Li_{1.1}M_{0.9}O_2$ compounds obtained by first principles calculations (VASP)

M in $Li_{2.1}M_{0.9}O_2/Li_{1.1}M_{0.9}O_2$	Average potential/V
Al	0.65
Ti	0.89
Cr	1.43
Mn	2.10
Fe	2.48
Co	2.23
Ni	3.45
Cu	2.87
Zn	5.24
Zr	1.27
Nb	1.30
Mo	1.08

The calculation results reveal that the average potential for lithium insertion was significantly affected by the M element in $Li_{1.1}M_{0.9}O_2$. Lithium ions were inserted within $Li_{1.1}M_{0.9}O_2$ at a relatively low potential (<1.5 V) when M was Al, Ti, Cr, Zr, Nb, or Mo. On the other hand, the potential was relatively high (>2.0 V) when M was Mn, Fe, Co, Ni, Cu, or Zn. Since low potential is favorable in the view of energy density, the former compounds are considered better candidates as active materials for the negative electrode than are the latter compounds. On the basis of the theoretical calculation data, we have tried to synthesize $Li_{1.1}M_{0.9}O_2$ (M = Al, Ti, Cr, Zr, Nb, and Mo) compounds by conventional solid state reaction method in order to investigate their electrochemical properties. Consequently, the $Li_{1.1}Nb_{0.9}O_{2-y}$ compound was obtained successfully through the synthetic procedure shown in the above experimental section. However, attempts to synthesize this compound by using the other M elements such as Al, Ti, Cr, Zr, and Mo, were unsuccessful. The details of the failures are not clear at present.

To the best of our knowledge, $Li_{1.1}Nb_{0.9}O_{2-y}$ was synthesized for the first time. Prior to the electrochemical measurements, Rietveld refinement was employed to determine the structural properties of the synthesized $Li_{1.1}Nb_{0.9}O_{2-y}$ compound. This is one of the most popular techniques used to characterize crystal structures. A perfect crystal structure model is very helpful for theoretical calculations, reaction mechanism analysis, and physical analysis of certain properties such as conductivity, magnetic susceptibility, chemical potential, etc. Powder XRD (or neutron diffraction) Rietveld refinement is one of the most popular methods used to characterize crystal structure.

The Rietveld refinement of the $Li_{1.1}Nb_{0.9}O_{2-y}$ model gave a satisfactory fit to the overall profile. The refinement processes were carried out with several variations such as baseline correction, cell parameters, zero correction, scale factor, temperature factor, atomic coordinates, etc. The best-fit of the XRD patterns are shown in Fig. 11.1. $Li_{1.1}Nb_{0.9}O_{2-y}$ is assigned to the hexagonal system with the *P63/mmc* (No. 194) space group. The cell parameters are $a = 2.904$ Å and $c = 10.456$ Å. The calculated cell volume is $V = 103.62A^3$. About 10 % of the Nb^{3+} (2a) sites are substituted for by

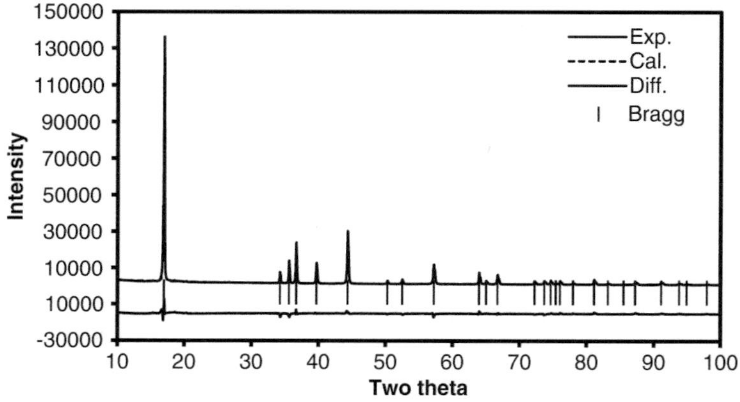

Fig. 11.1 The Rietveld refinement of $Li_{1.1}Nb_{0.9}O_{2-y}$ (Kim et al. 2012)

Table 11.2 Atomic coordinates of $Li_{1.1}Nb_{0.9}O_{2-y}$ which is obtained from Rietveld refinement (Kim et al. 2012)

	Site	x	y	z	Occupancy
Li	2a	0	0	0	1
Li	2d	0.66667	0.33333	0.25	0.1
Nb	2d	0.66667	0.33333	0.25	0.9
O	4f	0.33333	0.66667	0.1263	0.95

lithium ions. The Nb^{3+} is coordinated by six oxygen atoms and located in the octahedral site. Some Li^{+} is also located in the Nb^{3+} (2b) site. The fractional coordinates of the $Li_{1.1}Nb_{0.9}O_{2-y}$ at room temperature are listed in Table 11.2.

In this structure, one Nb plane is sandwiched by two O planes; the stacking of the layers forms the crystal. Thus, the crystal exhibits a layered structure due to the stacking of the planes (O-Nb-O)/Li/(O-Nb-O).

The calculation of the electronic states of lithium niobium oxide was carried out using cluster models taken from the bulk crystal. The model clusters, which are composed of (a) 51 atoms $(Li_{13}V_{12}O_{26})^{3-}$ cluster model for $Li_{1.1}V_{0.9}O_2$ (Fig. 11.2a) (b) 87 atoms $(Li_{13}Nb_{12}O_{62})^{73-}$ for $Li_{1.1}Nb_{0.9}O_{2-y}$ (Fig. 11.2b) and (c) 85 atoms $(Li_{17}Nb_{20}O_{48})^{14-}$ compared to $Li_{1.1}V_{0.9}O_{2-y}$ in R-3m (Fig. 11.2c), are employed for these purposes.

In order to discuss the valence electronic state and chemical bonding of lithium vanadium oxide, we made calculations using cluster models. The energy diagram density of $Li_{1.1}V_{0.9}O_{2-y}$ is obtained in this study as shown in Fig. 11.3. The filled band located from −8 to −3 eV is mainly composed of O 2p orbitals. The partially filled band located around −2 to 4 eV is mainly composed of V 3d orbitals. The unoccupied band located above 5 eV are made up of Li 2sp and V 4sp orbitals. Significant amounts of V 3d and 4sp states in the O 2p band were found. Strong covalent interaction between V and O was found to be present.

The energy diagram and the bond overlap population (BOP) diagrams or crystal overlap population (COP) for V–O(2.012 Å) and Li–O(2.096 Å) bonds were plotted,

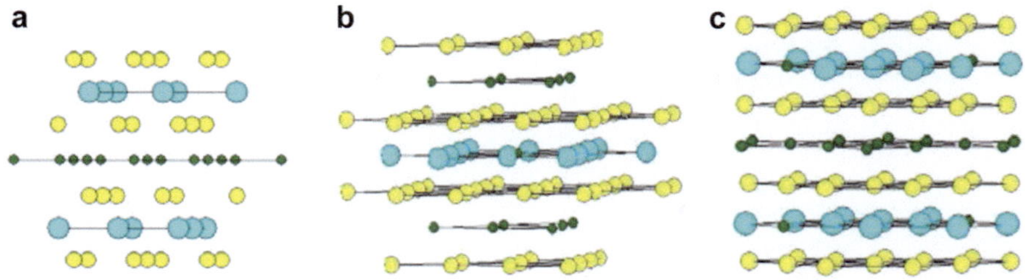

Fig. 11.2 (**a**) $(Li_{13}V_{12}O_{26})^{3-}$cluster model for $Li_{1.1}V_{0.9}O_2$ (R-3m), (**b**) $(Li_{13}Nb_{12}O_{62})^{73-}$ cluster model for $Li_{1.1}Nb_{0.9}O_2$(P63/mmc), and (**c**) $(Li_{17}Nb_{20}O_{48})^{14-}$ cluster model for $Li_{1.1}Nb_{0.9}O_2$ (P63/mmc)

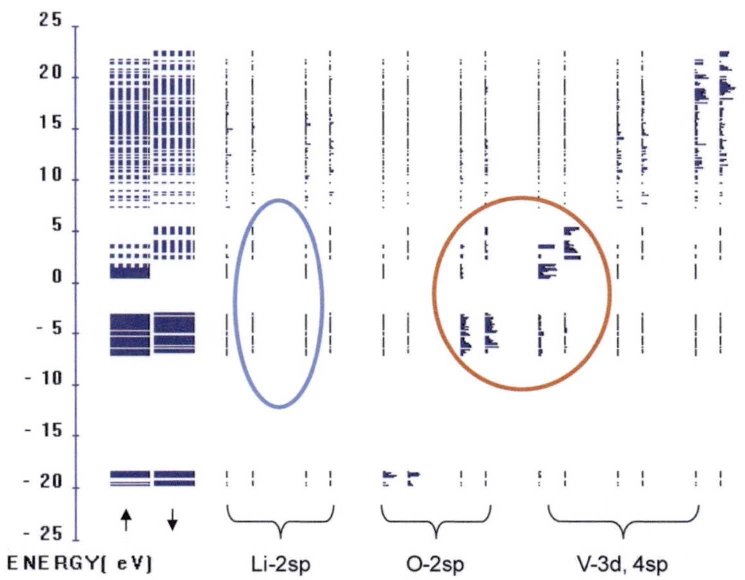

Fig. 11.3 Energy diagram of $Li_{1.1}V_{0.9}O_2$ with $(Li_{13}V_{12}O_{26})^{3-}$ cluster model by DV-Xα method

with results shown in Fig. 11.4. The interactions between V and O, and the Li–O bonds have been investigated individually. The value given in each diagram is the bond overlap population, which is given by the integration of the diagram with respect to the occupied bands. As has been qualitatively mentioned, strong covalent bonding can be noted between V and O. On the other hand, the Li–O bond overlap population is small, but not zero (0.063), as shown in Fig. 11.4. This provides evidence for the strong covalency between M (M = transition metal) and X (X = S and O), as discussed in our previous reports (Kim et al. 1998; Koyama et al. 1999a, b).

Table 11.3 shows BOP between M-O and Li-O bonds and the ionicity of the Li atom of $Li_{1.1}Nb_{0.9}O_{2-y}$ and $Li_{1.1}V_{0.9}O_2$, which belong to P63/mmc and R-3m respectively. The BOP of V-O shows a larger value than that of the BOP of Nb-O, even though these two showed similar bond lengths between metal oxygen.

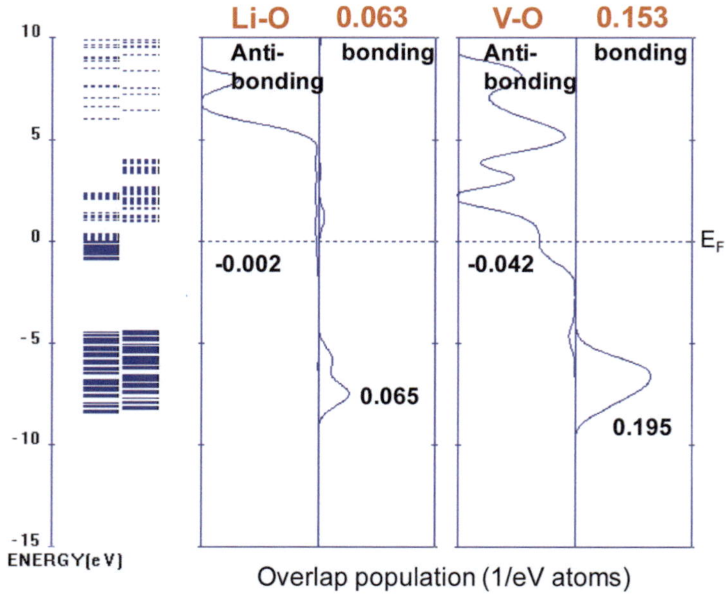

Fig. 11.4 Overlap population of $Li_{1.1}V_{0.9}O_2$ with $(Li_{13}V_{12}O_{26})^{3-}$ cluster model (Yin et al. 2008) by DV-Xα method

Table 11.3 Comparisons between $Li_{1.1}Nb_{0.9}O_{2-y}$ and $Li_{1.1}V_{0.9}O_2$ using DV-Xα method. (V-base[a] means, Vanadium simply replace to Niobium and remain *P63/mmc* crystal structure)

		BOP between Li-O	BOP between M-O	Ionicity of Li atom
$Li_{1.1}Nb_{0.9}O_{2-y}$ (Fig. 11.2b)	Nb-Base	0.0869 (2.1175 Å)	0.166 (2.1175 Å)	0.5265
	V-base[a]	0.0967	0.2051	0.534
$Li_{1.1}Nb_{0.9}O_{2-y}$ (Fig. 11.2c)	Nb-Base	0.0908 (2.1342 Å)	0.183 (2.1175 Å)	0.599
	V-base[a]	0.098	0.233	0.561
$Li_{1.1}V_{0.9}O_2$ (Fig. 11.2a)	R-3m	0.063 (2.096 Å)	0.153	0.73 (2.096 Å R-3m)

The BOP value of V-O in the $Li_{1.1}V_{0.9}O_2$ (R-3m) is smaller than that of the V substituted Nb-based system (P63/mmc). The ionicity of Li in the P63/mmc system is smaller than that of the Li in the R-3m system.

The BOP between the 3d transition metal-O and the Li-O, and the ionicity of the Li atoms, are listed in Table 11.4. For all of these, the BOP values of M-O and Li-O are larger than that of $Li_{1.1}V_{0.9}O_2$ (R-3m). For all of these samples, the iconicity of $Li_{1.1}M_{0.9}O_2$ (P63/mmc) is smaller than that of $Li_{1.1}V_{0.9}O_2$ (R-3m). This provides

Table 11.4 Comparisons of BOP and Ionicity of Li atom between $Li_{1.1}M_{0.9}O_2$ (P63/mmc) and $Li_{1.1}V_{0.9}O_2$ (R-3m) using DV-Xα method

	BOP between Li-O	BOP between M-O	Ionicity of Li atom
$Li_{1.1}Ti_{0.9}O_2$	0.106	0.156	0.51
$Li_{1.1}V_{0.9}O_2$	0.097	0.2051	0.53
$Li_{1.1}Cr_{0.9}O_2$	0.095	0.189	0.51
$Li_{1.1}Mn_{0.9}O_2$	0.106	0.182	0.52
$Li_{1.1}Fe_{0.9}O_2$	0.103	0.184	0.53
$Li_{1.1}Co_{0.9}O_2$	0.102	0.182	0.52
$Li_{1.1}Ni_{0.9}O_2$	0.101	0.177	0.53
$Li_{1.1}Cu_{0.9}O_2$	0.111	0.1713	0.49
$Li_{1.1}V_{0.9}O_2$(R-3m)	0.063	0.153	0.73

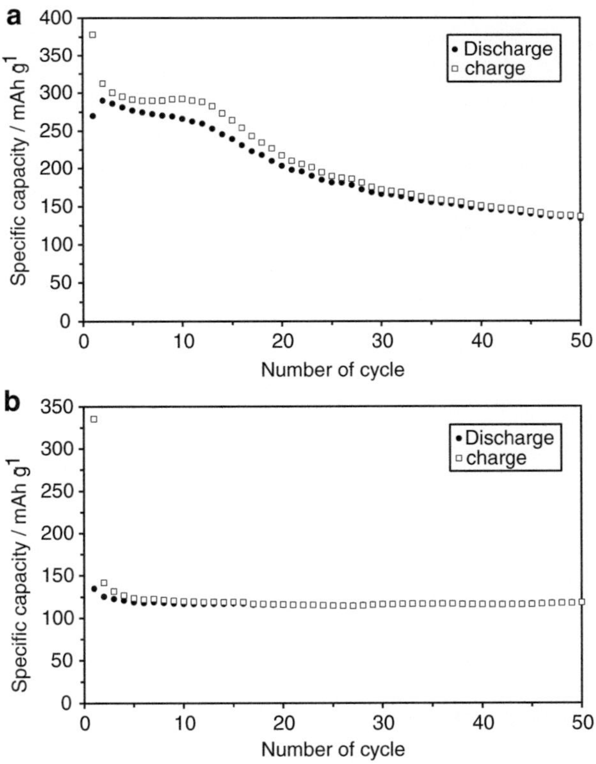

Fig. 11.5 Charge and discharge capacity retention with cycle number for (**a**) $Li_{1.1}V_{0.9}O_2$ and (**b**) $Li_{1.1}Nb_{0.9}O_{2-y}$ (Kim et al. 2012) electrode

evidence for the idea that the BOP of M-O and the ionicity of Li are strongly affected by the crystal structure.

Additionally, Fig. 11.5 shows the cycle performance of $Li_{1.1}V_{0.9}O_{2-y}$ (R-3m) and $Li_{1.1}Nb_{0.9}O_{2-y}$ (P63/mmc), respectively. $Li_{1.1}Nb_{0.9}O_{2-y}$ (P63/mmc) has strong covalent bonding compared to that of $Li_{1.1}V_{0.9}O_{2-y}$ (R-3m) in our cluster calculation. The stronger covalent bond maintains a stable structure during the charge and discharge; this stability can lead to good cycle life performance. The chemical bonding provides evidence of the necessary material parameters required to retain the crystal structure and stability of the cycle performance.

Finally, we do not at present have experimental evidence for the structural transformation of $Li_{1.1}Nb_{0.9}O_{2-y}$ during charge and discharge. Therefore more detailed analysis is necessary to clarify the nature of the irreversible reaction.

Conclusion

The DV-Xα method, which is a sort of first principles method using density functional theory and full potential molecular orbitals, was used to understand the electronic structure and to obtain a basis for new material design in lithium secondary batteries. Mulliken's population analysis has been thoroughly conducted to examine the net charge as well as the magnitude of the covalent bondings. $Li_{1.1}Nb_{0.9}O_{2-y}$ (P63/mmc) has shown covalent bonding stronger than that of $Li_{1.1}V_{0.9}O_2$ (R-3m) in our cluster calculation. $Li_{1.1}Nb_{0.9}O_{2-y}$ (P63/mmc), and the BOP value for Li–O and V–O are different in their structure. The chemical bonding provides evidence of the necessary material parameters required to maintain the crystal structure and good cycle performance. These results are contrary to the widely accepted picture of Li-intercalated compounds, and should be a very important consideration for the determination of battery properties such as OCV. The information about the chemical bonding and iconicity of these compounds should be helpful investigating the possibility of new electrode active materials.

Acknowledgments This work was supported by KBSI grant (D34808) to Y.-S. Kim.

References

Adachi H, Tsukada M, Satoko C (1978) Discrete variational Xα cluster calculations. I. application to metal clusters. J Phys Soc Jpn 45:875

Anani A, Crouch-Baker S, Huggins RA (1987) Measurement of lithium diffusion in several binary lithium alloys at ambient temperature. In: Dey AN (ed) Proceedings of the ECS symposium on lithium batteries, Pennington, p 365

Averill FW, Ellis DE (1973) An efficient numerical multicenter basis set for molecular orbital calculation: application to FeC_{14}. J Chem Phys 59:6412

Ceder G, Aydinol MK, Kohan AF (1997) Application of first-principles calculations to the design of rechargeable Li-batteries. Comput Mater Sci 8:161

Choi W, Lee JY, Jung BH, Lim HS (2004) Microstructure and electrochemical properties of a nanometer-scale tin anode for lithium secondary batteries. J Power Sources 136:154

Denis S, Baudrin E, Touboul M, Tarascon J-M (1997) Synthesis and electrochemical properties vs. Li of amorphous vanadates of general formula RVO4 (R = In, Cr, Fe, Al, Y). J Electrochem Soc 144:4099

Ellis DE, Adachi H, Averill FW (1976b) Molecular cluster theory for chemisorptions of first row atoms on nickel (100) surfaces. Surf Sci 58:496

Idota Y, Kubota T, Matsufuji A, Maekawa Y, Miyasaka T (1997) Tin- based amorphous oxide: a high-capacity lithium-ion-storage material. Science 276:1395
Kim Y-S, Koyama Y, Tanaka I, Adachi H (1998) Chemical bondings around intercalated Li atoms in $LiTiX_2$(X = S, Se, and Te). Jpn J Appl Phys 37:6440
Kim Y-S, Na K, Lee M-H, Back C, Baik Y, Jeong E-D, Hwang S-H, Lee S-R, Yin R-Z, Jeong S-K (2012) Synthesis and electrochemical characterization of $Li_{1.1}Nb_{0.9}O_{2-x}$ as a novel active material for the negative electrode of lithium secondary batteries. Mater Lett 83:14
Koyama Y, Kim Y-S, Tanaka I, Adachi H (1999a) Changes in chemical bondings by Li deintercalation in $LiMO_2$(M = Cr, V, Co, and Ni). Jpn J Appl Phys 38:2024
Koyama Y, Tanaka I, Kim Y-S, Nishitani SR, Adachi H (1999b) First principles study on factors determining battery voltages of $LiMO_2$ (M D Ti–Ni). Jpn J Appl Phys 38:4804
Koyama Y, Tanaka I, Adachi H, Makimura Y, Ohzuku T (2003) Crystal and electronic structures of superstructural $Li_{1-x}[Co_{1/3}Ni_{1/3}Mn_{1/3}]O_2$ ($0 \leq x \leq 1$). J Power Sources 119–121:664
Kresse G, Furthmuller J (1996a) Efficiency of ab-initio total energy calculations for metal and semiconductors using a plane-wave basis set. Comput Mater Sci 6:15
Kresse G, Furthmuller J (1996b) Efficiency interactive schemes for ab initio total- energy calculations using a plane-wave basis set. Phys Rev B 54:11669
Kresse G, Hafner J (1993) *Ab initio* molecular dynamics for open-shell transition metals. Phys Rev B 48:13115
Linden D (ed) (1995) Handbook of batteries. McGraw-Hill, New York
Mishra SK, Ceder G (1999) Structural stability of lithium manganese oxides. Phys Rev B 59:6120
Mulliken RS (1955) Electronic population analysis on LCAO-MO molecular wave functions. I. J Chem Phys 23:1833
Murphy TC, Cason-Smith DM (1990) Characterization of AA size lithium rechargeable cells. In: Proceedings of the 34th international power source symposium, p 176
Reed J, Ceder G (2002) Charge, potential, and phase stability of layered $Li(Ni_{0.5}Mn_{0.5})O_2$. Electrochem Solid-State Lett 5:A145
Sony lithium ion battery performance summary (1994) JEC Batt Newslett 2: 31
Yin RZ, Kim Y-S, Choi W, Kim SS, Kim H (2008) Structural analysis and first-principles calculation of lithium vanadium oxide for advanced Li-ion batteries. Adv Quantum Chem 54:23

Chapter 12
Chemical Bonding, Point Defects and Positron Lifetimes in $FeSi_2$ from First-Principles Calculations

Masataka Mizuno and Hideki Araki

The chemical bonding and formation energy of Fe and Si vacancies in α, β and γ-$FeSi_2$ have been investigated using first-principles calculations. The positron lifetimes of those $FeSi_2$ are also calculated. The bonding and antibonding bands in β-$FeSi_2$ are completely separated, which leads to a gap at the Fermi level in β-$FeSi_2$. In γ-$FeSi_2$, the separation is not perfect and the Fermi level lines in the bonding band. Whereas the formation energy of a Fe vacancy in β-$FeSi_2$ is higher than that of the other phases, it steeply decreases going from the Fe-rich to the Si-rich condition and shows the lowest value, about 0.4 eV, at the Si-rich limit. In α and γ-$FeSi_2$, the Fe vacancy has a shorter positron lifetime than the Si vacancy. However, in β-$FeSi_2$, positron lifetimes of the Fe and Si vacancies are close to each other. This result suggests that it may be difficult to identify vacancy sites in β-$FeSi_2$ by positron lifetime measurement.

12.1 Introduction

The $FeSi_2$ has three polymorphs, α, β and γ phases. In particular β-$FeSi_2$ has been attracting much interest as thermoelectronic conversion materials or optoelectronic devices. Many growth methods, such as chemical vapor transport, ion-beam synthesis, molecular beam epitaxy and sputtering, have been attempted for the formation of β-$FeSi_2$ films. Although β-$FeSi_2$ seems to be a line-compound in the equilibrium phase diagram, the Si/Fe ratio is an important factor to control various

M. Mizuno (✉) • H. Araki
Center for Atomic and Molecular Technologies, Osaka University, 2-1 Yamadaoka, Suita 565-0871, Osaka, Japan

Division of Materials and Manufacturing Science, Osaka University, 2-1 Yamadaoka, Suita 565-0871, Osaka, Japan
e-mail: mizuno@mat.eng.osaka-u.ac.jp

T. Ishii et al. (eds.), *The DV-Xα Molecular-Orbital Calculation Method*,
DOI 10.1007/978-3-319-11185-8_12

properties of β-$FeSi_2$. It has been reported that the Si/Fe ratios of β-$FeSi_2$, namely point defects, affect its structural, electronic and optical properties (Tani and Kido 2000; Seki et al. 2003; Liu et al. 2008).

Positron annihilation spectroscopy is a sensitive tool for vacancy-type defects in solids. In a perfect crystal, the positron wave function is delocalized in the interstitial region because of the repulsion from the ion cores. If vacancy-type defects exist in solids, the positron wave function is localized at the defects where the electron density is lower than in the other regions. The localization of the positron at the defects results in a longer lifetime of the positron compared to the bulk lifetime because the positron lifetime is inversely proportional to the electron density where the positron annihilated. In addition, positron lifetimes can be calculated by solving the positron wave function. A combination of positron lifetime measurements and its theoretical calculations allows us to identify more precisely the defects. We plan to do positron annihilation measurements on β-$FeSi_2$ in order to identify lattice defects. In this work, chemical bonding, positron lifetimes and vacancy formation energies are evaluated using first-principles calculations.

12.2 The Structure of $FeSi_2$

Space group, lattice parameters and atomic positions of $FeSi_2$ are listed in Table 12.1. α-$FeSi_2$ is a high-temperature phase and β-$FeSi_2$ is stable at room temperature. There is no available data for lattice parameters of γ-$FeSi_2$ because γ-$FeSi_2$ is formed only on Si substrate as thin film (Onda et al. 1992). The crystal structures of α, β and γ-$FeSi_2$ are shown in Fig. 12.1. In α and γ-$FeSi_2$, the Si atoms form a simple cubic sub-lattice and a half of body-centered sites are occupied by the Fe atoms. The Si atoms occupy the body-centered sites layer by layer in the α-$FeSi_2$. In γ-$FeSi_2$, the Fe atoms and the unoccupied sites are arranged next to each other. While the arrangement of atoms in β-$FeSi_2$ looks very complicated, the

Table 12.1 Space group, lattice parameters and atomic positions of $FeSi_2$ and FeSi

	Space group	Lattice parameter (Å)	Atomic position
α-$FeSi_2$	*P4/mmm* (No. 123)	$a=2.696$	Fe $x, y, z=0$
		$c=5.154$	Si 0.5, 0.5, 0.27
β-$FeSi_2$	*Cmca* (No. 64)	$a=9.863$	Fe1 0.2146, 0, 0
			Fe2 0.5, 0.3086, 0.1851
		$b=7.791$	Si1 0.1286, 0.2746, 0.0512
		$c=7.833$	Si2 0.3727, 0.0450, 0.2261
γ-$FeSi_2$	$Fm\bar{3}m$ (No. 225)	–	Fe $x, y, z=0$
			Si $x, y, z=0.25$
FeSi	$P2_13$ (No. 198)	$a=4.495$	Fe $x, y, z=0.3865$
			Si $x, y, z=0.09262$

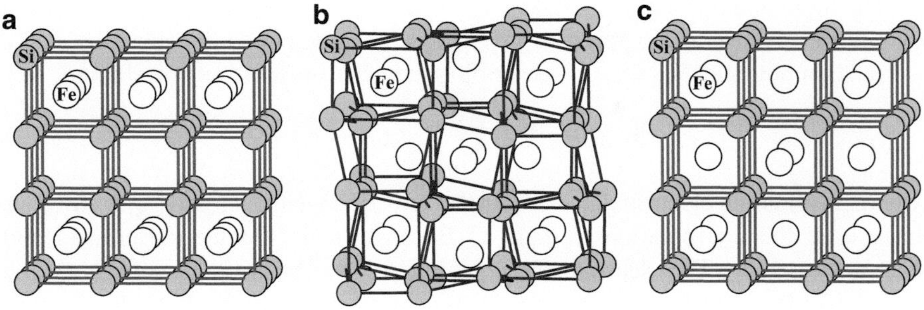

Fig. 12.1 Crystal structures of (**a**) α, (**b**) β and (**c**) γ-$FeSi_2$

Table 12.2 Average bond length and number of neighboring atoms in $FeSi_2$

	Average bond length (Å)/Number of neighboring atoms		
	Fe-Si at Fe	Si-Si	Fe-Fe
α-$FeSi_2$	2.360/8	2.371/1	2.696/4
		2.713/5	
β-$FeSi_2$	2.371/8	2.525/5	2.968/2
γ-$FeSi_2$	2.351/8	2.715/6	–/0

crystal structure of β-$FeSi_2$ can be considered as distorted γ-$FeSi_2$. Average bond lengths and the number of neighboring atoms are listed in Table 12.2. The Fe atom in all of $FeSi_2$ has eight neighboring Si atoms and average bond lengths of Fe-Si bonds are very similar to each other. The bond length of Si-Si bonds between the Si layers in α-$FeSi_2$ is 2.371 Å which is close to that in metallic Si, 2.35 Å. The unoccupied body-centered sites in β-$FeSi_2$ are compressed by the distortion of Si sub-lattice, which leads to the shorter average Si-Si bond length compared to γ-$FeSi_2$. The Fe-Fe bonds are also formed by the distortion in β-$FeSi_2$.

12.3 Computational Method

12.3.1 Chemical Bonding and Positron Lifetime

In order to evaluate the chemical bonding and obtain the electron density and potential for the positron lifetime calculations, the discrete-variational (DV) Xα cluster method using a program code SCAT (Adachi et al. 1978; Ellis and Painter 1970) is employed. We use the model clusters composed of about 150 atoms. The bonding and antibonding characters are evaluated using overlap population diagrams which are obtained by convoluting overlap populations, calculated based on the Mulliken population analysis (Mulliken 1955), at each molecular orbital with Gaussian function of 1.0 eV FWHM.

The three dimensional potential $V_+(r)$ sensed by the positron is constructed as a sum of the electron static potential $V_c(r)$ due to the nuclei and the electrons, and a correlation potential $V_{corr}(n_-(r))$ describing the electron-positron correlation:

$$V_+(r) = V_c(r) + V_{corr}(n - (r)), \quad (12.1)$$

where $n_-(r)$ is the electron density. For the density dependence of $V_{corr}(n_-(r))$, we have used the interpolation formula of Boroński and Nieminen (Boroński and Nieminen 1986). The Schrödinger equation for the positron is solved by the finite-difference method (Puska and Nieminen 1983). The positron lifetime τ is calculated as the reciprocal value of the positron annihilation rate λ by

$$\lambda = \pi r_0^2 c \int n_+(r) n_-(r) \gamma(r) dr, \quad (12.2)$$

where r_0 is the classical electron radius, c is the speed of the light, $n_+(r)$ is the positron density and γ(r) is the enhancement factor describing the short-range pileup of the electron at the positron. For the enhancement factor, we have used the interpolation formula (Boroński and Nieminen 1986). Fractions of positron annihilation for each orbital are obtained using the Mulliken population analysis.

12.3.2 Vacancy Formation Energy

In order to obtain energies needed for calculations of vacancy formation energies, we employed the plane-wave pseudopotential method as implemented in the Vienna *ab initio* simulation package (VASP) (Kresse and Furthmuller 1996a, b) with generalized gradient approximation proposed by Perdew and Wang (1992). Potential based upon the all-electron projector augmented wave (PAW) method were used (Kresse and Joubert 1999; Blöchl 1994). First, we calculated the equilibrium lattice parameters of α, β and γ-$FeSi_2$ using the kinetic energy cutoff of 350 eV and 21 × 21 × 11, 6 × 8 × 8 and 11 × 11 × 11 k-meshes in the Monkhorst-Pack scheme, respectively. We obtained the equilibrium lattice parameters of $a = 2.7065$ Å and $c = 5.1334$ Å for α-$FeSi_2$, and $a = 9.8942$ Å, $b = 7.7700$ Å and $c = 7.8212$ Å for β-$FeSi_2$ which well reproduce the experimental values shown in Table 12.1. The equilibrium lattice parameter of $a = 5.3913$ Å for γ-$FeSi_2$ is in good agreement with the previous theoretical work, $a = 5.389$ Å (Christensen 1990).

β-$FeSi_2$ virtually exists only on the stoichiometric composition in the equilibrium phase diagram. In this case, processes of vacancy formations inevitably accompany formations of other phases. Tani and kido calculated formation energies of point defects in β-$FeSi_2$ using first-principles pseudopotential calculations (Tani and Kido 2003). They employed the energy of Fe and Si crystals as the chemical potentials of atoms subtracted as a vacancy. However, a wide range of the Si/Fe ratio is employed for the growth of β-$FeSi_2$ to control its properties, indicating that deviations from the stoichiometric composition are compensated by introducing vacancies or antisite atoms in β-$FeSi_2$. The process of vacancy formation without formation of other phases, therefore, needs to be considered.

In the process of vacancy formation without formation of other phases, two or more defects are introduced to maintain the stoichiometric composition. In order to determine the process of vacancy formation, we have calculated formation energies of non-stoichiometric α, β and γ-$FeSi_2$ using supercells including a vacancy or an antisite atom. In the present work, we employed supercells composed of 32 Fe and 64 Si lattice sites for α, β and γ-$FeSi_2$. The supercells for β-$FeSi_2$ are the $2 \times 2 \times 2$ primitive cells whose cell vectors are (0, 0.5*b*, 0.5*c*), (0.5*a*, 0, 0.5*c*) and (0.5*a*, 0.5*b*, 0). The supercells for α and γ-$FeSi_2$ are the $4 \times 4 \times 2$ and $2 \times 2 \times 2$ unit cells, respectively. Relaxations of the atomic positions and lattice parameters were allowed. In addition to the supercells composed of 96 lattice sites, the $3 \times 3 \times 2$ and $1 \times 1 \times 1$ unit cells were used for the supercells of α and β-$FeSi_2$, respectively. For the supercells of γ-$FeSi_2$, the $2 \times 2 \times 2$ primitive cells composed of 8 Fe and 16 Si sites were also employed. The formation energies were calculated by subtracting the total energies of the constituent elemental solids from the total energy of the compounds. The calculated formation energies of α, β and γ-$FeSi_2$ are plotted as a function of Si composition in Fig. 12.2. Whereas there are two sites for Fe and Si in β-$FeSi_2$ as shown in Table 12.1, the difference in formation energies between two sites is less than 0.003 eV/atom for both Fe and Si site. The hierarchy of formation energies of α, β and γ-$FeSi_2$ agree with their stabilities: β-$FeSi_2$ is low-temperature phase and γ-$FeSi_2$ does not appear in Fe-Si equilibrium phase diagram. In the Fe-rich side of α and β-$FeSi_2$, the Fe antisite atom shows lower formation energies than the Si vacancy, which indicates that Fe antisite atoms are introduced in the Fe-rich side by the deviation from the stoichiometric composition. On the other hand, in the Si-rich side of α-$FeSi_2$, the Fe vacancy is more stable than the Si antisite atom, whereas the Si antisite atom shows slightly lower formation energies than the Fe vacancy in the Si-rich side of β-$FeSi_2$. In the case of γ-$FeSi_2$, the Fe and Si vacancies are introduced around the stoichiometric composition.

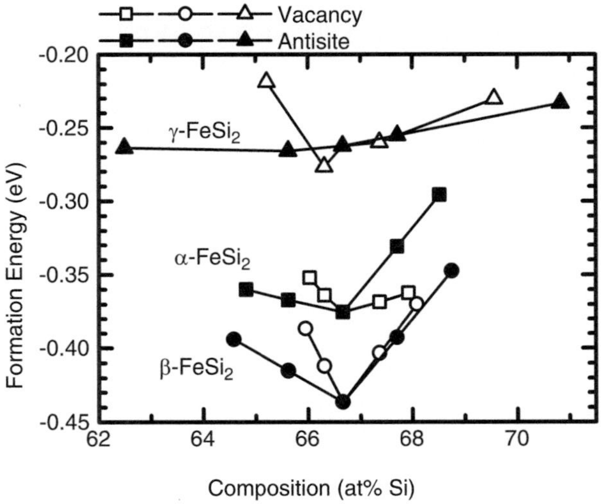

Fig. 12.2 Formation energies of α, β and γ-$FeSi_2$ as a function of Si composition

In the case of the Fe-rich side of β-$FeSi_2$ in which Fe antisite atoms are introduced by the deviation from the stoichiometric composition, the formation of Fe and Si vacancies proceeds via reactions:

$$\text{Fe vacancy}: 0 \rightarrow 3V_{\text{Fe}} + 2\text{Fe}_{\text{Si}} + \text{FeSi}_2, \tag{12.3}$$
$$\text{Si vacancy}: \text{Fe}_{\text{si}} \rightarrow 3V_{\text{Si}} + \text{FeSi}_2. \tag{12.4}$$

Here V_{Fe} and V_{si} denote Fe and Si vacancy, respectively. Fe_{Si} is an antisite atom: a Fe atom located at a Si site. The last term, $FeSi_2$, is the additional lattice site occupied by atoms subtracted as vacancies. The compositional deviation caused by the vacancy formation is compensated by the creation or annihilation of Fe antisite atoms. The processes of the formation of Fe and Si vacancies in the Si-rich side can be expressed in the same manner.

$$\text{Fe vacancy}: 2\text{Si}_{\text{Fe}} \rightarrow 3V_{\text{Fe}} + \text{FeSi}_2, \tag{12.5}$$
$$\text{Si vacancy}: 0 \rightarrow 3V_{\text{Si}} + \text{Si}_{\text{Fe}} + \text{FeSi}_2. \tag{12.6}$$

The process of the vacancy formation in the Fe rich side of α-$FeSi_2$ is the same as that of β-$FeSi_2$. In the Si rich side of α-$FeSi_2$, Fe vacancies are introduced by the deviation from the stoichiometric composition. The formation energy of the Fe vacancy is, therefore, zero. The formation of a Si vacancy proceeds via reactions:

$$\text{Si vacancy}: 0 \rightarrow V_{\text{Fe}} + 2V_{\text{Si}} + \text{FeSi}_2. \tag{12.7}$$

In γ-$FeSi_2$, the formation energy of a Fe vacancy in the Si-rich side and a Si vacancy in the Fe rich-side is zero because these vacancies exist for the compensation of compositional deviations. The formation of a Fe vacancy in the Fe-rich side and a Si vacancy in the Si-rich side proceed via the same reaction as (12.7). At the stoichiometric composition, possible processes of vacancy formations are the reactions expressed by (12.3), (12.6) and (12.7), in which there are no initial defects needed for vacancy formations.

The β-$FeSi_2$ is usually deposited on Si substrates as a thin film. In this case, the vacancy formation in β-$FeSi_2$ proceeds with Si from substrates via reactions:

$$\text{Fe vacancy}: 2\text{Si} \rightarrow V_{\text{Fe}} + \text{FeSi}_2, \tag{12.8}$$
$$\text{Si vacancy}: 0 \rightarrow V_{\text{Si}} + \text{Si}. \tag{12.9}$$

The vacancy formation energies can be obtained by calculating the energy difference of these reactions. In the case of the Fe vacancy expressed by (12.3), the formation energy, $\widetilde{E}^{\text{Fe}}_{vac}$, can be calculated as

$$\begin{aligned}\widetilde{E}^{\text{Fe}}_{vac} &= \left[E\left(N_{\text{Fe}} - 1, N_{\text{Si}}, V^{\text{Fe}}\right) - E(N_{\text{Fe}}, N_{\text{Si}}, 0)\right] \\ &+ \frac{2}{3}[E(N_{\text{Fe}} + 1, N_{\text{Si}} - 1, \text{Fe}_{\text{Si}}) - E(N_{\text{Fe}}, N_{\text{Si}}, 0)] \\ &+ \frac{1}{N_{\text{Fe}} + N_{\text{Si}}} E(N_{\text{Fe}}, N_{\text{Si}}, 0).\end{aligned} \tag{12.10}$$

Here $E(N_{Fe}-1,N_{Si},V^{Fe})$ in the first term denotes the energy of a supercell with $(N_{Fe}-1)+N_{Si}$ atoms including a Fe vacancy. $E(N_{Fe}+1,N_{Si}-1,\ Fe_{Si})$ in the second term is the energy of a supercell with $(N_{Fe}+1)+(N_{Si}-1)$ atoms including a Fe antisite atom. The energy of $FeSi_2$ in the last term is also calculated using a supercell composed of $N_{Fe}+N_{Si}$ atoms in order to reduce the error arising from the supercell size. For calculating these energies, we employed supercells composed of 96 atoms: $N_{Fe}=32$ and $N_{Si}=64$.

Vacancy formation energy can be described as a sum of two terms: the difference in energy between the defective and perfect supercells, and the chemical potential of atoms subtracted as defects. The first term of (12.10) is the difference in energy between the supercell including a Fe vacancy and the perfect supercell. The sum of the second and third terms of (12.10), therefore, corresponds to the chemical potential of the Fe atom subtracted as a vacancy.

In the Si-lean condition, it was reported that FeSi phase is probably formed first (Liu et al. 2008). The formation energies of Fe and Si vacancies in FeSi are also calculated in the same manner as $FeSi_2$. FeSi crystallizes in the cubic type structure, which belongs to the space group $P2_13$, with lattice parameters of $a=4.495$ Å. We have calculated formation energies of non-stoichiometric FeSi using supercells composed of 64 lattice sites. The calculated formation energies of FeSi are plotted as a function of Si composition in Fig. 12.3. The defects introduced by the deviation from the stoichiometric composition in FeSi are the same as those in α-$FeSi_2$: Fe antisite atoms in the Fe-rich side and Fe vacancies in the Si-rich side. The formation of Fe and Si vacancies proceeds via reactions:

Fe-rich side:

$$\text{Fe vacancy}: 0 \rightarrow 2V_{Fe} + Fe_{Si} + FeSi, \tag{12.11}$$
$$\text{Si vacancy}: Fe_{Si} \rightarrow 2V_{Si} + FeSi. \tag{12.12}$$

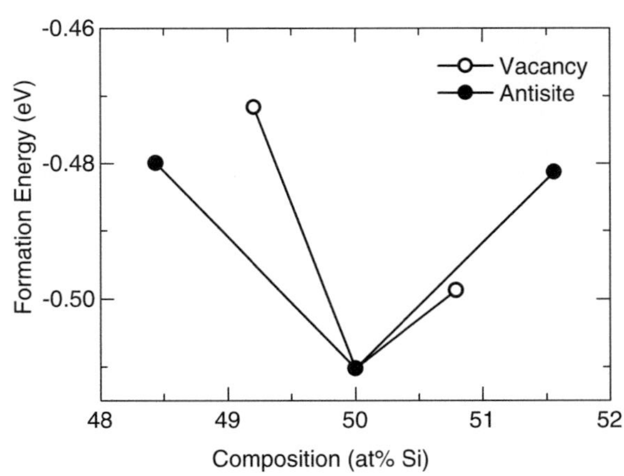

Fig. 12.3 Formation energies of FeSi as a function of Si composition

Si-rich side:

$$\text{Si vacancy}: 0 \rightarrow V_{\text{Fe}} + V_{\text{Si}} + \text{FeSi}. \tag{12.13}$$

At the stoichiometric composition, the following reaction is also possible in addition to the reactions of (12.11) and (12.13).

$$\text{Si vacancy}: 0 \rightarrow 2V_{\text{Si}} + \text{Si}_{\text{Fe}} + \text{FeSi}. \tag{12.14}$$

12.4 Results and Discussion

12.4.1 Chemical Bonding in $FeSi_2$

The total and partial density of states (DOS) and overlap population diagrams of perfect crystals of α, β and γ-$FeSi_2$ are shown in Figs. 12.4, 12.5 and 12.6, respectively. The DOS curves are shifted so as to set the highest occupied molecular orbital at 0 eV. The general shapes of the density of states are similar all the cases. The large peaks near the Fermi level are mainly composed of the Fe 3d states interacting with the Si 3s and 3p states. The Si 3s and 3p states widely spread from -14 eV to the conduction band because of the formation of

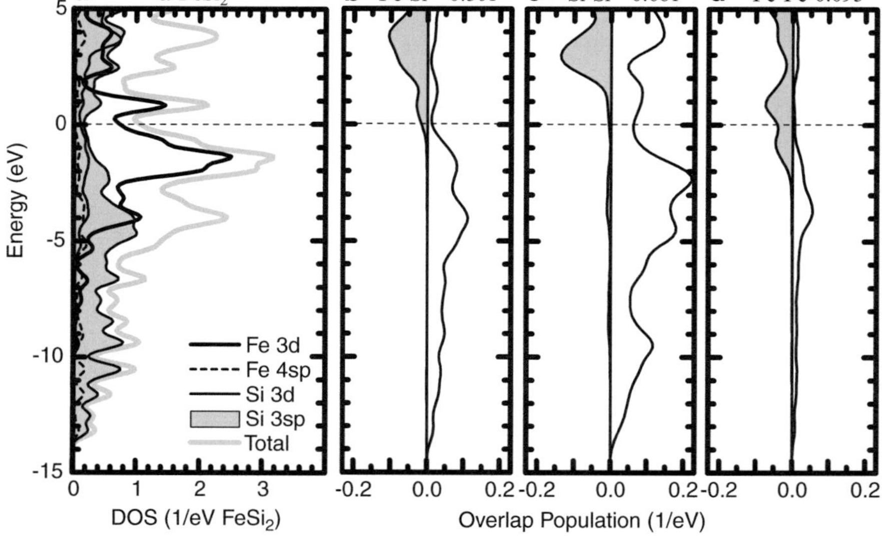

Fig. 12.4 (**a**) Total and partial density of states and overlap population diagrams of (**b**) Fe-Si bond, (**c**) Si-Si bond and (**d**) Fe-Fe bond in α-$FeSi_2$. Bond overlap populations are given on the right top of each panel

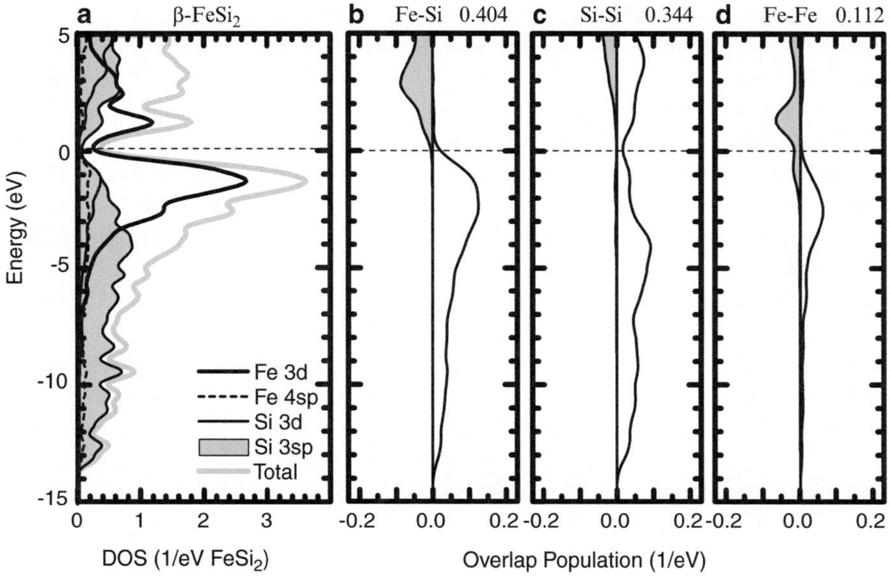

Fig. 12.5 (**a**) Total and partial density of states and overlap population diagrams of (**b**) Fe-Si bond, (**c**) Si-Si bond and (**d**) Fe-Fe bond in β-$FeSi_2$. Bond overlap populations are given on the right top of each panel

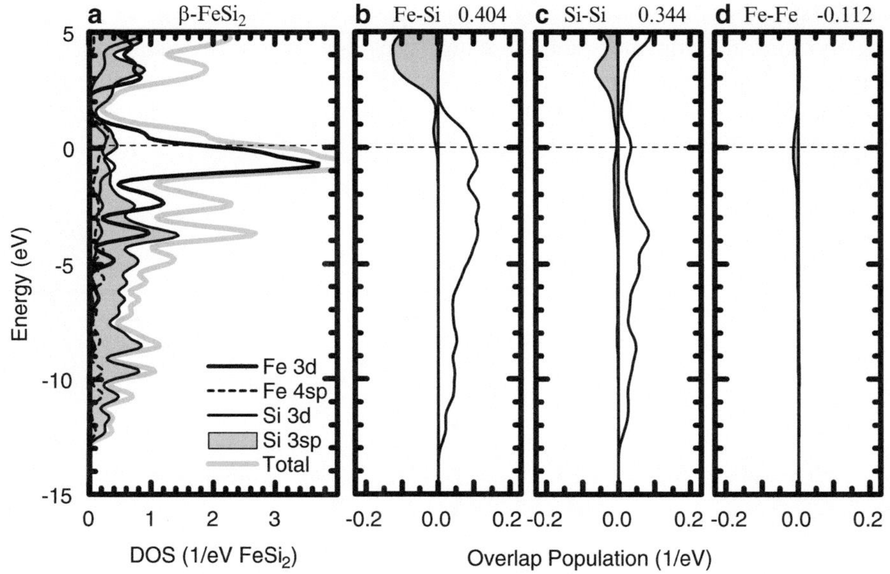

Fig. 12.6 (**a**) Total and partial density of states and overlap population diagrams of (**b**) Fe-Si bond, (**c**) Si-Si bond and (**d**) Fe-Fe bond in γ-$FeSi_2$. Bond overlap populations are given on the right top of each panel

the Si-Si bonds. α and γ-$FeSi_2$ are metallic phases and the Fermi level lines in the middle of the Fe 3d bands. On the other hand, the Fermi level in β-$FeSi_2$ lies at the deep minimum in the Fe 3d bands. This feature well reproduces the semiconducting behavior of β-$FeSi_2$.

The bonding and antibonding bands of the Fe-Si bond in β-$FeSi_2$ are completely separated by the interaction between the Fe 3d and the Si 3s and 3p states. This interaction leads to a gap at the Fermi level in β-$FeSi_2$. In γ-$FeSi_2$, the separation is not perfect and the Fermi level lines in the bonding band. The distortion in the crystal structure of β-$FeSi_2$ can be, therefore, considered as a Jahn-Teller distortion which lowers the energy and forms a band gap. There are two types of Si-Si bonds in α-$FeSi_2$ as shown in Table 12.2. The overlap population diagram of the Si-Si bond in α-$FeSi_2$ is plotted for shorter Si-Si bonds connecting the Si layers. The Si-Si bonds in α-$FeSi_2$ are shorter than those in β and γ-$FeSi_2$ because there are Si-Si layers in which the Fe atoms are not occupied. The bonding states of the Si-Si bond in α-$FeSi_2$ are, therefore, larger than those in β and γ-$FeSi_2$. The Fe-Fe bonds in α-$FeSi_2$ are also shorter than those in β and γ-$FeSi_2$ because the adjacent body-centered sites are occupied by the Fe atoms in α-$FeSi_2$. However the Fermi level lies in the middle of the antibonding bands of the Fe-Fe bond. The Fe-Fe bond is much weaker than the Fe-Si and Si-Si bond because of the antibonding contribution.

The bond overlap populations (BOP) of the Fe-Fe, Fe-Si and Si-Si bonds are shown in Fig. 12.7. In α-$FeSi_2$, the Si-Si bonds between the Si layers are very strong as described above. Compared to γ-$FeSi_2$, the BOPs of the Si-Si and Fe-Fe bonds in β-$FeSi_2$ are larger. This is due to the difference in bond length. Local structures around the Fe site are shown in Fig. 12.8. In γ-$FeSi_2$, the bond lengths of the Fe-Fe are too long to form the stable bonding. The Si-Si bonds are also longer than those in β-$FeSi_2$. The bond lengths of the Si-Si bonds in β-$FeSi_2$ are adjusted to the suitable length by breaking two Si-Si bonds around the Fe site. The Fe-Fe distances are also shortened to form the Fe-Fe bonds in β-$FeSi_2$.

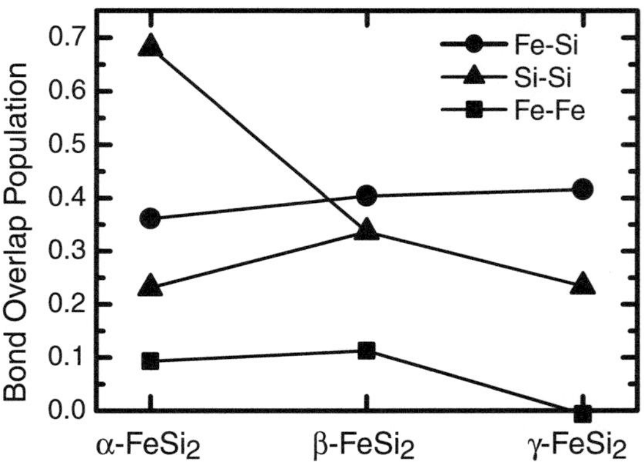

Fig. 12.7 Bond overlap populations of the Fe-Si, Si-Si and Fe-Fe bonds in $FeSi_2$

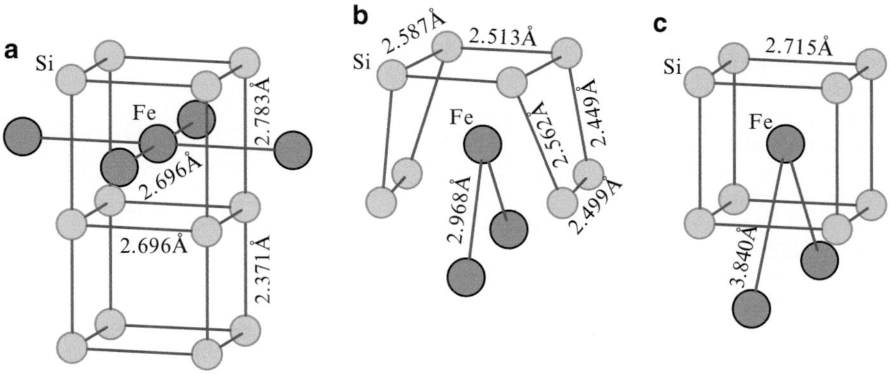

Fig. 12.8 Local structures around the Fe site in (**a**) α, (**b**) β and (**c**) γ-$FeSi_2$

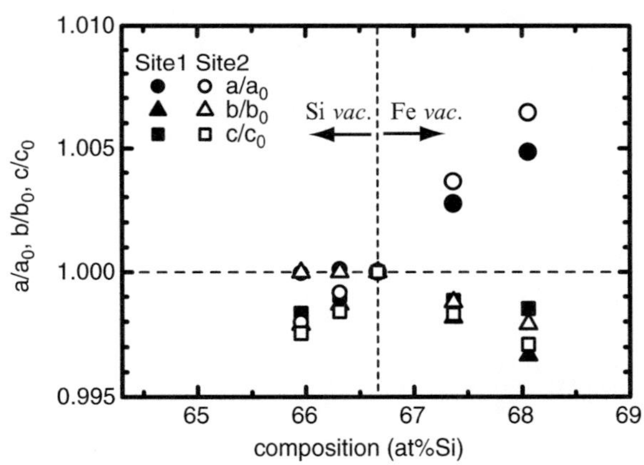

Fig. 12.9 Calculated lattice parameters of β-$FeSi_2$, in which compositional deviations are compensated by the Fe or Si vacancies, as a function of Si composition. The lattice parameters are normalized by the equilibrium lattice parameter of the stoichiometric composition

12.4.2 *Compositional Dependence of Lattice Parameters of β-$FeSi_2$*

The Si/Fe ratio is an important factor for both sintered β-$FeSi_2$ and β-$FeSi_2$ thin films. Figures 12.9 and 12.10 show the calculated lattice parameters of β-$FeSi_2$ including vacancies and antisite atoms as a function of Si composition, respectively. In the case of vacancies, the Fe vacancies both on Fe1 and Fe2 sites increase the lattice parameter of *a*-axis. This is due to anisotropic atomic relaxations around the Fe vacancy as described below. Whereas the Si antisite atoms also increase the lattice parameters, the Si1 and Si2 sites show different trends: the Si1 antisite atoms increase the lattice parameters of the *a* and *c*-axes; the Si2 antisite atoms increase those of *b* and *c*-axes. The influence of the Fe antisite atoms on the lattice parameters is smaller than the other defects. In the case of sintered $FeSi_2$, it was reported that the lattice parameter of the *a*-axis increases with increasing Si/Fe ratio, but those of the *b* and *c*-axes remain almost constant (Kobayashi et al. 1990).

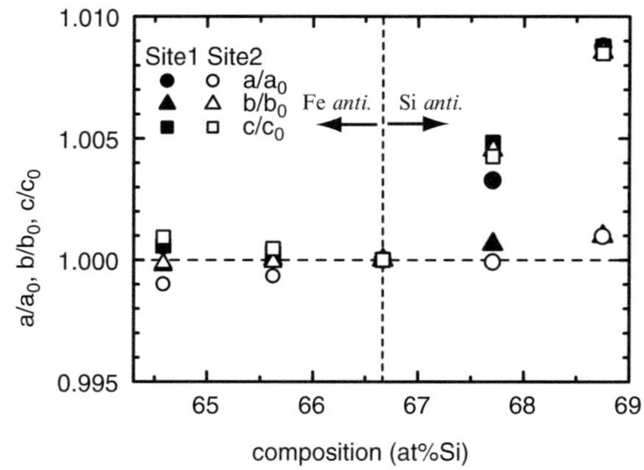

Fig. 12.10 Calculated lattice parameters of β-$FeSi_2$, in which compositional deviations are compensated by the Fe or Si antisite atoms, as a function of Si composition. The lattice parameters are normalized by the equilibrium lattice parameter of the stoichiometric composition

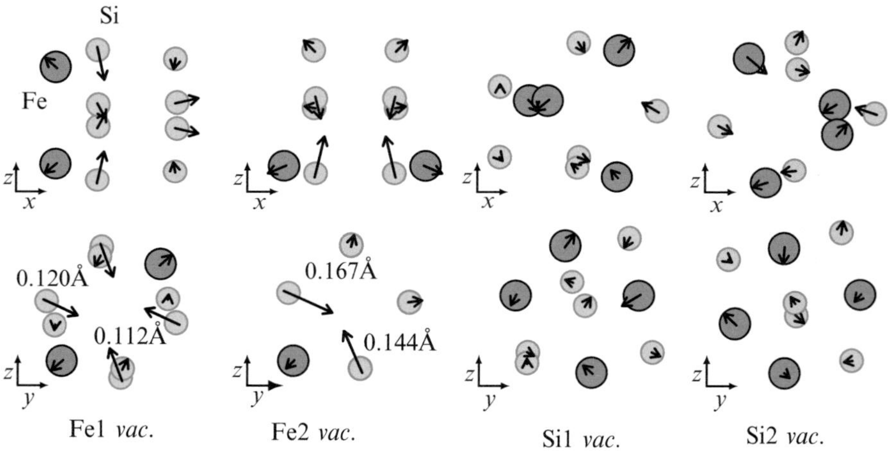

Fig. 12.11 Structures around the Fe and Si vacancies in β-$FeSi_2$. *Arrows* with a scale factor of 10 show the displacements from the ideal position in β-$FeSi_2$

This trend cannot be explained by the change in the lattice parameters arising from single defect studied here.

The arrangements of the neighboring atoms around the vacancy and antisite atom are shown in Figs. 12.11 and 12.12, respectively. With respect to the vacancy, the lattice parameter of the *a*-axis shows an increase with increasing Fe vacancy, which indicates that anisotropic relaxations are induced around the Fe vacancy. As shown in Fig. 12.11, around the Fe vacancy, the Si atoms near the neighboring Fe atoms significantly displace inward towards the Fe vacancy and the neighboring Fe atoms move outward from the Fe vacancy, which leads to the increase of the lattice parameter of the *a*-axis. Compared to the Fe vacancy, the neighboring atoms around the Si vacancy show mild relaxations. The lattice parameters are almost constant or decreased by the Si vacancy. In the case of the antisite atom, the Si antisite atom significantly increases the two lattice parameters while the change in lattice

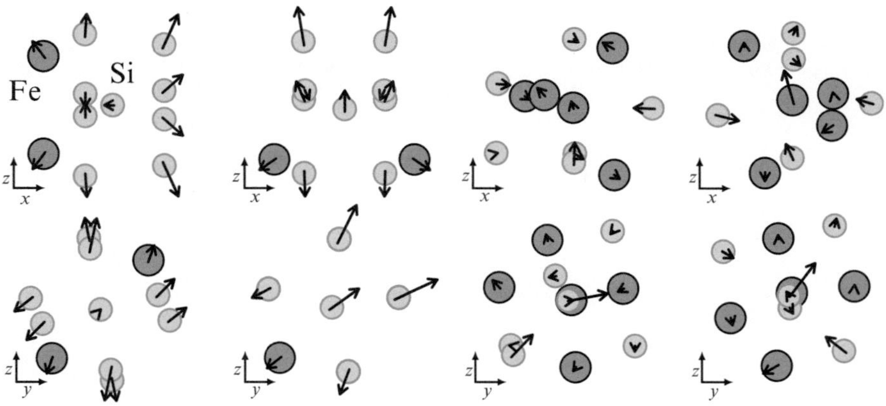

Fig. 12.12 Structures around the Fe and Si antisite atoms in β-$FeSi_2$. *Arrows* with a scale factor of 10 show the displacements from the ideal position in β-$FeSi_2$

parameters induced by the Fe antisite atom is gentle. The average bond length of the Si-Si bonds is longer than that of the Fe-Si bond as shown in Table 12.2. The Si antisite atom, therefore, induces outward relaxations of the neighboring Si atoms to form the Si-Si bonds, which leads to the increase of the lattice parameters.

12.4.3 Vacancy Formation Energy

The calculated formation energies of Fe and Si vacancies for all of the possible reactions are listed in Tables 12.3 and 12.4, respectively. The lowest energies of Fe and Si vacancies at each condition in $FeSi_2$ are shown in Figs. 12.13 and 12.14, respectively. In all of the conditions, β-$FeSi_2$ has the highest vacancy formation energy both of the Fe and Si sites. In γ-$FeSi_2$, the vacancy formation energies are zero or negative. These trends arise from the difference in the formation energies of $FeSi_2$ shown in Fig. 12.2. The energetic penalty of vacancy formation, namely the energy of breaking the bonds between the neighboring atoms expressed by the first term of Eq. (12.10), is the highest in β-$FeSi_2$ having the lowest formation energies at any compositions. On the other hand, the vacancies are introduced at around the stoichiometric composition of γ-$FeSi_2$. In particular, the non-stoichiometric γ-$FeSi_2$ including the Si vacancies possesses lower formation energies than the stoichiometric γ-$FeSi_2$. These features lead to the negative formation energies of the Fe and Si vacancy in γ-$FeSi_2$.

The formation energy of the Fe vacancy steeply decreases going from the Fe-rich to the Si-rich condition. The formation energy of the Si vacancy has an opposite trend and exhibits gradual changes compared to the Fe vacancy. This is because the sum of the chemical potential of one Fe and two Si atoms is equal to the

Table 12.3 Formation energies of a Fe vacancy in $FeSi_2$ and FeSi

	Reaction	Formation energy (eV)
α-$FeSi_2$	$0 \rightarrow 3V_{Fe} + 2Fe_{Si} + FeSi_2$	1.19
	$0 \rightarrow V_{Fe} + 2V_{Si} + FeSi_2$	0.95
	$V_{Fe} \rightarrow V_{Fe}$	0
	$2Si \rightarrow V_{Fe} + FeSi_2$	−0.08
β-$FeSi_2$	$0 \rightarrow 3V_{Fe} + 2Fe_{Si} + FeSi_2$	4.64, 4.53
	$0 \rightarrow V_{Fe} + 2V_{Si} + FeSi_2$	2.60, 2.63
	$2Si_{Fe} \rightarrow 3V_{Fe} + FeSi_2$	0.47, 0.35
	$2Si \rightarrow V_{Fe} + FeSi_2$	2.40, 2.29
γ-$FeSi_2$	$0 \rightarrow 3V_{Fe} + 2Fe_{Si} + FeSi_2$	−2.48
	$0 \rightarrow V_{Fe} + 2V_{Si} + FeSi_2$	−0.83
	$V_{Fe} \rightarrow V_{Fe}$	0
	$2Si \rightarrow V_{Fe} + FeSi_2$	−0.30
FeSi	$0 \rightarrow 2V_{Fe} + Fe_{Si} + FeSi$	1.70
	$0 \rightarrow V_{Fe} + V_{Si} + FeSi$	1.58
	$V_{Fe} \rightarrow V_{Fe}$	0
	$Si \rightarrow V_{Fe} + FeSi$	0.22

Table 12.4 Formation energies of a Si vacancy in $FeSi_2$ and FeSi

	Reaction	Formation energy (eV)
α-$FeSi_2$	$Fe_{Si} \rightarrow 3V_{Si} + FeSi_2$	0.82
	$0 \rightarrow V_{Fe} + 2V_{Si} + FeSi_2$	0.95
	$0 \rightarrow 3V_{Si} + Si_{Fe} + FeSi_2$	1.42
	$0 \rightarrow V_{Si} + Si$	1.46
β-$FeSi_2$	$Fe_{Si} \rightarrow 3V_{Si} + FeSi_2$	1.64, 1.69
	$0 \rightarrow V_{Fe} + 2V_{Si} + FeSi_2$	2.60, 2.63
	$0 \rightarrow 3V_{Si} + Si_{Fe} + FeSi_2$	3.73, 3.77
	$0 \rightarrow V_{Si} + Si$	2.76, 2.80
γ-$FeSi_2$	$V_{Si} \rightarrow V_{Si}$	0
	$0 \rightarrow V_{Fe} + 2V_{Si} + FeSi_2$	−0.83
	$0 \rightarrow 3V_{Si} + Si_{Fe} + FeSi_2$	−1.24
	$0 \rightarrow V_{Si} + Si$	−1.09
FeSi	$Fe_{Si} \rightarrow 2V_{Si} + FeSi$	1.46
	$0 \rightarrow V_{Fe} + V_{Si} + FeSi$	1.58
	$0 \rightarrow 2V_{Si} + Si_{Fe} + FeSi$	3.36
	$0 \rightarrow V_{Si} + Si$	2.94

energy of $FeSi_2$: $\Delta\mu_{Fe} = -2\Delta\mu_{Si}$. In the Si-rich condition of β-$FeSi_2$, the formation energy of the Fe vacancy is about 0.4 eV, indicating that the Fe vacancies are introduced in this condition. The change in lattice parameters associated with the Si/Fe ratio observed in sintered β-$FeSi_2$, described in the previous section, can be ascribed to the formation of the Fe vacancies and Si antisite atoms: the Fe vacancies increase the lattice parameter of *a*-axis and the decrease in the lattice parameters of the *b* and *c*-axis is reduced by the Si antisite atoms.

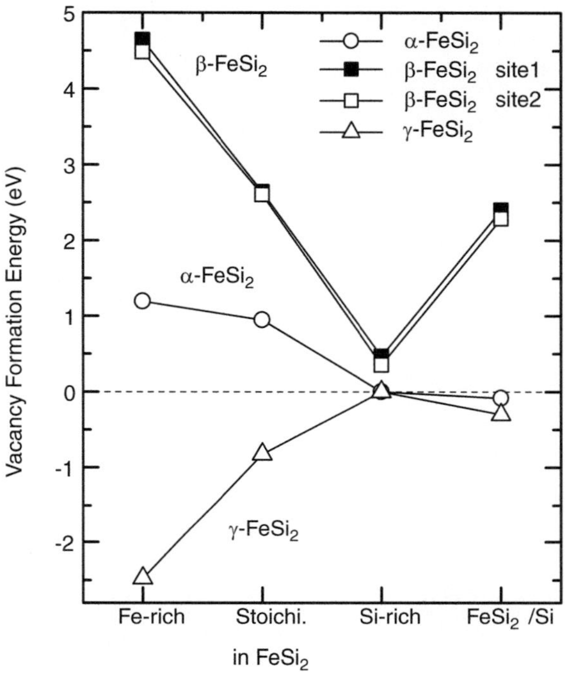

Fig. 12.13 Calculated formation energies of a Fe vacancy in $FeSi_2$

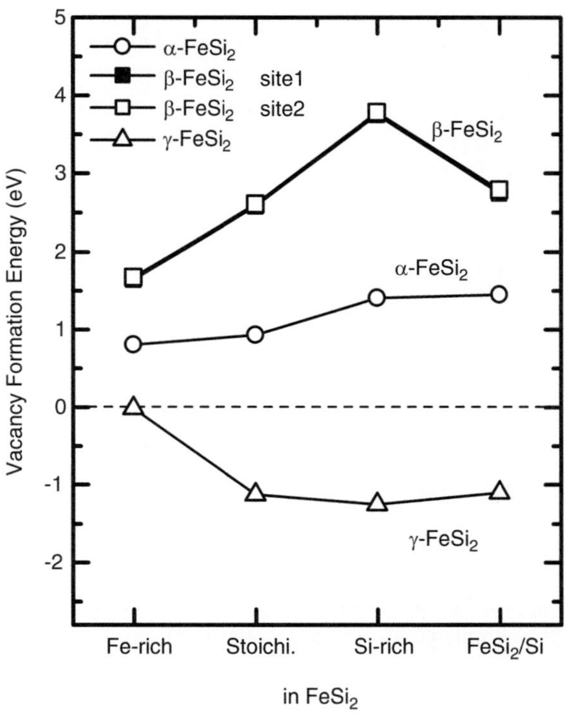

Fig. 12.14 Calculated formation energies of a Si vacancy in $FeSi_2$

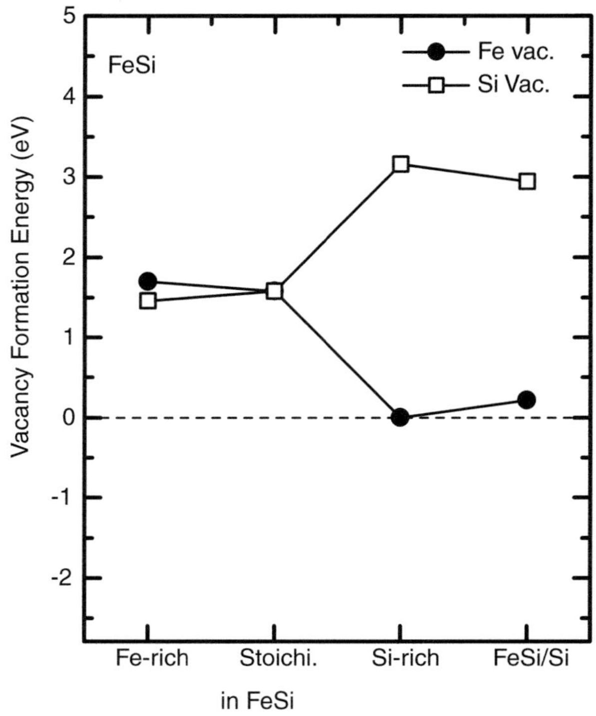

Fig. 12.15 Calculated formation energies of a Fe and a Si vacancy in FeSi

The formation energies of Fe and Si vacancies in FeSi are shown in Fig. 12.15. Compared to β-$FeSi_2$, the formation energies of Fe vacancies in FeSi are lower at each condition. This trend mainly originates from the difference in the chemical bonding of the Fe-Si bond. The energetic penalty of a Fe vacancy formation in FeSi is about 2 eV lower than that in β-$FeSi_2$. Whereas the Fe atom in $FeSi_2$ has eight neighboring Si atoms at the average distance of 2.37 Å, the Fe atom in FeSi has seven Si atoms at the average distance of 2.41 Å. In addition, in the Si-rich condition, the Fe vacancies are introduced in FeSi by the deviation from the stoichiometric composition. These results suggest that more vacancies are introduced in the case FeSi is formed first in the Si-lean condition for the growth of β-$FeSi_2$.

12.4.4 Positron Lifetimes of Bulk State

The distributions of positron density in the bulk state of $FeSi_2$ are shown in Fig. 12.16. The positron lifetimes and annihilation fractions are listed in Table 12.5. In general, positrons in a perfect crystal are delocalized in the interstitial region because of the repulsion from the ion cores. In the structure of $FeSi_2$, the Si atoms form a simple cubic sub-lattice and a half of the body centered sites are still

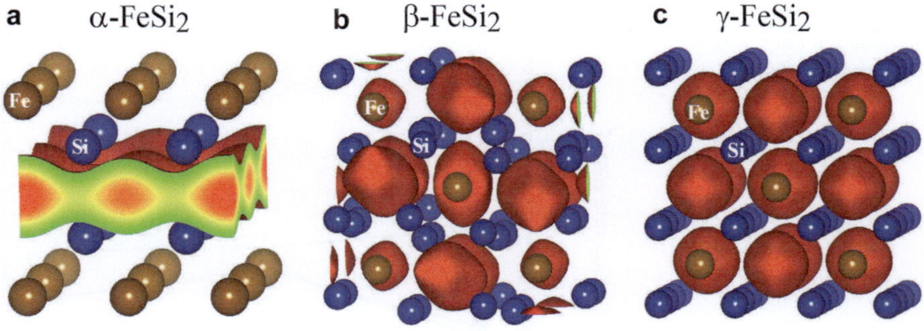

Fig. 12.16 Positron density distributions of the bulk states in (**a**) α, (**b**) β and (**c**) γ-$FeSi_2$

Table 12.5 Positron lifetime and annihilation fraction of Fe in $FeSi_2$

	Positron lifetime (ps)/Annihilation fraction of Fe (%)		
	Bulk	Fe vacancy	Si vacancy
α-$FeSi_2$	147/14.6	165/10.3	194/22.6
β-$FeSi_2$	137/30.7	172/19.1	175/39.1
γ-$FeSi_2$	148/24.2	171/14.3	188/41.3

unoccupied. Positrons, therefore, tend to localize at unoccupied body centered sites. In α-$FeSi_2$, positrons are distributed between the Si layers in which the Fe atoms are not occupied. The annihilation fraction of Fe atoms is, therefore, lower than that of the other phases. In β-$FeSi_2$, the interstitial sites are distorted and the free volume becomes smaller. As a result, the annihilation fraction of the Fe atoms increases and positron lifetime becomes shorter than the other phases.

The annihilation rates in the bulk state of $FeSi_2$ are shown in Fig. 12.17. The reciprocal of the sum of annihilation rates is equivalent to positron lifetime as shown in Eq. (12.2). Annihilation rates are proportional to annihilation fractions in the case distributions of positrons are similar such as the bulk state of $FeSi_2$. Changes induced by a vacancy show different trends for annihilation rates and fractions, because the annihilation rates reflect not only the fractions of orbitals but also the electron densities.

12.4.5 Positron Lifetimes of Fe and Si Vacancies

The distribution of positron density at the Fe and Si vacancies in $FeSi_2$ are shown in Fig. 12.18. The positron annihilation rates of the Fe and Si vacancies are also shown in Figs. 12.19 and 12.20, respectively. The Fe vacancy in each phase has a similar positron lifetime of about 170 ps. This is because the Fe atom in each phase has eight neighboring Si atoms and the free volume of the Fe vacancy is comparable. In the case of bcc-Fe, the positron lifetime of a Fe vacancy is about 70 ps longer

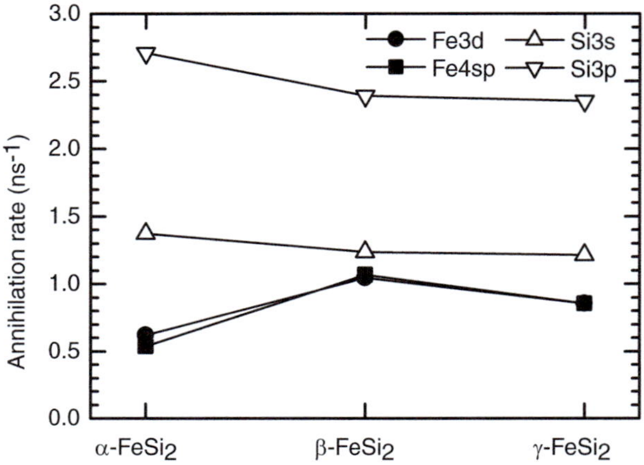

Fig. 12.17 Positron annihilation rates of the bulk states in $FeSi_2$

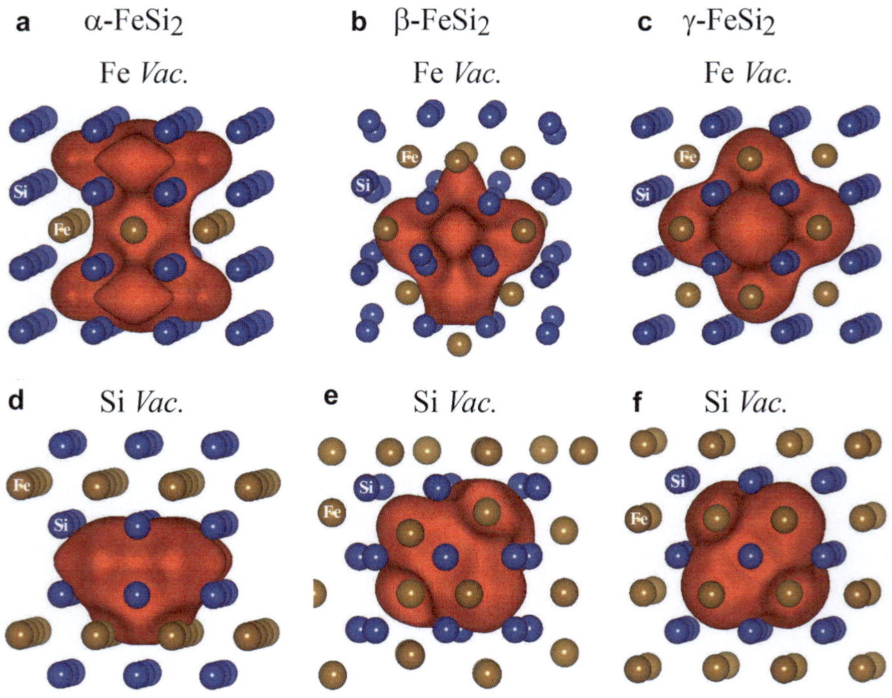

Fig. 12.18 Positron density distributions at a Fe vacancy in (**a**) α, (**b**) β and (**c**) γ-$FeSi_2$ and a Si vacancy in (**d**) α, (**e**) β and (**f**) γ-$FeSi_2$

than that of the bulk state. The increase in the positron lifetime induced by a Fe vacancy in $FeSi_2$ is less than half of that in bcc-Fe. This is because positrons in the bulk state of $FeSi_2$ are mainly distributed at the unoccupied body-centered sites. Compared to the bulk state, the positron annihilation rates of the Si 3s and 3p orbitals at the Fe vacancy remain almost unchanged, indicating that the unoccupied

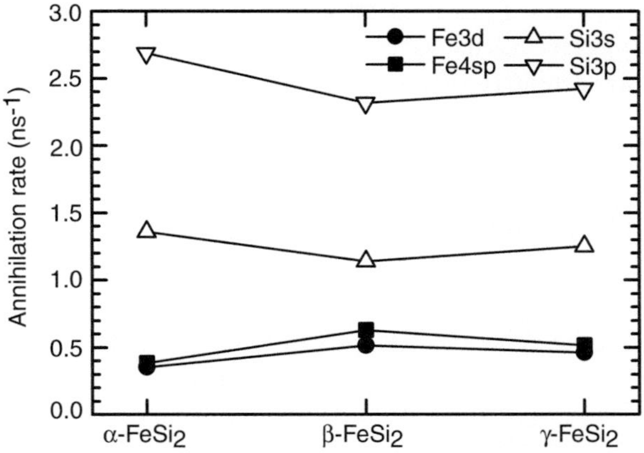

Fig. 12.19 Positron annihilation rates at a Fe vacancy in $FeSi_2$

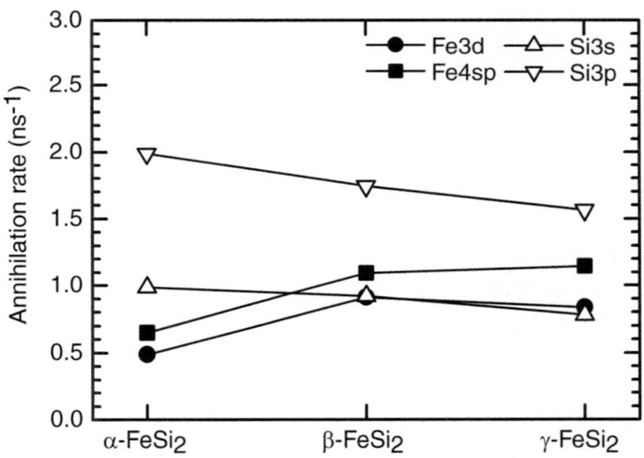

Fig. 12.20 Positron annihilation rates at a Si vacancy in $FeSi_2$

body-centered sites as well as the Fe vacancy contribute to the positron annihilation. This trend can be seen at the distribution of positron density shown in Fig. 12.18. In general, positrons tend to be strongly localized at a vacancy. Positrons at the Fe vacancy in $FeSi_2$ spread to the adjacent unoccupied sites.

On the other hand, the Si vacancy in α and γ-$FeSi_2$ has a longer positron lifetime than the Fe vacancy. The increase in the positron lifetime originates from the decrease in the annihilation rates of the Si 3s and 3p orbitals. The annihilation rates of the Fe 3d and Fe 4sp orbitals show only small changes, even though the annihilation fractions of the Fe atoms significantly increase. These results indicate that positrons are localized at the Si vacancy. The Si vacancy in β-$FeSi_2$ has a shorter positron lifetime than the other Si vacancies because the free volume around the Si atom is decreased by the Jahn-Teller distortion.

Whereas the Fe vacancy in α and γ-$FeSi_2$ has a shorter positron lifetime than the Si vacancy, positron lifetimes of the Fe and Si vacancies in β-$FeSi_2$ are close to each other. This result suggests that it may be difficult to identify vacancy sites in β-$FeSi_2$ by positron lifetime measurement. However, the annihilation fraction of the

Fe atoms of the Si vacancy is about twice as large as that of the Fe vacancy, which suggests that the vacancies in β-$FeSi_2$ can be identified using Coincidence Doppler broadening (CDB) technique (Asoka-Kumar et al. 1996).

12.5 Summary

Chemical bonding and positron lifetimes in α, β and γ-$FeSi_2$ are evaluated using the discrete-variational (DV) Xα cluster method. The DOS of β-$FeSi_2$ shows a deep minimum around the Fermi level, which corresponds to the semiconducting behavior. The deep minimum arises from the complete separation between the bonding and antibonding bands of the Fe-Si bonds. The calculated positron lifetimes of the Fe and Si vacancies are similar to each other in β-$FeSi_2$. It may be, therefore, difficult to identify vacancy sites in β-$FeSi_2$ by positron lifetime measurement. Vacancy formation energies are obtained using the plane-wave pseudopotential method. In β-$FeSi_2$, the formation energy of the Fe vacancy steeply decreases going from the Fe-rich to the Si-rich condition and shows the lowest value, about 0.4 eV, at the Si-rich limit. The Fe vacancy in β-$FeSi_2$ increases the lattice parameter of *a*-axis due to the anisotropic lattice relaxations around the Fe vacancy. These results suggest that the increase of the lattice parameter of *a*-axis with the Si/Fe ratio can be ascribe to the formation of Fe vacancies in β-$FeSi_2$. In the Si-lean condition for the growth of β-$FeSi_2$, more vacancies are introduced in the case FeSi is formed first because the formation energy of the Fe vacancy in FeSi is lower than that in β-$FeSi_2$.

Acknowledgements This work supported by 'Priority Assistance of the Formation of Worldwide Renowned Centers of Research – The Global COE Program (Project: Center of Excellence for Advanced Structural and Functional Materials Design)' from the Ministry of Education, Culture, Sports, Science and Technology of Japan, and in part by a Grant-in-Aid for Scientific Research from Ministry of Education, Culture, Sports, Science and Technology of Japan.

References

Adachi H, Tsukada M, Satoko C (1978) Discrete variational Xα cluster calculations. I. Application to metal clusters. J Phys Soc Jpn 45:875–883

Asoka-Kumar P, Alatolo M, Ghosh VJ, Kruseman AC, Nielsen B, Lynn KG (1996) Increased elemental specificity of positron annihilation spectra. Phys Rev Lett 77:2097–2100

Blöchl PE (1994) Projector augmented-wave method. Phys Rev B 50:17953–17979

Boroński E, Nieminen RM (1986) Electron-positron density-functional theory. Phys Rev B 34:3820–3831

Christensen NE (1990) Electronic structure of β-$FeSi_2$. Phys Rev B 42:7148–7153

Ellis DE, Painter GS (1970) Discrete variational method for the energy-band problem with general crystal potentials. Phys Rev B 2:2887–2898

Kobayashi M, Hijikata K, Ido S (1990) The composition dependence of some electrical properties of $FeSi_x$ thin films. Jpn J Appl Phys 29:1118–1121

Kresse G, Furthmuller J (1996a) Efficiency of ab-initio total energy calculations for metals and semiconductors using a plane-wave basis set. Comput Mater Sci 6:15–50
Kresse G, Furthmuller J (1996b) Efficient iterative schemes for ab initio total-energy calculations using a plane-wave basis set. Phys Rev B 54:11169–11186
Kresse G, Joubert D (1999) From ultrasoft pseudopotentials to the projector augmented-wave method. Phys Rev B 59:1758–1775
Liu Z, Tanaka M, Kuroda R, Osamura M, Makia Y (2008) Influence of Si/Fe ratio in multilayer structures on crystalline growth of β-$FeSi_2$ thin film on Si substrate. Appl Phys Let 93:021907-1-3
Mulliken RS (1955) Electronic population analysis on LCAO-MO molecular wave functions. I. J Chem Phys 23:1833–1840
Onda N, Henz J, Müller E, Mäder KA, von Känel H (1992) Epitaxy of fluorite-structure silicides: metastable cubic $FeSi_2$ on Si(111). Appl Surf Sci 56:421–426
Perdew JP, Wang Y (1992) Accurate and simple analytic representation of the electron-gas correlation energy. Phys Rev B 45:13244–13249
Puska MJ, Nieminen RM (1983) Defect spectroscopy with positrons: a general calculational method. J Phys F Met Phys 13:333–346
Seki N, Takakura K, Suemasu T, Hasegawa F (2003) Conduction type and defect levels of β-$FeSi_2$ films grown by MBE with different Si/Fe rations. Mater Sci Semicond Process 6:307–309
Tani J, Kido H (2000) Hall effect and thermoelectric properties of $FeSi_x$. Jpn J Appl Phys 39:1054–1057
Tani J, Kido H (2003) First-principles study of native point defects in β-$FeSi_2$. J Alloys Comp 352:153–157

Chapter 13
Structural Analysis of Al_2TiO_5 at Room Temperature and at 600 °C by DV-Xα Approach (II)

Myung Chul Chang

13.1 Introduction

The Al_2TiO_5 ceramics have advantages such as high melting point, non-wetting behavior with aluminum alloys, non-reactivity with aluminum alloys, extremely good thermal shock resistance, and very low thermal expansion coefficient (Miyauchi et al. 2007; Kim et al. 2003). However, instead of low thermal expansion coefficient it has a low Young's modulus, indicating the low mechanical strength.

It is known (Morosin and Lynch 1972) that there is a complete disorder in the metal sites for Al_2TiO_5. The differences in the structure at room temperature and 600 °C appear to support that the coordination polyhedra about the metal ions tend towards more regular configurations with the temperature increase, and the importance of cleavage planes, resulting from edge-shared octahedra in the hysteresis of the thermal expansion of the ceramic. As particular interest materials; when heated, it contracts along at least one crystallographic direction, meaning the thermal expansion anisotropy and the distortion of atomic coordination polyhedra. Bond length and angles vary markedly about the cations. DV-Xα analysis is investigated for the cluster models at room temperature and 600 °C.

The excellent thermal shock-resistance of Al_2TiO_5 ceramics results from the combination of low thermal expansion and Young's modulus, but Al_2TiO_5 ceramics has a low mechanical strength because of micro-cracks induced by the high anisotropy of thermal expansion coefficients along the crystallographic axes. The preparation of stabilized Al_2TiO_5 phase (Morosin and Lynch 1972; Tsetsekou 2005; Ghiorso et al. 1999) is intended during the sintering of Al_2O_3 and TiO_2 above 1,300 °C, but unstabilized Al_2TiO_5 phase decomposes into Al_2O_3 and TiO_2 between 800 and 1,300 °C. The decomposition results from the collapse of adjacent octahedra of Al and Ti, because of too large occupancy of the lattice site occupied

M.C. Chang (✉)
Kunsan National University, 573-701 Miryong-Dong, Kunsan, Republic of Korea
e-mail: mcchang@kunsan.ac.kr

T. Ishii et al. (eds.), *The DV-Xα Molecular-Orbital Calculation Method*,
DOI 10.1007/978-3-319-11185-8_13

by Al atom. Al atom migrates from its position and the structural dissolution occurs into rutile and corundum because of the thermal energy induced by the octahedral collapse. The thermal stability of Al_2TiO_5 can be improved by the incorporation of MgO, Fe_2O_3, or TiO_2, which are isomorphous with the mineral pseudobrookite, causing a decrease in the decomposition temperature of isostructural compounds. As another stabilization the grain growth can be limited by the addition of SiO_2 and/or ZrO_2, restraining the tendency of Al_2TiO_5 toward decomposition (Miyauchi et al. 2007; Kim et al. 2003).

13.2 Calculations

The incorporation of Mg and Si into Al_2TiO_5 cluster structure was modeled in order to investigate the molecular orbital energy change for the Tialite structure at room temperature and 600 °C. The clusters were prepared by using pseudobrookite data (Morosin and Lynch 1972; Tsetsekou 2005; Ghiorso et al. 1999) as follows; *Four molecules* per cell in an orthorhombic unit; metal atoms on fourfold sites at $\pm$(x, ¼, 0) and $\pm$(x + ½,¼, ½), on eightfold sites at $\pm$(x, y, 0), $\pm$(x, ½ − y, 0), $\pm$(x + ½, y, ½), and $\pm$(x + ½,½ − y, ½), and the oxygen atoms on similar symmetry sites (one fourfold and two different eightfold sites). Al-centered clusters [$Al_5Ti_3O_6$] and Ti-centered clusters [$Ti_4Al_3O_6$] were prepared, and Mg and/or Si were incorporated into a metal center, named like as Al-centered-Mg [$MgAl_4Ti_3O_6$] or Ti-centered-Si [$SiTi_3Al_3O_6$] cluster (Fig. 13.1).

The total charges for ALTO [$Al_6Ti_3O_7$], ALTO-u [$Al_6Ti_3O_7$, 600 °C], and ALTO-BCT [$Al_6Ti_3O_7$] clusters are +16, +16, +15, respectively. For the disordered structure metal sites of Al and Ti were replaced and the clusters were coded as ALTO-r1, ALTO-r2, ALTO-BCT-r1, and ALTO-u-r1, etc. (Fig. 13.2).

13.3 Results and Discussion

The electronic structures of Al_2TiO_5 at room temperature and at 600 °C were investigated by using discrete variational Xα molecular orbital methods (Chang 2009). In aluminum titanate having pseudobrookite isomorphous structure (Morosin and Lynch 1972; Tsetsekou 2005; Ghiorso et al. 1999), there are of compositions $M_2^{3+}Ti^{4+}O_5$ (in which M^{3+} may be Al) or $M^{2+}Ti_2^{4+}O_5$ (in which M^{2+} may be Mg). These materials crystallize in space group Bbmm, with lattice constants similar to those for Al_2TiO_5. At room temperature, the reported lattice constants of Al_2TiO_5 are $a_0 = 9.429$, $b_0 = 9.636$, and $c_0 = 3.591$ Å. At 600 °C the lattice constants are $a_0 = 9.481$, $b_0 = 9.738$, and $c_0 = 3.583$ Å. By using the positional and isotropic thermal parameters for Al_2TiO_5 DV-Xα molecular orbital energy was calculated. With the addition of Mg and/or Si into Al site in Al_2TiO_5 cluster [$Al_{18}Ti_{12}O_{50}$], variations of the calculated chemical bonds were coordinated

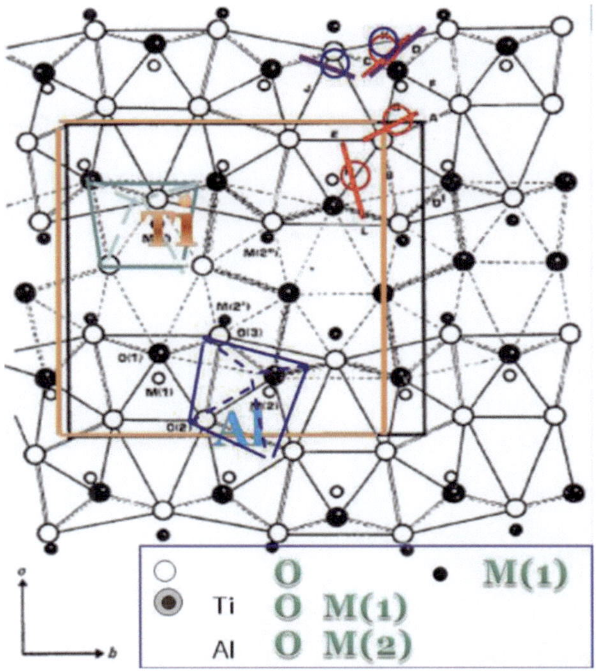

Fig. 13.1 The molecular structure of Al_2TiO_5 crystal; cleavage occurs across the weaker shared edges (edges type of C and D), shared edge is shorter than unshared edge, a cleavage plane – perpendicular to the a-axis (C and D edges), while that perpendicular to the b axis takes a rather crooked path (A, B, and D edges) (Morosin and Lynch 1972)

with the lattice structure in order to understand the single crystal thermal expansion and inter-atomic separations in Al_2TiO_5 at room temperature and 600 °C. The regular configurations of the coordination polyhedral about the metal ions with the temperature increase were calculated with the degree of the disorder in the metal sites. The real data of crystalline anisotropy as the degree of distortion of atomic coordination polyhedra were compared with the marked variation of the bond lengths and angles about the cations and the resulting total energy (Figs. 13.3 and 13.4).

The strength of the ceramic sharply increases at temperatures above 550 °C and the chemical recombination above 550 °C is accomplished by tetrahedral-to-octahedral interaction of surface metal ions on the opposite sides of a given microfracture (Morosin and Lynch 1972). The ease of which such interactions accomplish recombination accounts for the rather remarkable ability for Al_2TiO_5 ceramics to sustain repeated temperature cycling without degradation of the fracture-recombination mechanism. In the room temperature structure the metal ions are completely disordered in Al_2TiO_5. On the other hands in 600 °C crystal the metal ions are ordered in Al_2TiO_5. The differences between room temperature and 600 °C structures explain the anisotropic thermal expansion behavior of the single crystal (Morosin and Lynch 1972). There is a tendency towards more symmetric configurations about the metal ions with increasing temperature for Al_2TiO_5.

The results of Mg^{2+} or Si^{4+} incorporation into the Al or Si sites in AlO_6 or TiO_6 octahedral coordination may be reported as follows (Morosin and Lynch 1972);

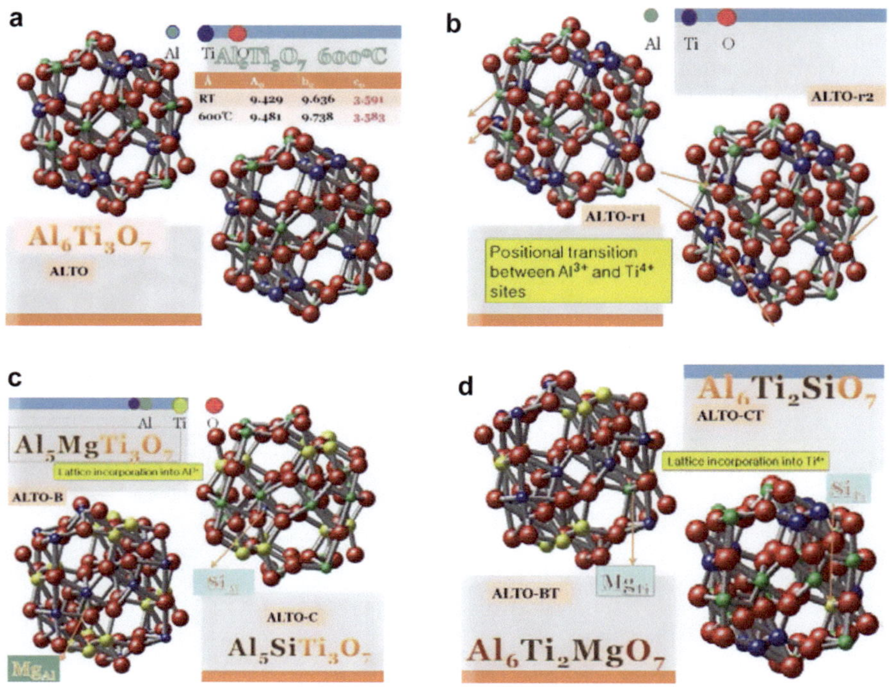

Fig. 13.2 The schematic diagrams of Al-Ti-O clusters with the positional change of Al and Ti sites, and the incorporation of Mg^{2+} or Si^{4+} into Al_2TiO_5; (**a**) $Al_6Ti_3O_7$ at 600 °C (**b**) positional transition of Al^{3+} and Ti^{4+} in $Al_6Ti_3O_7$ cluster (**c**) lattice incorporation of Si^{4+} into Al^{3+} site in $Al_6Ti_3O_7$ cluster (**d**) lattice incorporation of Mg^{2+} into Ti^{4+} site in $Al_6Ti_3O_7$ cluster

Structural differences at the two temperature result in the different coordination polyhedra about the metal ions. The cleavage planes, resulting from edge-shared octahedra in the thermal hysteresis of the ceramic cause binding energy change. The micro-cracking in polycrystalline Tialite ceramics is as follows; Single phase Aluminum Titanate is a low strength material. Thermal expansion anisotropy of three crystal directions is the reason of the grain boundary micro-cracking. These micro-cracks often result in negative thermal coefficients accompanied by pronounced hysteresis. Such anisotropy creates complicate internal stresses during cooling after sintering. The magnitude of internal stress is a direct function of anisotropic thermal expansion, which is sufficient to exceed the intrinsic fracture strength of the material, resulting in the severe micro-cracking at room temperature, the low mechanical strength, and low Young's modulus. The development of micro-cracks accounts for low thermal expansion coefficients, which are a bulk average value based on the behavior of individual grains.

The amounts of total energy for each model were compared individually from the DV-Xα calculation. As one of results the Al site incorporation was energetically preferable, but the Ti site incorporation was not. The positional transformation of Al^{3+} and Si^{4+} octahedral sites was preferable. It is known that the thermal stability

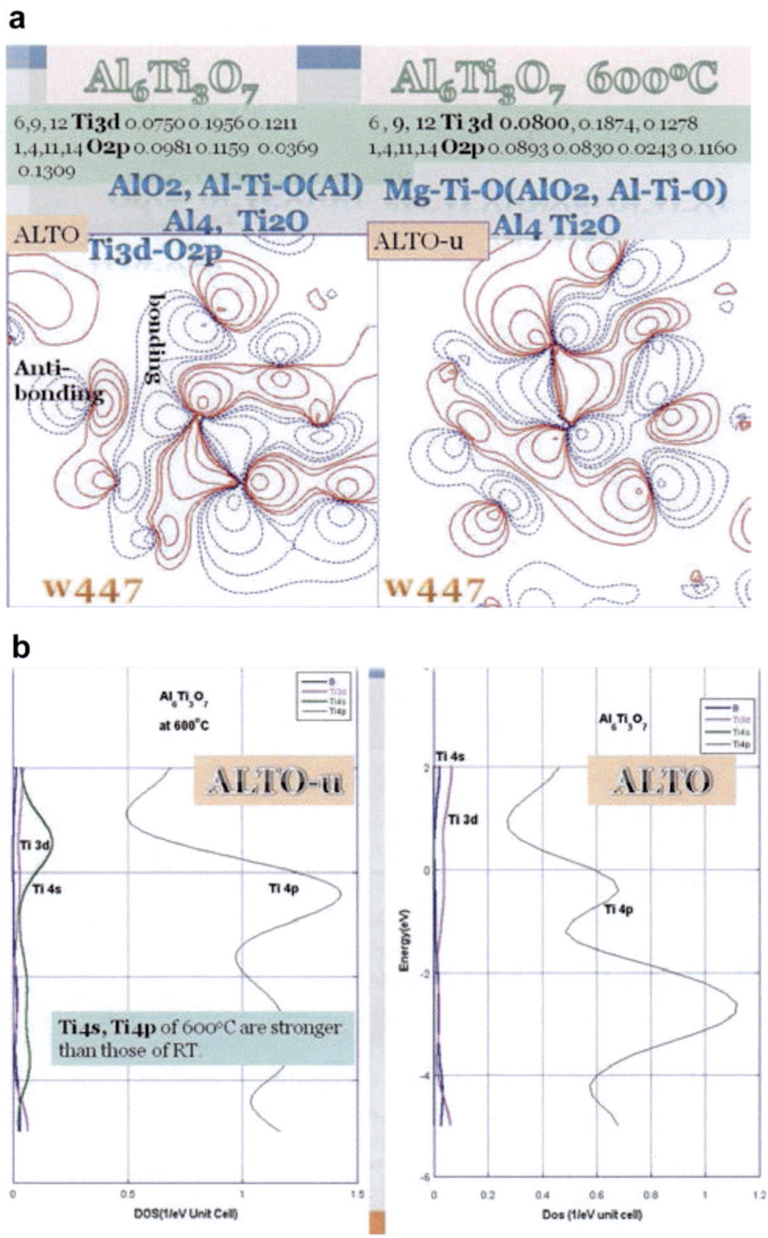

Fig. 13.3 The energy contour (**a**) and band energy population (**b**) in the clusters of ALTO and ALTO-U

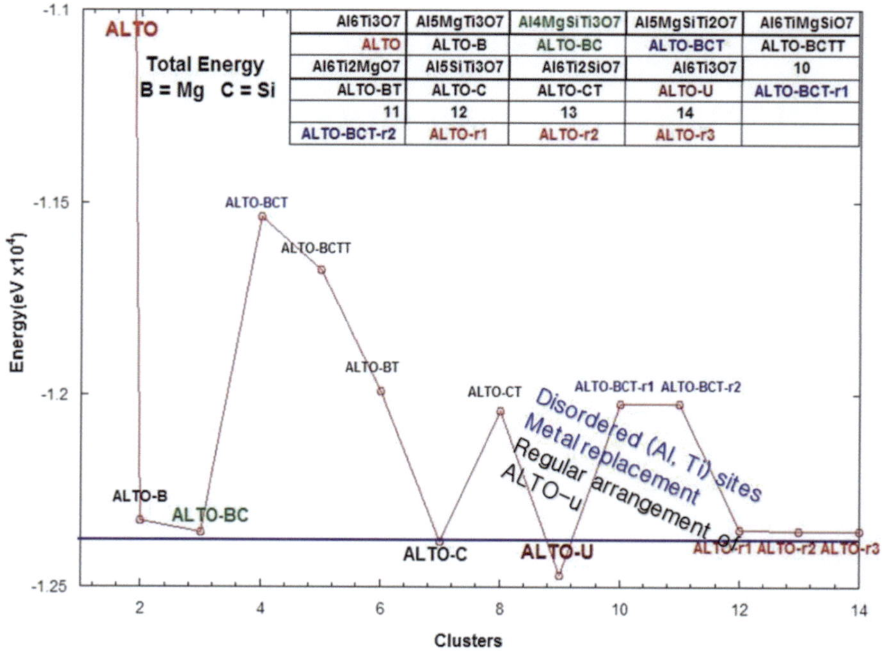

Fig. 13.4 The total energy in clusters: ALTO, ALTO-B, ALTO-BC, ALTO-BCT, ALTO-BCTT, AQLTO-BT, ALTO-C, ALTO-CT, and ALTO-U

of ALTO by the formation of solid solution with MgO is isomorphous with the mineral pseudobrookite and the mechanical stability by limiting its grain growth with the additives of SiO_2, or mulite (A_3S_2, $3Al_2O_3.2SiO_2$), not forming solid solution, but restrains the decomposition of Al_2TiO_5. The decomposition of Al_2TiO_5 is caused by the collapse of adjacent Al and Ti octahedra because the lattice site occupied by Al is too large. The microcracks show the crack opening, closing and reopening hysteresis with the temperature, indicating the mechanical healing.

The structural differences at the two temperatures result in the coordination polyhedra about the metal ions. The cleavage planes, resulting from edge-shared octahedra in the thermal hysteresis of the ceramic causes the change of binding energy. Tialite ceramic body contracts along c-axis with heating as follows; The thermal expansion coefficients for ALTO and ALTO-A_3S_2 are 9.70×10^{-6}/K, and $0.68-5.48 \times 10^{-6}$/K, respectively. Al_2O_3-SiO_2-TiO_2 (Tialite-mullite) ceramics; Composites with more than 10 vol.% mullite (A_3S_2) as a second phase, enhance the Young's modulus, thermal expansion coefficient and room temperature strength (Fig. 13.5).

Adjacent Al and Ti octahedra collapses because of the lattice site occupied by Al is too large, as shown in the polycrystalline ceramics of ALTO with 10 % mullite. In these simulation we used the MgO stabilized Tialite, and Tialite (Al_2TiO_5)-Mica

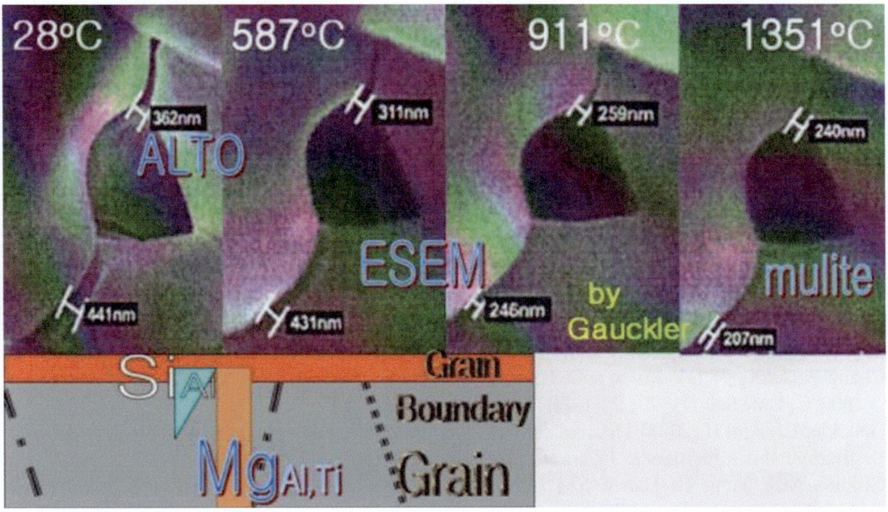

Fig. 13.5 The stress development concept in Tialite ceramics (Gauckler and Kim 2008)

($Mg_3Si_4O_{10}(OH)_2$), which may be the cluster model showing the microcracks (Gauckler and Kim 2008). In the structure of microcracks there will be open, closing and reopening hysteresis, which may be the cause of the mechanical healing.

The critical feature of crystalline anisotropy, leading to thermal expansion anisotropy, is the degree of distortion of the atomic coordination polyhedra. By distortion the bond lengths and angles vary markedly about the cations. Aluminum titanate (Tialite), Al_2TiO_5, as a ceramic body, exhibits a very low bulk thermal expansion coefficient, primarily because of a complex system of internal stresses and fractures. The stresses and fractures are developed upon cooling from the high sintering temperature, and are the result of the anisotropic structural properties of the individual grains which form the ceramic body. Successful application of the material has depended on the ability to control the micro-cracking phenomena.

Conclusion

Compared to other cluster Models, $(Al_5Mg)(Ti_2Si)O_7$ cluster shows the stable incorporation of Mg into Al site and Si into Ti site in Al_2TiO_5 crystal. In the Mg, Si incorporated cluster in ALTO-u system the DV-Xα calculation was not converged and the calculation for the disordered placement of metal sites in ALTO-u was also not converged. Conclusively the disordered structure of metal sites in Al_2TiO_5 is favorable in room temperature structure, but not favorable in 600 °C structure. The similar results were obtained for the incorporation of Mg and/or Si into Al and Ti sites as shown in ALTO-BCT cluster, respectively. The disordered metal structure for ALTO-BCT cluster showed the similar DV-Xα calculation results compared with ALTO clusters.

(continued)

In the incorporation of Mg, Si into ALTO-u the DV-Xα calculation was not converged. On the contrary ALTO-u shows regular distribution. There is the stable incorporation of Mg/Si into Al site in Al_2TiO_5 crystal. The site replacement using Mg^{2+} and Si^{4+} in AlO_6 octahedra induces the increase of total molecular energy, approaching the energy of ALTO-u and indicating the stability of the incorporated cluster structure.

References

Chang MC (2009) DV-Xα approach for the structure of Al_2TiO_5 at room temperature and at 600°C. Bull Soc DV-Xα 22(1–2):228–230

Gauckler LJ, Kim IJ (2008) Thermal durability of aluminum titanate–mullite composites with high thermal shock resistance. J Ceram Process Res 9(3):240

Ghiorso MS, Yang H, Hazen RM (1999) Thermodynamics of cation ordering in Karrooite ($MgTi_2O_5$). Am Mineral 84:1370

Kim IJ, Zhao F, Gong J, Lee KS, Han IS, Kuk WS (2003) J Ceram Process Res 4(2):71

Miyauchi Y, Kagomiya Y, Shimizu Y, Osato H (2007) Influence of TiO_2 particle sizes on the sintering and annealing of Al_2TiO_5-TiO_2 miocrowave dielectric ceramics. J Ceram Soc Jpn 115(11):797

Morosin B, Lynch RW (1972) Structure studies on Al_2TiO_5 at room temperature and at 600°C. Acta Cryst B 28:1040

Tsetsekou A (2005) A comparison study of tialite ceramics doped with various oxide material and tialite–mullite composites: microstructural, thermal and mechanical properties. J Eur Ceram Soc 25:335–348

Chapter 14
Wavelength of Luminescence and Energy Level Structure of Binuclear Copper(I) Complex

Tomohiko Ishii, Masahiro Kenmotsu, Kiyoshi Tsuge, Genta Sakane, Yoichi Sasaki, Masahiro Yamashita, and Brian K. Breedlove

14.1 Introduction

Metal complexes are very attractive compounds because their physical properties, such as electronic and magnetic properties and optical phenomena, can be explained straightforwardly in relation to the ligand field splitting of the d or f orbitals of the central metal atom (Cotton et al. 1994, 1999; Figgis and Hitchman 1999). Many coordination chemists are currently studying how to control the ligand field splitting by chemically modifying the ligands and the coordination environment of the central metal ion. Unfortunately, no rules for systematically changing the ligand field splitting have been reported up to now. Systematic control of the wavelength at which a complex absorbs light has been extensively studied. For example, Tsuchida et al. have developed a spectrochemical series (Tsuchida 1938a, b; Shimura and Tsuchida 1956; Schönherr 2004a, b; Jørgensen 1958),

T. Ishii (✉) • M. Kenmotsu
Department of Advanced Materials Science, Faculty of Engineering, Kagawa University, 2217-20 Hayashicho, Takamatsu 761-0396, Kagawa, Japan
e-mail: tishii@eng.kagawa-u.ac.jp

K. Tsuge
Department of Chemistry, Faculty of Science, University of Toyama, Gofuku 3190, Toyama-shi 930-8555, Toyama, Japan

G. Sakane
Department of Chemistry, Faculty of Science, Okayama University of Science, 1-1, Ridaicho, 700-0005 Okayama, Japan

Y. Sasaki
Division of Chemistry, Graduate School of Science, Hokkaido University, Sapporo 060-0810, Hokkaido, Japan

M. Yamashita • B.K. Breedlove
Department of Chemistry, Graduate School of Science, Tohoku University, Aoba, Aramaki, Aoba-ku, Sendai 980-8578, Miyagi, Japan

T. Ishii et al. (eds.), *The DV-Xα Molecular-Orbital Calculation Method*,
DOI 10.1007/978-3-319-11185-8_14

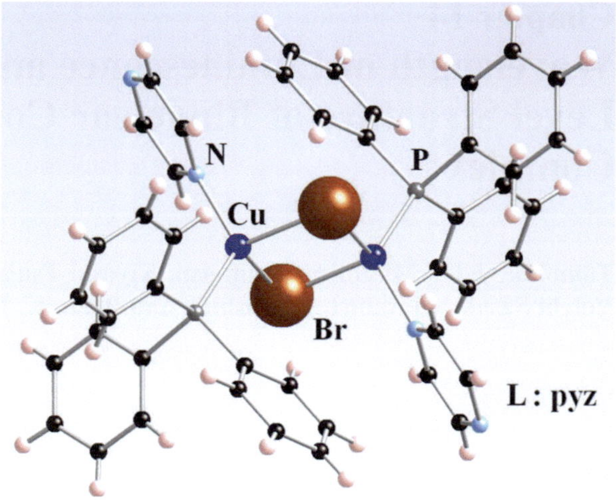

Fig. 14.1 Molecular structure of bromine-bridged copper(I) binuclear complex $[Cu_2(\mu\text{-}Br)_2(PPh_3)_2L]$ (L = pyz) exhibiting luminescence

$$CO > CN^- > PPh_3 > NO_2^- > SO_3^{2-} > phen > bpy > en > NH_3 > py > CH_3CN > ONO^- > NCS^- > OH_2 > C_2O_4^{2-} > ONO_2^- > OSO_3^{2-} > OCHO^- > OH^- > OCO_2^{2-} > OCOR^- > S_2O_3^{2-} > F^- > N_3^- > Cl^- > SCN^- > S^{2-} > Br^- > I^-,$$

showing the relationship between the first absorption band of the octahedral $[Co(NH_3)_5L]$ and ligand L. In addition, they have developed a spectrochemical series in relation to the metal ion (M) in $[M(NH_3)_6]^{n+}$ and $[M(H_2O)_6]^{n+}$:

$$Pt^{4+} > Ir^{3+} > Pd^{4+} > Ru^{3+} > Rh^{3+} > Mo^{3+} > Mn^{4+} > Co^{3+} > Fe^{3+} > V^{2+} > Fe^{2+} > Co^{2+} > Ni^{2+} > Mn^{2+}.$$

These spectrochemical series are independent from each other. Recently, we have reported a novel 2D universal spectrochemical series that can be used for any combination of central metal atom and ligand molecules (Ishii et al. 2009). Therefore, the physical properties arising from the ligand field splitting width of not only known complexes but also unknown ones can be predicted by means of molecular orbital calculations.

In the case of the ligand dependence on the emission spectrum, Araki and Tsuge et al. have reported (Araki et al. 2005) that the luminescence wavelength from red to blue light can be controlled systematically by changing the ligand (L) in halogen-bridged copper(I) binuclear complexes, $[Cu_2(\mu\text{-}X)_2(PPh_3)_2L]$ (X = Br or I) (L = N-heteroaromatic ligands) (Figs. 14.1 and 14.2).

The interatomic distances in the bromine-bridged copper(I) binuclear complexes change slightly in relation to the L used. The Cu-Br, Cu-P, Cu-N, and nearest Cu. . .Cu distances are summarized in Fig. 14.3.

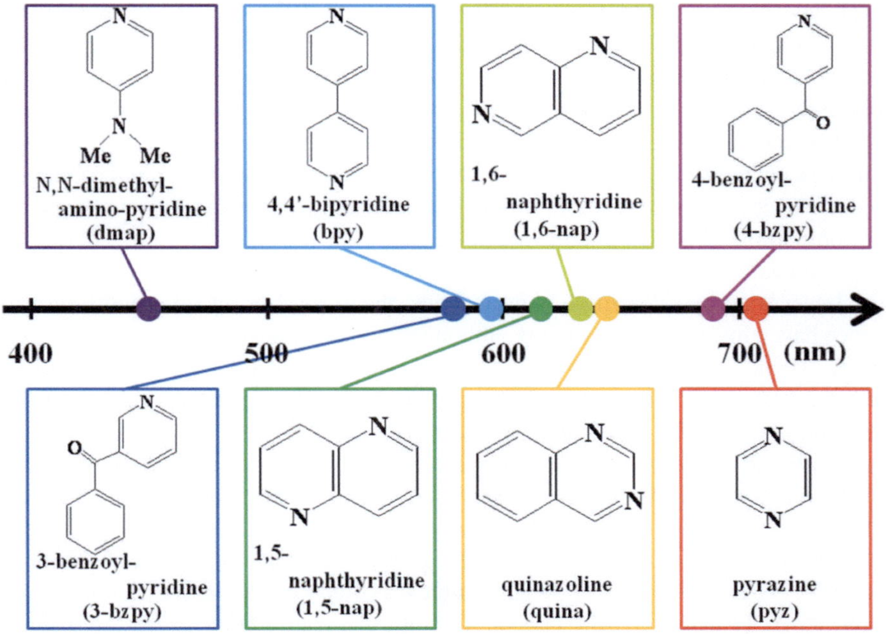

Fig. 14.2 Observed luminescence wavelength (nm) bromine-bridged copper(I) binuclear complex $[Cu_2(\mu\text{-}Br)_2(PPh_3)_2L]$ in relation to L (dmap, 3-bzpy, bpy, 1,5-nap, 1,6-nap, quina, 4-bzpy, and pyz)

Fig. 14.3 Cu-Br, Cu-P, Cu-N, and nearest Cu...Cu distances in the bromine-bridged copper(I) binuclear complex $[Cu_2(\mu\text{-}Br)_2(PPh_3)_2L]$ (L = dmap, 3-bzpy, bpy, 1,5-nap, 1,6-nap, quina, 4-bzpy, and pyz) versus experimental luminescence wavelengths

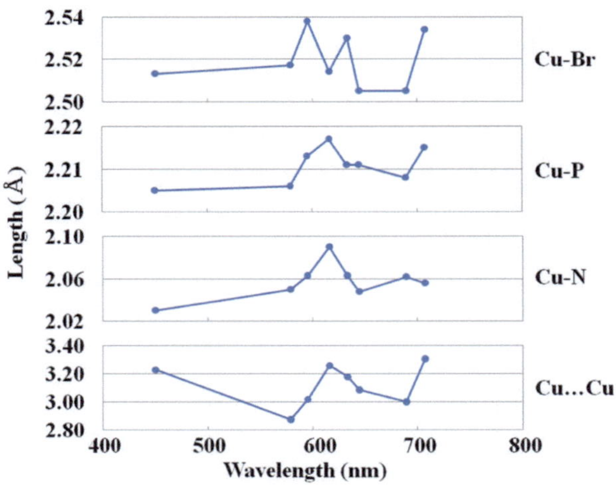

On the basis of Fig. 14.3, there appears to be no relation between the bond distances and the experimental luminescence wavelength. In other words, the experimental luminescence wavelength cannot be discussed on the basis of the molecular structure. Therefore, the electronic structures are the only way to explain the experimental luminescence wavelength of these metal complexes. Tsuge et al. have reported that the observed luminescence wavelength can be explained by the difference in the reduction potential. Of course, changing of the reduction potential is one of the most important ways to change the electronic structure. However, it is only possible for metal complexes whose reduction potentials are already known. Therefore, we investigated the relationship between the observed luminescence wavelength and the electronic structure of the halogen-bridged binuclear copper complexes in order to predict the luminescence wavelength of not only known but also unknown complexes.

In general, luminescence occurs when an excited electron decays in energy to the lower state by releasing a photon. Thus, the absorption of light and luminescence have almost the same energy, although they are opposite processes from each other. Therefore, if we accurately know the energies of the ground and excited states of the metal complexes, we can predict the experimental wavelength of the absorption and/or the luminescence correctly.

In order to discuss the luminescence mechanism, an accurate electronic structure of the excited states is needed. However, it can be difficult to calculate such an electronic structure. Although luminescence involves an electron decaying in energy from excited to ground states, the ground state structure cannot be used to describe the excited states of metal complexes. Many coordination chemists have been trying to determine the crystal structure of metal complexes in the excited state (Ozawa et al. 2003, 2008). Although, only a few percent of the surface structure of a single crystal becomes excited when the single crystal is irradiated with an ultraviolet lamp or CW laser during data acquisition, structural data of the complex in an excited state can be separated from the ground state crystal structural data. In this study, we calculated the electronic structures of the excited and ground states in order to compare the energy levels before and after irradiating with an ultraviolet lamp and to determine their relationship to the observed luminescence wavelength. In addition, we compared the electronic structures and the crystal structures of the ground and excited states in order to discuss the luminescence mechanism. On basis of our calculations, we found that the experimental luminescence wavelength was not due to differences in the crystal structures but due to the difference in the electronic structures of the metal complexes. In other words, the energy of the luminescence is proportional to the ligand field splitting width of the 3d metal orbitals when a ligand coordinates to a metal ion, which is similar to the case of absorption (Ishii et al. 2009).

14.2 Calculations

In order to determine the electronic structure of the halogen-bridged binuclear copper complexes, we carried out electronic structure calculations by using a discrete variational (DV)-Xα molecular orbital method, which has been described elsewhere (Adachi et al. 2005; Ishii et al. 2009; Rosén and Ellis 1975; Adachi et al. 1978a, b; Satoko et al. 1978; Adachi and Taniguchi 1978). The crystal

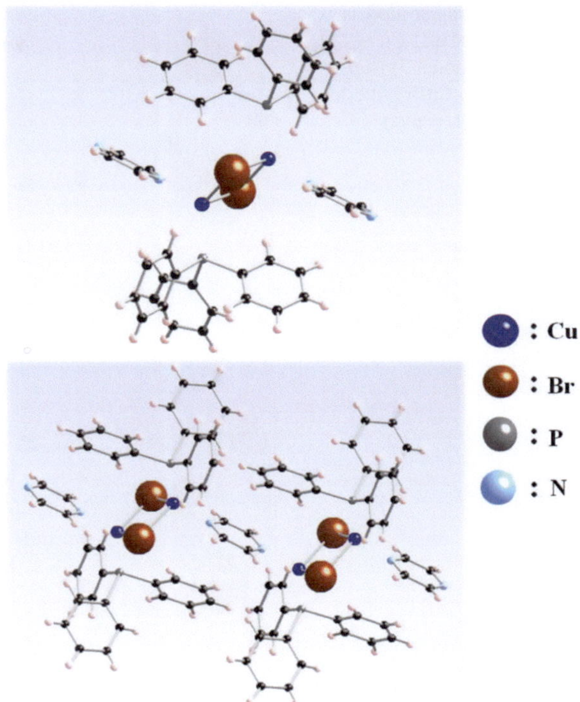

Fig. 14.4 Cluster models of halogen-bridged binuclear copper complexes, L-Cu(PPh_3)(μ-Br)$_2$Cu(PPh_3)-L (model 1) (*top*) and L-Cu(PPh_3)(μ-Br)$_2$Cu(PPh_3)-L-Cu(PPh_3)(μ-Br)$_2$Cu(PPh_3)-L (model 2) (*bottom*) (L = pyz)

structures of the metal complexes of halogen-bridged binuclear copper complexes were taken from the Cambridge Structure Database (Allen 2002), and the CIF data is included in the Electronic Supporting Information (ESI) (Araki et al. 2005). The chemical formulae used for the molecular orbital calculations of the metal complexes were L-Cu(PPh_3)(μ-Br)$_2$Cu(PPh_3)-L (L = dmap, 3-bzpy, bpy, 1,5-nap, 1,6-nap, quina, 4-bzpy, and pyz) (L-M-L) (model 1) and L-Cu(PPh_3)(μ-Br)$_2$Cu(PPh_3)-L-Cu(PPh_3)(μ-Br)$_2$Cu(PPh_3)-L (L-M-L-M-L) (model 2) (L = bpy, 1,5-nap, 1,6-nap, quina, and pyz), as shown in Fig. 14.4.

In order to compare the cluster model dependence and the size effect, we employed two more models: Cu(μ-Br)$_2$Cu(PPh_3)-L-Cu(PPh_3)(μ-Br)$_2$Cu (model 3) and Br_2Cu(PPh_3)-L-Cu(PPh_3)Br_2 (model 4), as shown in Fig. 14.5.

Preliminary crystal structural data of the excited state obtained by using X-ray diffraction while irradiating with CW laser were collected at low temperatures (104, 153, and 253 K) by Ozawa and Toriumi et al. (Ozawa et al. 2003, 2008). They have tried to observe direct geometrical distortion of the complex by photo exited state crystallographic technique. Single crystal X-ray diffraction experiment was performed by using the low temperature vacuum X-ray camera at SPring-8 BL02B1 beam-line. Full intensity data of both under light irradiated by CW laser, and non-irradiated conditions were collected by multiple-exposure IP method. In the excited state, the nearest Cu...Cu distance was almost the same, whereas the nearest Br...Br distance was about 10 % shortened after irradiating with CW

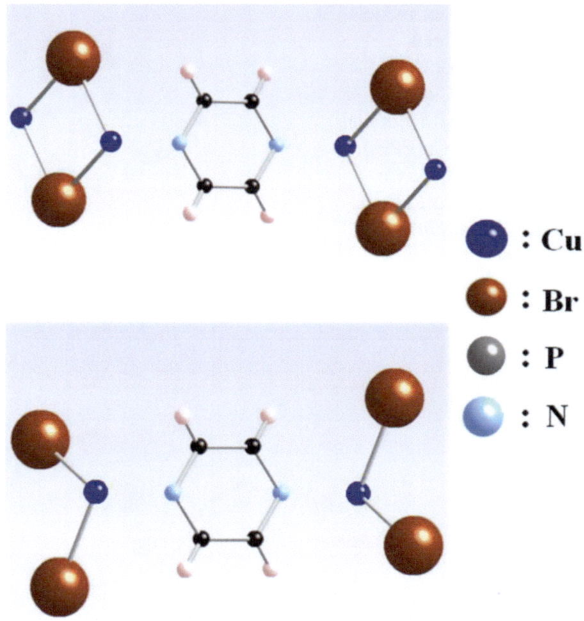

Fig. 14.5 Cluster models of Cu(μ-Br)$_2$Cu-L-Cu (μ-Br)$_2$Cu (model 3) (*top*) and $CuBr_2$-L-$CuBr_2$ (model 4) (*bottom*) (L = pyz)

laser, by photo-difference Fourier technique. Therefore, we developed a molecular structure model of the excited state (model 5, Fig. 14.19 in the ESI) and calculated the electronic structure in order to compare the electronic structures between the ground and excited states.

Non-relativistic DV-Xα calculations were performed by using a Slater exchange parameter (α) of 0.7 for all atoms and up to a million DV sampling points, which gave a precision of better than 0.001 eV for the valence electron energy eigenvalues. We employed the basic functions of the copper ions up to 4p orbital and those of the halogen and atoms of the ligand molecules up to the 2p, 3p, 4p or 5p orbitals, depending of the kind of atom, in order to optimize the electronic structure. The calculations were carried out self-consistently until the difference in orbital populations between the initial and final states of the iteration was less than 0.0001 electrons. Detailed results of the calculations are summarized in the ESI.

14.3 Results and Discussions

14.3.1 Energy Level Structure

In the case of organic compounds, absorption involves a π-π* transition and/or the energy difference between the HOMO and the LUMO energy levels. On the other hand, in the case of metal complexes, there is a relationship between the observed wavelength of the absorption and the energy difference among the d orbitals

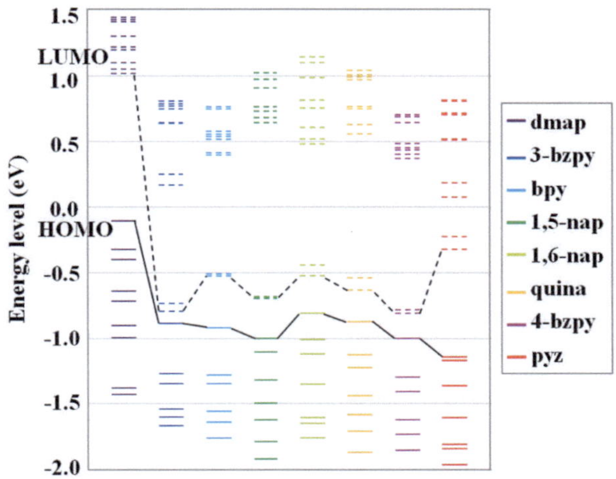

Fig. 14.6 Ligand (L) dependence of the energy level diagrams of L-Cu(PPh_3)(μ-Br)$_2$Cu(PPh_3)-L (L = dmap, 3-bzpy, bpy, 1,5-nap, 1,6-nap, quina, 4-bzpy, and pyz) (L-M-L) (model 1). *Solid* and *dashed lines* denote the occupied and unoccupied molecular orbitals, respectively. The colors used in this figure correspond to the observed luminescence wavelengths

separated due to ligand field splitting. However, the molar extinction coefficient (ε) of the d-d orbital transition is in the range from 10^0 to 10^2 (M^{-1} cm^{-1} (1 M = 1 mol dm^{-3})) and is much smaller than those of the intervalence transition (10^2–10^3), charge transfer transition (10^3–10^4), and the ligand-based transition (10^3–10^5). Since ε corresponds to the intensity of the absorption band, in this study, it is of no concern because we were interested in the luminescence wavelength (photon energy). So, among analogous metal complexes, the wavelength should change in relation to a combination of a d-d transition, metal to ligand charge transfer (MLCT), halide to ligand charge transfer (XLCT), ligand to ligand charge transfer (LLCT), ligand to metal charge transfer (LMCT), and so on.

Energy level diagrams of the cluster model 1(L-M-L) with several ligands are plotted in Fig. 14.6.

Further details concerning the energy level diagrams are summarized in the ESI (Fig. 14.20). On the basis of the energy level diagrams, there is no relationship between the observed luminescence wavelength and the energy difference between the HOMO and the LUMO levels. In addition, there is no relationship involving the energy dispersion among the occupied and the unoccupied energy levels. The molecular orbitals of the metal complexes are a complex mixture of σ, π, and π* orbitals from the copper 3d, bromine 4p, and phosphorus 3p orbitals. Therefore, it is very difficult to discuss the luminescence mechanism directly in relation to specific molecular orbitals. In the next section, we discuss the atomic orbitals rather than the molecular orbitals, which are linear combinations of atomic orbitals (LCAOs).

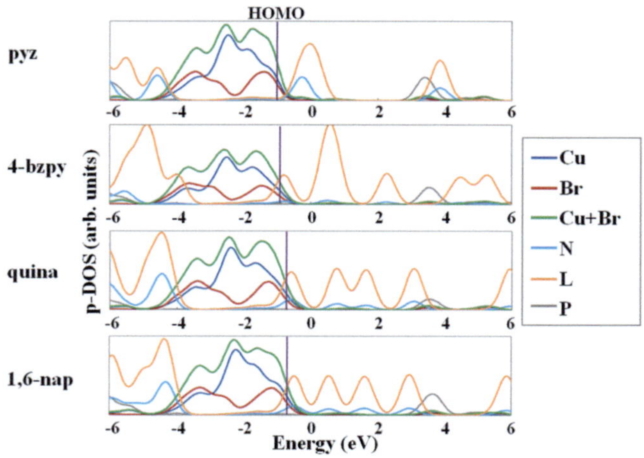

Fig. 14.7 Partial density of states (p-DOS) near the Fermi energy level of L-Cu(PPh_3)(μ-Br)$_2$Cu (PPh_3)-L (L-M-L) (model 1) (L = pyz, 4-bzpy, quina, and 1,6-nap). *Solid vertical line* denotes the energy level of the HOMO

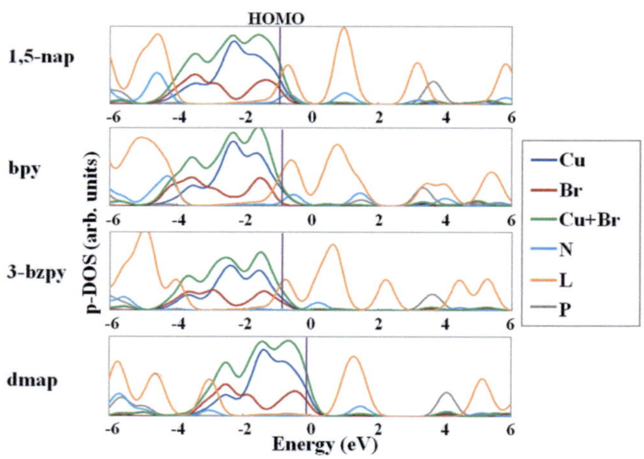

Fig. 14.8 Partial density of states (p-DOS) near the Fermi energy level of L-Cu(PPh_3)(μ-Br)$_2$Cu (PPh_3)-L (L-M-L) (model 1) (L = 1,5-nap, bpy, 3-bzpy, and dmap). *Solid vertical line* denotes the energy level of the HOMO

14.3.2 Partial Density of States (p-DOS)

Each LCAO can be separated into the atomic orbitals by means of the Mulliken's population analysis (Mulliken 1955a, b, c, d). From this analysis, a partial density of states (p-DOS), which is a percentage of each atomic orbital of the density of states, can be obtained. The p-DOSs of the cluster model 1 (L-M-L) with several bidentate ligand molecules are shown in Figs. 14.7 and 14.8.

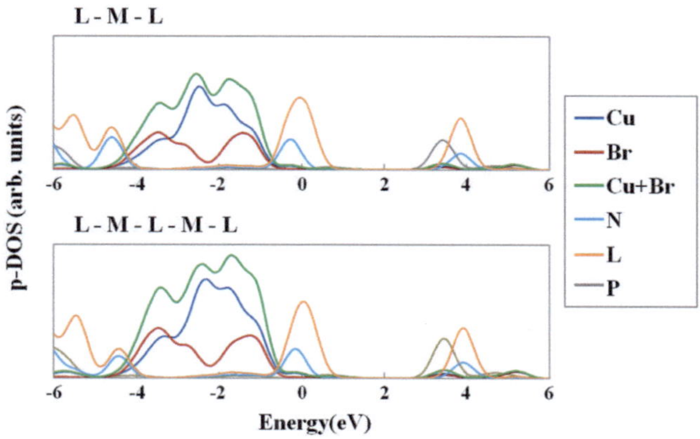

Fig. 14.9 Partial density of states (p-DOS) near the Fermi energy level of L-Cu(PPh_3)(μ-Br)$_2$Cu(PPh_3)-L (L-M-L) (model 1) (*top*) and L-Cu(PPh_3)(μ-Br)$_2$Cu(PPh_3)-L-Cu(PPh_3)(μ-Br)$_2$Cu(PPh_3)-L (L-M-L-M-L) (model 2) (*bottom*) (L = pyz)

Near the Fermi energy in the occupied orbital region, the molecular orbitals were mainly a mixture of copper 3d and bromine 4p orbitals. On the other hand, in the unoccupied region, they are mainly a combination of ligand σ, π, and π^* orbitals. On the basis of the p-DOSs, the luminescence occurs via a transition from an excited state comprised mainly of the ligand orbitals to a ground state comprised of copper 3d and bromine 4p orbitals, after an MLCT-type transition caused by irradiation with an ultraviolet lamp.

Now, we should discuss the validity of the cluster size in the molecular orbital calculations. The original metal complex has a one-dimensional chain structure, in which a bidentate ligand molecule bridges Cu_2(μ-Br)$_2$ units. Therefore, the cluster size of model 1 could be too small for discussing the electronic structures. So, we used an additional model with a larger cluster size having the chemical formula L-Cu(PPh_3)(μ-Br)$_2$Cu(PPh_3)-L-Cu(PPh_3)(μ-Br)$_2$Cu(PPh_3)-L (L-M-L-M-L) (model 2). The p-DOSs of both models 1 and 2 are summarized in Fig. 14.9.

In this figure, the p-DOSs of models 1 and 2 are similar, indicating that the size of the model 1 is large enough to discuss the electronic structure of the metal complex.

Another reason why the two p-DOSs are almost the same is due the simple nature of the metal complex. In other words, the physical properties of metal complexes can be explained straightforwardly by means of the ligand field splitting of the d or f orbitals of the central metal atom, as mentioned in the Introduction. The one dimensional chain structure can be used for discussing electrical conductivity and single-chain magnets (SCMs) (Miyasaka et al. 2009), whereas the electronic structure of the binuclear unit can only be used for discussing optical properties, such as absorption and emission. Therefore, we can discuss the electronic structures of the metal complexes if we can create model of the metal complex with a central metal ion and a minimum number of ligands of an appropriate size. Luminescence from analogous binuclear copper complexes with one-dimensional chain as well as

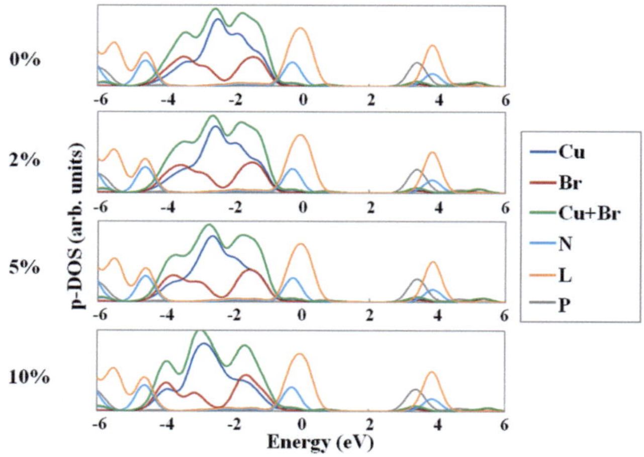

Fig. 14.10 Partial density of states (p-DOS) near the Fermi energy level after 0, 2, 5, and 10 % structural deformation of L-Cu(PPh_3)(μ-Br)$_2$Cu(PPh_3)-L (L-M-L) (model 1) (L = pyz)

non-chain discrete structures has been observed. Therefore, model 2, which has a larger size, is not required. Details of the calculations with bpy, quina, 1,6-nap, 1,5-nap, and pyz are summarized in Figs. 14.21, 14.22, 14.23, and 14.24 in the ESI.

From Fig. 14.9, the character of the ligand field splitting of the copper 3d orbitals could be obtained. The ligand field splitting is caused by the bridging bromine, terminal triphenylphosphine and L. The mechanism for luminescence involves an MLCT transition between the copper 3d and the ligand molecular orbitals. Therefore, the observed luminescence wavelength is closely related to the energy splitting widths of the copper 3d and the ligand molecular orbitals. In the case of the copper 3d orbitals, the 4p orbitals of the bridging bromine atoms bind tightly to the copper 3d orbitals on the basis of p-DOS. We can compare the energy splitting widths more clearly by means of the sum of the copper 3d and bromine 4p orbitals, as shown in Fig. 14.9. Similar calculations involving other ligand molecules are summarized in Figs. 14.21, 14.22, 14.23, and 14.24 in the ESI.

Ozawa et al. have performed crystal structural analyses on these binuclear copper complexes in the excited state while irradiating the crystal with CW laser (Ozawa et al. 2003, 2008). In their experiment, the interatomic distance between the two copper atoms is almost the same, whereas that between the two bromine atoms is shortened in the excited state. The structural change is reported to be about 10 %. In the excited state, the electron in the copper 3d orbital slightly moved outside from the Cu_2Br_2 unit plane due to MLCT. Thus, the intermolecular distance between the two bromine atoms is shortened due to the expansion of the two copper atomic orbitals. Luminescence can be explained by the electron in the ligand molecular orbital dropping into the copper 3d and bromine 4p orbitals.

We calculated the electronic structures using extremely deformed cluster models in order to compare the effect from the structure changing between the ground and the excited states. The p-DOSs of the deformed cluster models with pyz as a ligand are shown in Fig. 14.10.

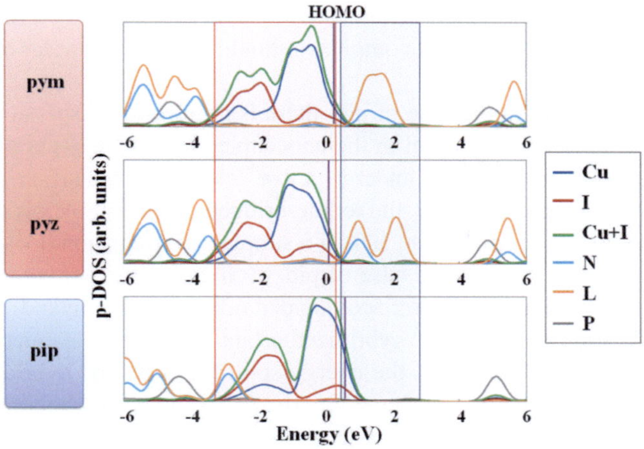

Fig. 14.11 Partial density of states (p-DOS) near the Fermi energy level of L-Cu(PPh_3)(μ-I)$_2$Cu (PPh_3)-L (*top*; L = pym, *middle*; L = pyz, *bottom*; L = pip). *Red* and *blue* rectangle regions denote occupied (mainly by Cu 3d orbital) and unoccupied (mainly by ligand molecular orbital) p-DOSs in relation to an MLCT-type electron transfer, respectively

We did not find much difference in the electronic structures with 0, 2, 5, and 10 % deformation. Similar results were obtained when L = 1,5-nap, bpy, and 3-bzpy, as shown in Fig. 14.25, 14.26, and 14.27 in the ESI.

Taking into account that Ozawa et al. observed about 10 % structural difference between the ground and excited states, the molecular orbital calculations using cluster models based on the ground state CIF data can be used to discuss the electronic structure of the excited state in order to determine the luminescence mechanism.

Luminescence has been observed not only from the bromine-bridged binuclear copper complex but also from the iodine-bridged ones (Araki et al. 2005). In the case of the iodine-bridged analogues, the metal complexes with bpy, pym, and pyz show luminescence, whereas the one with piperidine (pip) does not. The luminescence mechanism is due to an electron dropping after MLCT excitation. Therefore, luminescence should not occur if there is no p-DOS of the ligand orbital near the Fermi energy level in the unoccupied region. Among the analogous binuclear copper complexes, only in the case of the iodine-bridged complex with pip, luminescence is not observed. However, when L has a conjugated π system, luminescence is observed from both the iodine- and bromine-bridged complexes. Therefore, since luminescence appears to be closely related to the conjugated π character of L, we calculated the electronic structure of the complex with pip in order to compare the luminescence mechanisms, and p-DOSs of the iodine-bridged complexes with pym, pyz, and pip are shown in Fig. 14.11.

On the basis of the p-DOSs, the copper 3d and the iodine 5p orbitals are dispersed around the Fermi energy level in the occupied region. In addition, similar electronic dispersions of the ligand orbitals were obtained near the Fermi energy level in the unoccupied region in the cases of pym and pyz. Therefore, both the iodine- and the bromine-bridged complexes should exhibit luminesce via the same mechanism involving MLCT.

On the other hand, in the case of pip, there is no electron dispersion of p-DOS just above the Fermi energy level in the unoccupied region. In general, an organic compound needs to be able to undergo a π-π^* electronic transition in order for luminescence to occur. However, the iodine-bridged metal complex is not a simple organic material, meaning that the lack of a simple π-π^* transition does not explain the lack of luminescence. In the case of pip, each carbon or nitrogen atom is sp^3 hybridized. So, there are no π- or π^*-conjugated orbitals in this ligand. The overlap of a metal dz^2 orbital with an sp^3 hybridized orbital and orbital is much better than that with an sp^2 one. Therefore, the energy splitting of an sp^3 hybridized orbital between the bonding and the anti-bonding orbitals should be much more larger than those of the pym and pyz, which have sp^2 hybridized orbitals. As a result, in the case of pip, the energy difference between the copper 3d orbital and the ligand sp^3 molecular orbital is too large for luminescence in the visible region. In Sect. 3.3, we discuss the relationship between the experimental luminescence wavelength of the bromine-bridged binuclear metal complex and L.

14.3.3 Relationship Between the Experimental Luminescence Wavelength and the Energy Splitting Width

The luminescence mechanism of binuclear copper complexes involves an electron in an excited state due to MLCT returning to the ground state. Strictly speaking, the exact energy splitting width of the copper 3d and the halogen 4p (bromine) or 5p (iodine) orbitals in the ground state and the energy splitting width of the ligand molecular orbital in the excited state must be calculated with high accuracy in order to discuss the luminescence wavelength quantitatively.

Now, we consider the reason for the energy splitting. Needless to say, the copper 3d orbitals split in energy when a ligand coordinates to a copper ion. In a metal complex, the metal d orbitals and the ligand σ, π, and π^* orbitals mix to form the bonding and antibonding orbitals. Therefore, the energy level splitting of the ligand molecular orbitals is not independent from the energy level splitting of the metal d orbitals. So, in this section, we compare the electronic structure and the observed luminescence wavelength by means of the energy level splitting of the copper 3d and the halogen 4p or 5p orbitals.

Although we can obtain accurate crystal structural data in the ground state, it is impossible to obtain accurate crystal structural data in the excited state. So, it is impossible to calculate the luminescence wavelength accurately by means of any molecular orbital calculations. However, in order to discuss the same structure, similar chemical structures or analogous metal complexes, it is more effective to compare

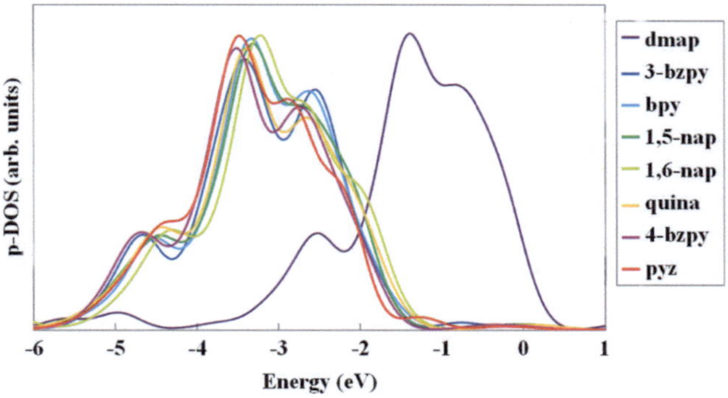

Fig. 14.12 Partial density of states (p-DOS) of the 3d orbitals of the Cu^+ ions near the Fermi energy level of L-Cu(PPh_3)(μ-Br)$_2$Cu(PPh_3)-L (L-M-L) (model 1) (L = dmap, 3-bzpy, bpy, 1,5-nap, 1,6-nap, quina, 4-bzpy, and pyz). The colors used in this figure correspond to the experimentally observed luminescence wavelength

Fig. 14.13 Plot of the energy splitting width of Cu 3d orbital versus the experimental luminescence wavelength. The colors used in this figure correspond to the experimentally observed wavelength

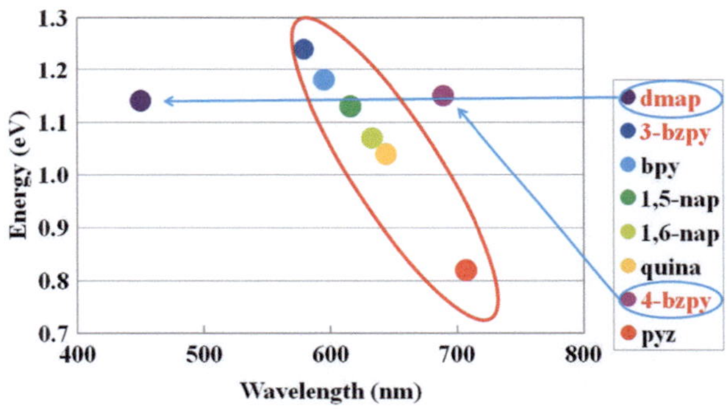

qualitative energy level diagrams. In this study, we compared the experimental luminescence wavelength on the basis of the electronic structures because the binuclear copper complexes have similar structures with similar bidentate ligands. Now, we discuss the energy splitting width of the copper 3d orbitals.

P-DOSs of the copper 3d orbitals of the binuclear complexes are plotted in Fig. 14.12.

The shape of the copper 3d orbitals shown in Fig. 14.12 is due to the distorted tetrahedral crystal field splitting between e_g and t_{2g} groups of the Cu^+ ion by the two bromine ions, triphenylphosphine, and L. The shape of each p-DOS is almost the same, making it possible to compare the energy level splittings to each other.

The relationship between the luminescence wavelength and the energy splitting width of the p-DOS is plotted in Fig. 14.13.

In the region indicated with a red ellipse in Fig. 14.13, a clear linear correlation between the experimental wavelengths and the calculated splitting energies was

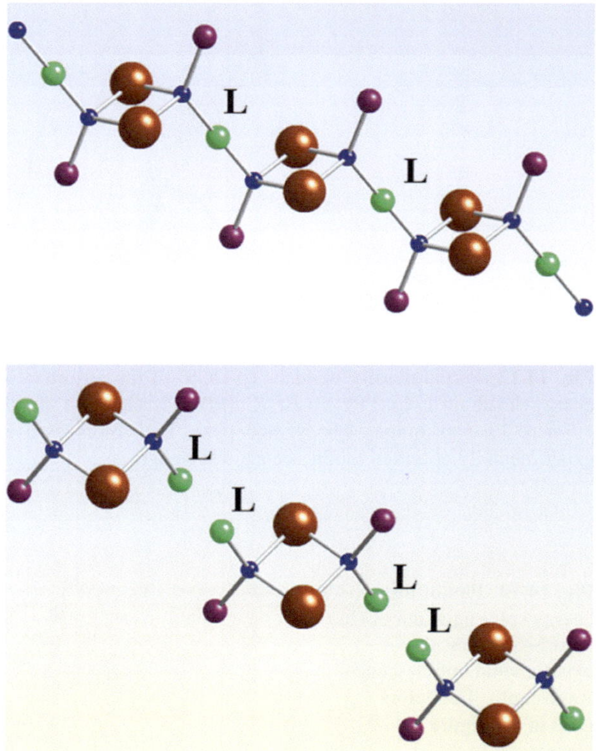

Fig. 14.14 One-dimensional chain structure (L = bpy, 1,5-nap, 1,6-nap, quina, and pyz) (*top*) and a discrete structure (L = dmap, 3-bzpy, and 4-bzpy) (*bottom*) of a binuclear copper complex $Cu_2(\mu\text{-}Br)_2(PPh_3)_2L$

observed. In this figure, the value for dmap and 4-bzpy do not correlate with the other data. One of the reasons for the disagreement comes from the difference in the structure of the metal complex. The structure of binuclear copper complex with dmap, 3-bzpy, or 4-bzpy is not a one-dimensional chain but a discrete structure, as shown in Fig. 14.14.

Dmap, 3bzpy, or 4bzpy are not bidentate ligands but monodentate ones in the discrete structure. In other words, the different coordination environment could give rise to the distance from the linear correlation.

Finally, we employed an another type of cluster model, $CuBr_2(PPh_3)$-L-$CuBr_2(PPh_3)$ (M-L-M) (Fig. 14.5) in order to compare the cluster model dependence with model 1 (L-M-L). The p-DOSs of the M-L-M cluster models 3 ($Cu(\mu\text{-}Br)_2Cu$-L-$Cu(\mu\text{-}Br)_2Cu$) and 4 ($Cu(\mu\text{-}Br)_2Cu$-L-$Cu(\mu\text{-}Br)_2Cu$) with pyz is shown in Fig. 14.15.

On the basis of the p-DOSs of the two models, there are scarcely any differences in the electronic structures. Therefore, the electronic structure of the metal complex is dependent on the coordination environment and the coordination ability of L. Similar results were obtained in the cases of quina (Fig. 14.28), 1,6-nap (Fig. 14.29), 1,5-nap (Fig. 14.30), and bpy (Fig. 14.31) in the ESI.

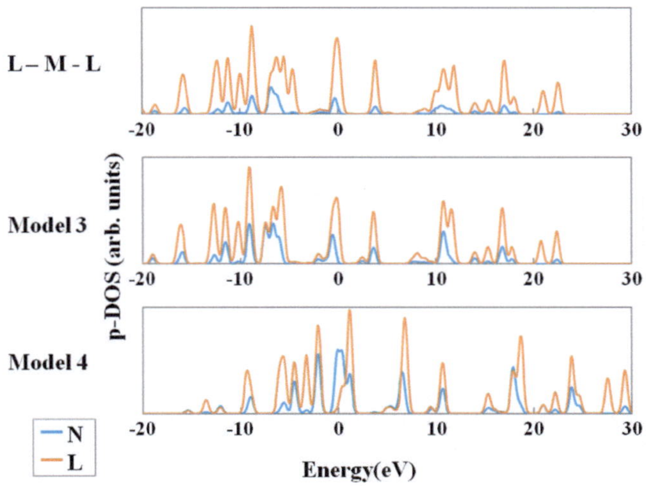

Fig. 14.15 Partial density of states (p-DOS) near the Fermi energy level of L-Cu(PPh_3)(μ-Br)$_2$Cu (PPh_3)-L (L-M-L) (model 1) (*top*), Cu(μ-Br)$_2$Cu-L-Cu(μ-Br)$_2$Cu (model 3) (*middle*) and ç (model 4) (*bottom*) (L = pyz)

14.3.4 Relationship Between the Experimental Wavelength and the Energy Level Difference Between the Metal 3d and Ligand Molecular Orbitals

We have discussed that there is a relationship between the experimental luminescence wavelength of the ligand field splitting width of the copper 3d orbitals caused by the surrounding ligands. As mentioned in Sect. 3.3, the splitting width and luminescence wavelength are linearly correlated except when L = dmap and 4-bzpy. It should be noted that a d-d electronic transition is not involved because the Cu^+ ion has a d^{10} system. Since luminescence from the binuclear copper complexes occurs when an electron returns to the ground state after being excited via MLCT, a discussion involving only the energy splitting of the copper 3d orbitals is insufficient. In addition, because luminescence is observed from both the one-dimensional chain and discrete complexes, other methods of analysis are needed in order to understand the luminescence mechanism even if it is sufficient for the complexes with dmap and 4-bzpy. Therefore, we now introduce a new way for discussing the energy level splitting between the copper 3d and ligand molecular orbitals, as shown in Fig. 14.16.

In Fig. 14.16, near the Fermi energy level in the occupied region, there is an electron dispersion created by the copper 3d and the bridging bromine 4p orbitals. On the other hand, in the unoccupied region, the electron dispersion mainly involves the

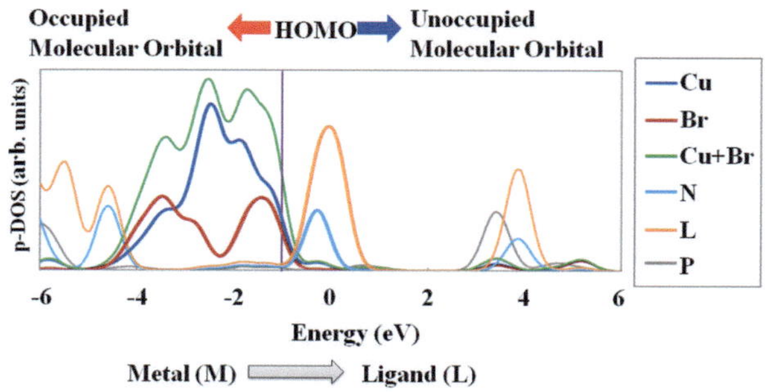

Fig. 14.16 Strategy for discussing the experimental luminescence wavelength of the binuclear copper metal complex using the energy difference between the metal's 3d orbitals in the occupied region and the ligand molecular orbitals in the unoccupied region of the partial density of states (p-DOS) near the Fermi energy level of L-Cu(PPh_3)(μ-Br)$_2$Cu(PPh_3)-L (model 1) (L = pyz). Solid vertical line denotes the HOMO energy level

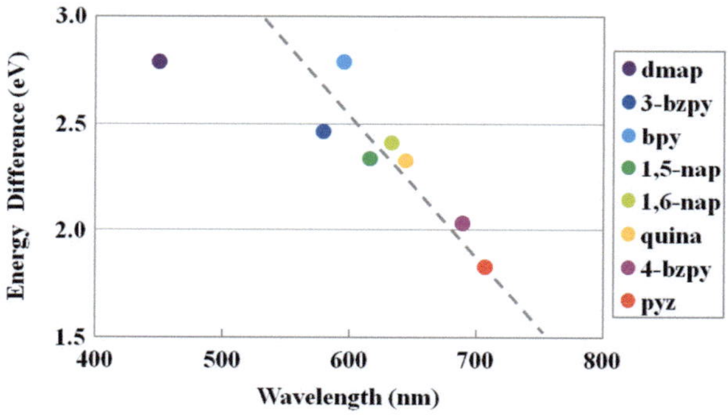

Fig. 14.17 Plot of the energy level difference between the Cu 3d and ligand orbitals versus the experimental luminescence wavelength of L-Cu(PPh_3)(μ-Br)$_2$Cu(PPh_3)-L (L-M-L) (model 1). *Dashed line* indicates a rough linear correlation

ligand molecular π-conjugated orbitals. Therefore, after irradiating with an ultraviolet lamp, an electron in the copper 3d and the bromine 4p orbitals is excited into the unoccupied ligand molecular orbitals via MLCT. After a certain lifetime, the excited electron returns to the original ground state emitting a photon. So, we determined the excitation energy, which is the difference between the numerical averages of the occupied copper 3d and unoccupied ligand molecular orbitals. P-DOSs of the binuclear copper complexes with pyz, 4-bzpy, quina, and 1,6-nap are plotted in Fig. 14.7, and with 1,5-nap, bpy, 3-bzpy, and dmap in Fig. 14.8.

The excitation energy is the energy difference between the copper 3d and ligand molecular orbitals in the p-DOS diagrams, and its relationship to the observed luminescence wavelength is plotted in Fig. 14.17.

From this plot, the tendency of the relationship between the calculated excitation energy and the experimental luminescence wavelength can be described with a linear function. The coordination strength can change not only the energy splitting width of

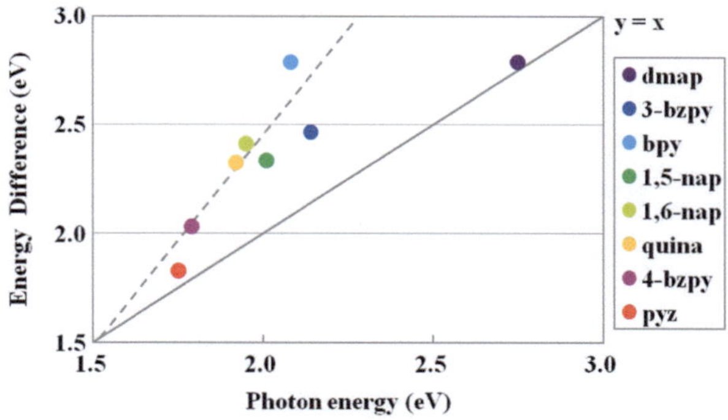

Fig. 14.18 Plot of the energy level difference between Cu 3d and the ligand orbital versus the experimental photon energy of emission of L-Cu(PPh_3)(μ-Br)$_2$Cu(PPh_3)-L (L-M-L) (model 1). *Dashed line* indicates the rough linear correlation between them

the copper 3d orbitals but also that of the ligand molecular orbitals. Therefore, in the case of a luminescence mechanism involving MLCT, the observed wavelength can be explained by the relationship between the excitation energy determined by using molecular orbital calculations and the experimental wavelength. The relationship in Fig. 14.17 is reproduced with the unit of eV, as shown in Fig. 14.18.

Each point is located in the area ($y > x$) above the diagonal of the plot, meaning that the energy of the luminescence should be lower than that of the excitation.

Conclusion

From the result of our calculation, it is revealed that there is the possibility to predict the luminescence wavelength of not only the well-known but also the unknown binuclear copper complexes by means of the strong relationship between the observed and the calculated photon energies. Such a strong relationship had been already confirmed very suitable for any metal complexes. In this mechanism, the luminescence of a light can be represented by an electronic charge transfer between a metal and a ligand (MLCT), after the electron were excited after irradiating by a UV light. The luminescence wavelength was linearly correlated to the energy difference between the metal's 3d orbitals below the HOMO level and the ligand's π-conjugated molecular orbitals above the LUMO level.

Acknowledgement The authors express their thanks to Drs. F. Izumi and K. Momma (Nat'l Inst. for Materials Sci., Japan) for permission to use the 3D visualization program "VESTA". The authors express their gratitude to Prof. H. Adachi (Kyoto Univ.) for permission to use his computational program. The authors also thank Drs. R. Sekine (Shizuoka Univ.), Y. Kowada (Hyogo Univ. of Teacher Edu.), and M. Mizuno (Osaka Univ.) for fruitful discussions and useful technical advice. The authors also thank to Profs. K. Toriumi and Y. Ozawa (Univ. of Hyogo) for fruitful discussions about the molecular structure in the excited state.

Electronic Supporting Information (ESI)

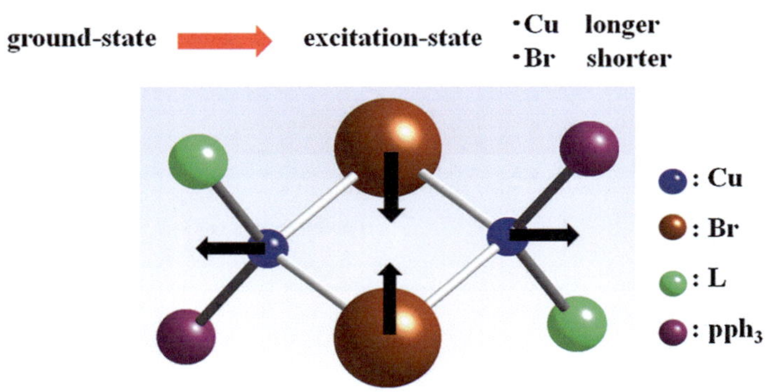

Cu 2%·5%·10% longer

Br 2%·5%·10% shorter

ligand	atom	amount of change (Å) 0%	2%	5%	10%
pyz	Cu-Cu	3.30	3.37	3.46	3.63
	Br-Br	3.84	3.76	3.65	3.46

Fig. 14.19 Expected molecular structural deformation of $Cu_2(\mu\text{-}Br)_2(PPh_3)_2L_2$ after irradiating with an ultraviolet light (model 5)

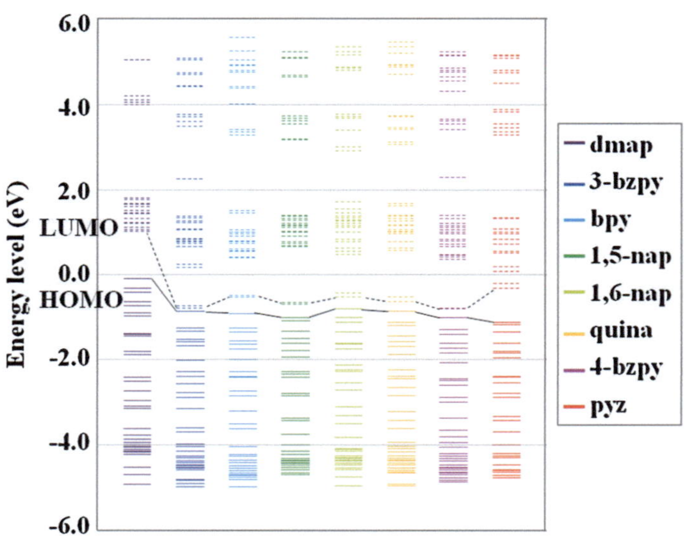

Fig. 14.20 Ligand (L) dependence of the energy level structures of L-$Cu(PPh_3)(\mu\text{-}Br)_2Cu(PPh_3)$-L (L = dmap, 3-bzpy, bpy, 1,5-nap, 1,6-nap, quina, 4-bzpy, and pyz) (model 1). *Solid* and *dashed lines* denote the occupied and the unoccupied molecular orbitals, respectively. The *colors* used in this figure correspond to the observed luminescence wavelengths

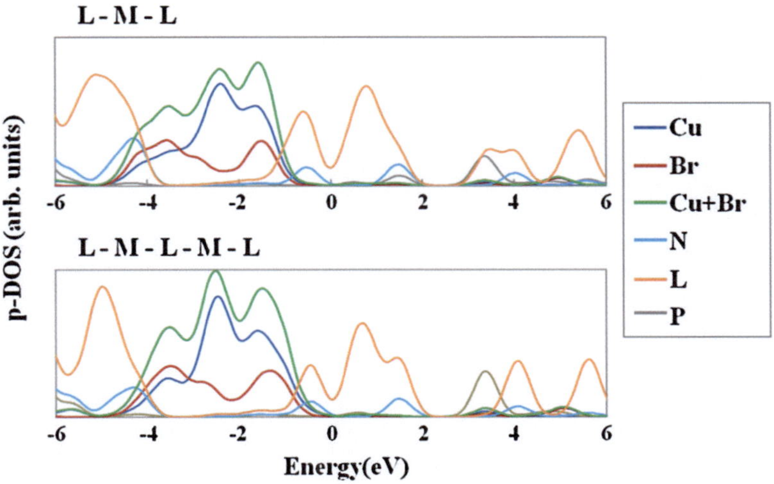

Fig. 14.21 Partial density of states (p-DOS) near the Fermi energy level of L-Cu(PPh_3)(μ-Br)$_2$Cu(PPh_3)-L (L-M-L) (model 1) (*top*) and L-Cu(PPh_3)(μ-Br)$_2$Cu(PPh_3)-L-Cu(PPh_3)(μ-Br)$_2$Cu(PPh_3)-L (L-M-L-M-L) (model 2) (*bottom*) (L = bpy)

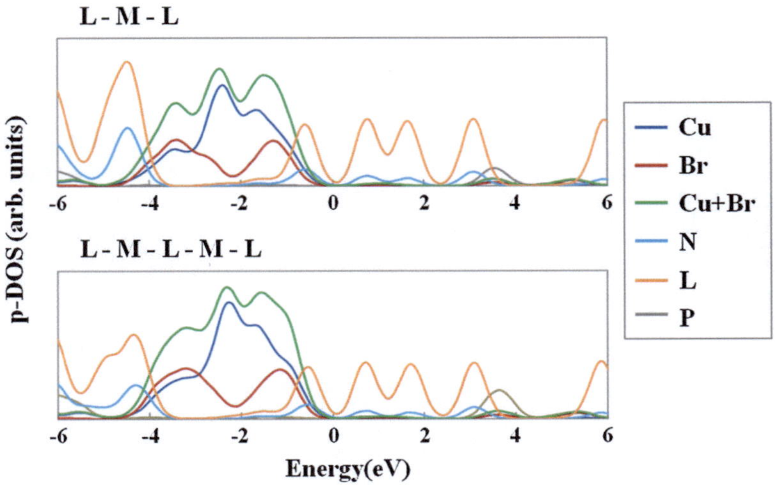

Fig. 14.22 Partial density of states (p-DOS) near the Fermi energy level of L-Cu(PPh_3)(μ-Br)$_2$Cu(PPh_3)-L (L-M-L) (model 1) (*top*) and L-Cu(PPh_3)(μ-Br)$_2$Cu(PPh_3)-L-Cu(PPh_3)(μ-Br)$_2$Cu(PPh_3)-L (L-M-L-M-L) (model 2) (*bottom*) (L = quina)

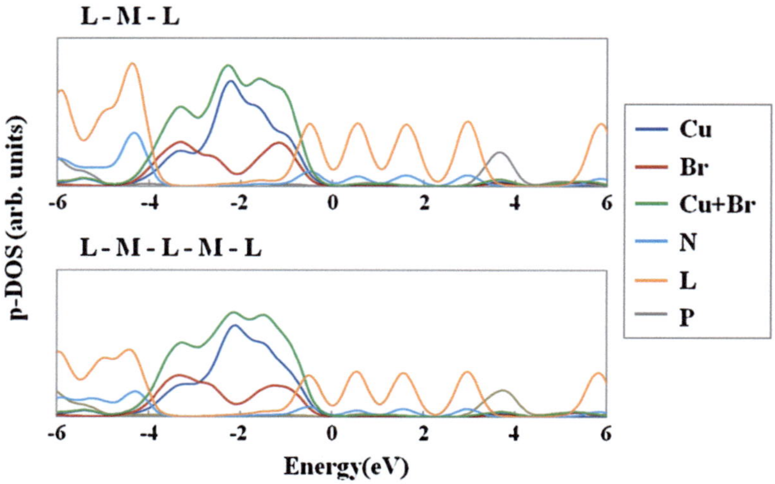

Fig. 14.23 Partial density of states (p-DOS) near the Fermi energy level of L-Cu(PPh_3)(μ-Br)$_2$Cu(PPh_3)-L (L-M-L) (model 1) (*top*) and L-Cu(PPh_3)(μ-Br)$_2$Cu(PPh_3)-L-Cu(PPh_3)(μ-Br)$_2$Cu(PPh_3)-L (L-M-L-M-L) (model 2) (*bottom*) (L = 1,6-nap)

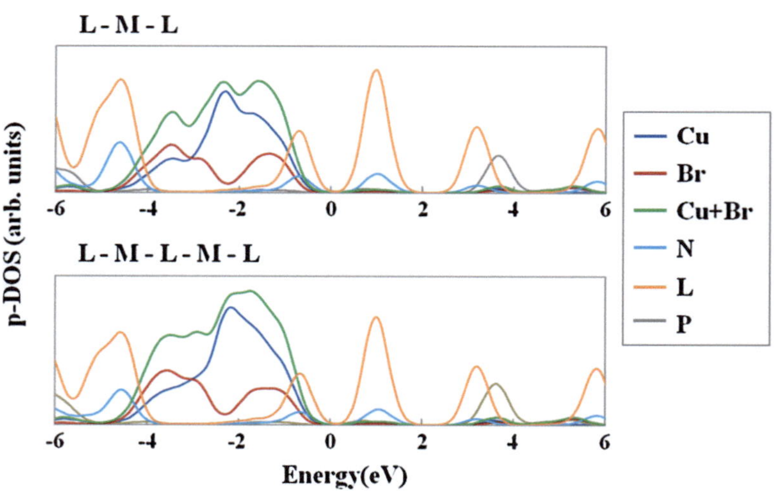

Fig. 14.24 Partial density of states (p-DOS) near the Fermi energy level of L-Cu(PPh_3)(μ-Br)$_2$Cu(PPh_3)-L (L-M-L) (model 1) (*top*) and L-Cu(PPh_3)(μ-Br)$_2$Cu(PPh_3)-L-Cu(PPh_3)(μ-Br)$_2$Cu(PPh_3)-L (L-M-L-M-L) (model 2) (*bottom*) (L = 1,5-nap)

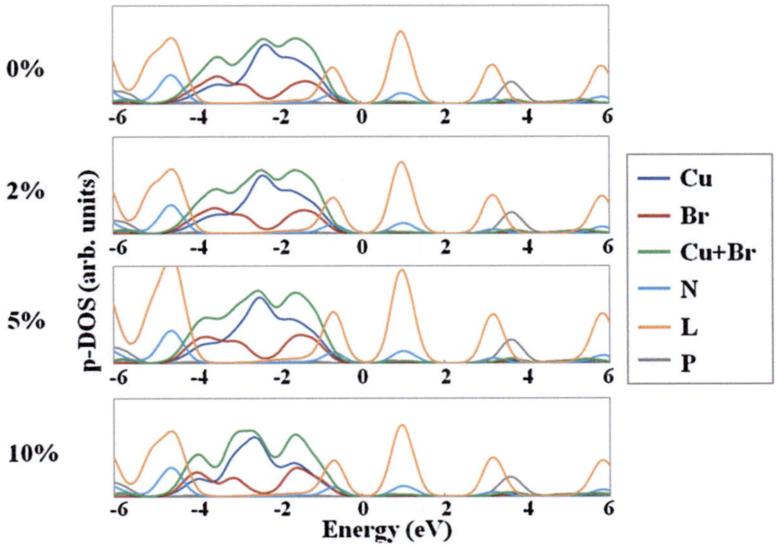

Fig. 14.25 Partial density of states (p-DOS) near the Fermi energy level after 0, 2, 5, and 10 % structural deformation of L-Cu(PPh_3)(μ-Br)$_2$Cu(PPh_3)-L (L-M-L) (model 1) (L = 1,5-nap)

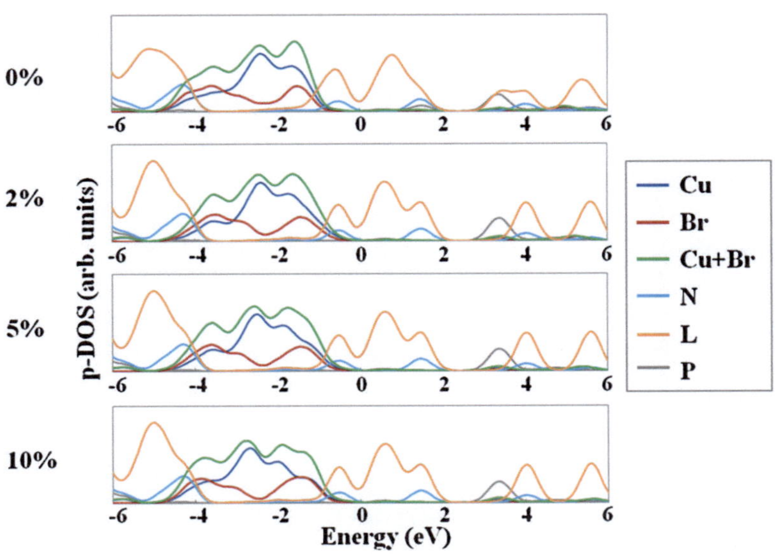

Fig. 14.26 Partial density of states (p-DOS) near the Fermi energy level after 0, 2, 5, and 10 % structural deformation of L-Cu(PPh_3)(μ-Br)$_2$Cu(PPh_3)-L (L-M-L) (model 1) (L = bpy)

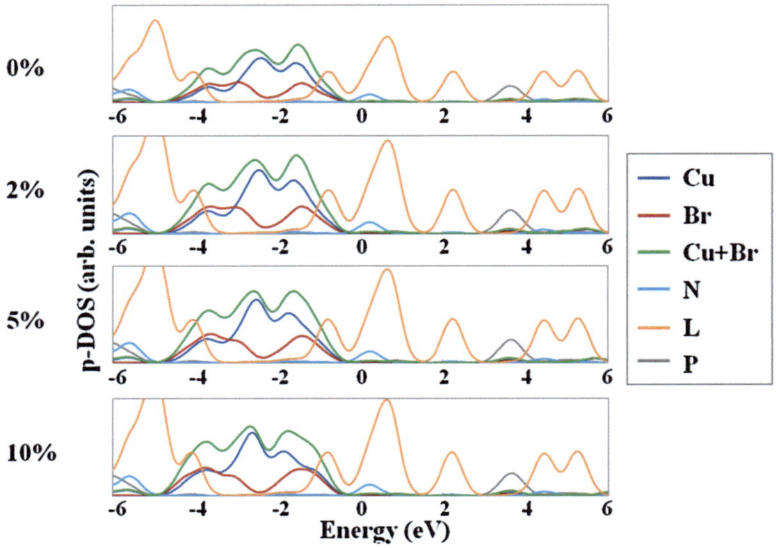

Fig. 14.27 Partial density of states (p-DOS) near the Fermi energy level after 0, 2, 5, and 10 % structural deformation of L-Cu(PPh_3)(μ-Br)$_2$Cu(PPh_3)-L (L-M-L) (model 1) (L = 3-bzpy)

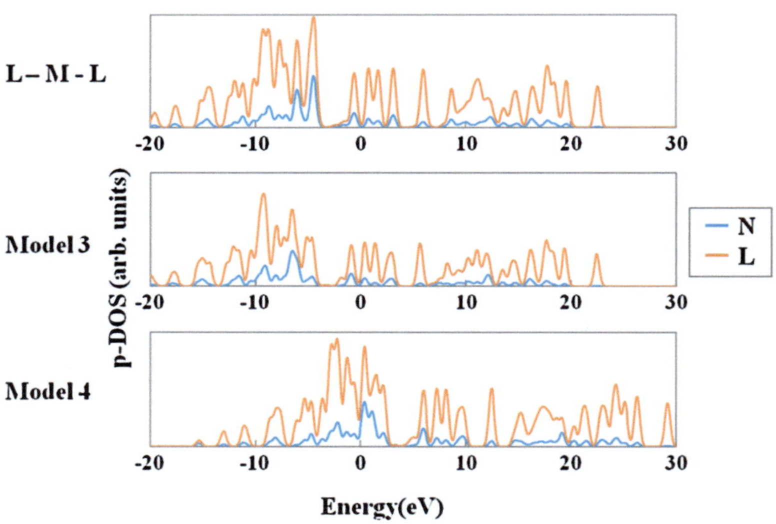

Fig. 14.28 Partial density of states (p-DOS) near the Fermi energy level of L-Cu(PPh_3)(μ-Br)$_2$Cu (PPh_3)-L (L-M-L) (model 1) (*top*), Cu(μ-Br)$_2$Cu-L-Cu(μ-Br)$_2$Cu (model 3) (*middle*), and $CuBr_2$-L-$CuBr_2$ (model 4) (*bottom*) (L = quina)

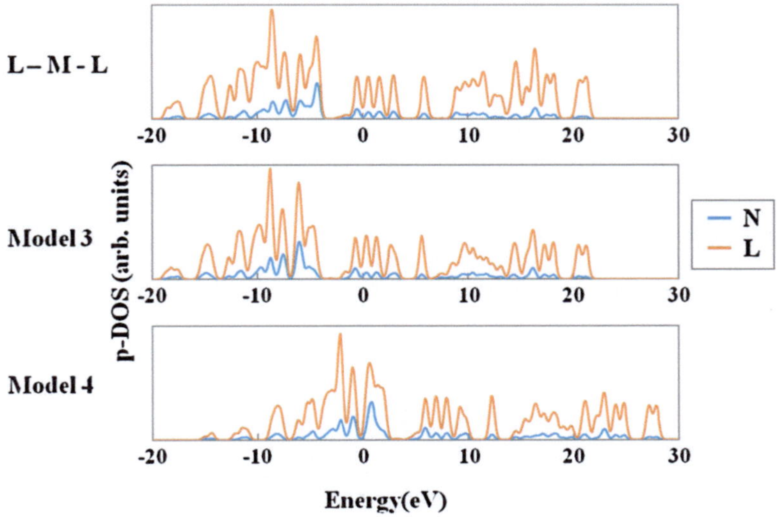

Fig. 14.29 Partial density of states (p-DOS) near the Fermi energy level of L-Cu(PPh_3)(μ-Br)$_2$Cu (PPh_3)-L (L-M-L) (model 1) (*top*), Cu(μ-Br)$_2$Cu-L-Cu(μ-Br)$_2$Cu (model 3) (*middle*), and $CuBr_2$-L-$CuBr_2$ (model 4) (*bottom*) (L = 1,6-nap)

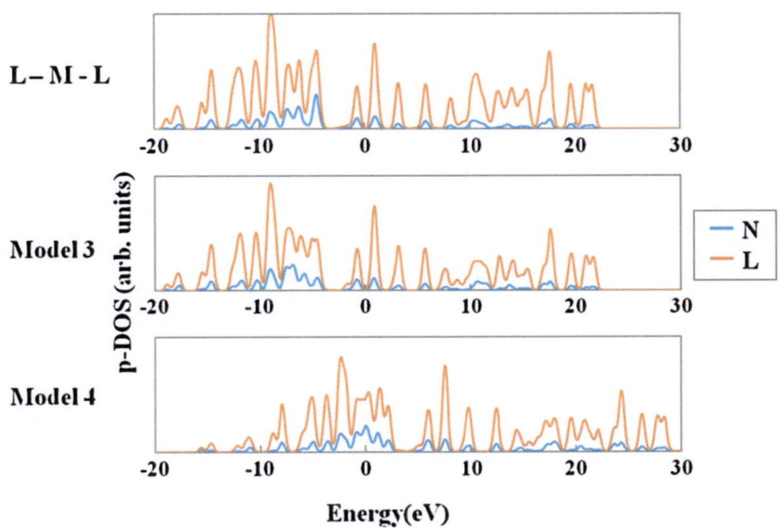

Fig. 14.30 Partial density of states (p-DOS) near the Fermi energy level of L-Cu(PPh_3)(μ-Br)$_2$Cu (PPh_3)-L (L-M-L) (model 1) (*top*), Cu(μ-Br)$_2$Cu-L-Cu(μ-Br)$_2$Cu (model 3) (*middle*), and $CuBr_2$-L-$CuBr_2$ (model 4) (*bottom*) (L = 1,5-nap)

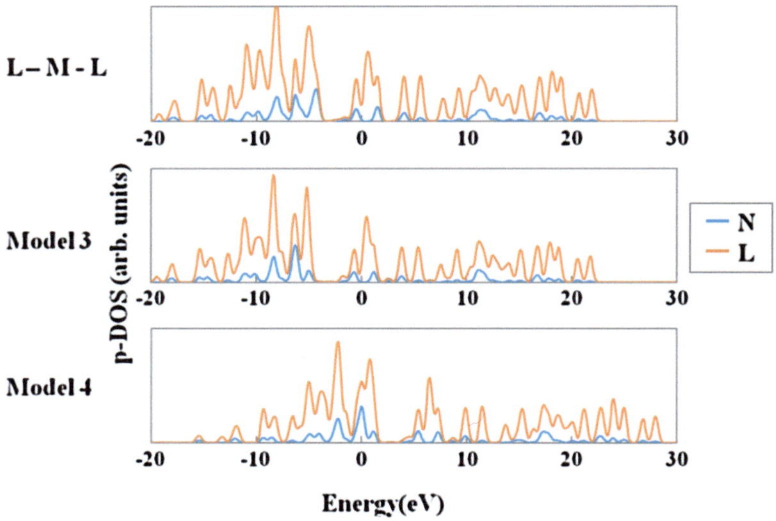

Fig. 14.31 Partial density of states (p-DOS) near the Fermi energy level of L-Cu(PPh_3)(μ-Br)$_2$Cu (PPh_3)-L (L-M-L) (model 1) (*top*), Cu(μ-Br)$_2$Cu-L-Cu(μ-Br)$_2$Cu (model 3) (*middle*), and $CuBr_2$-L-$CuBr_2$ (model 4) (*bottom*) (L = bpy)

References

Adachi H, Taniguchi K (1978) Discrete variational Xα cluster calculations. IV. Application to X-ray emission study. J Phys Soc Jpn 49(5):1944–1953

Adachi H, Tsukada M, Satoko C (1978a) Discrete variational Xα cluster calculations. I. Application to metal clusters. J Phys Soc Jpn 45(3):875–883

Adachi H, Shiokawa S, Tsukada M, Satoko C, Sugano S (1978b) Discrete variational X α cluster calculations. III. Application to transition metal complexes. J Phys Soc Jpn 47(5):1528–1537

Adachi H, Mukoyama T, Kawai J (2005) Hartree–Fock–Slater method for materials science, the DV-Xα method for design and characterization of materials, vol 84, Springer series in materials science. Springer, Berlin/Heidelberg/New York. ISBN 978–3540245087

Allen FH (2002) The Cambridge structural database, a quarter of a million crystal structures and rising. Acta Crystallogr Sect B Struct Sci 58:380–388

Araki H, Tsuge K, Sasaki Y, Ishizaka S, Kitamura N (2005) Luminescence ranging from red to blue: a series of copper(I) – halide complexes having rhombic {Cu2(μ-X)2} (X = Br and I) units with N-heteroaromatic ligands. Inorg Chem 44:9667–9675

Cotton FA, Wilkinson G, Gaus PL (1994) Basic inorganic chemistry, 3rd edn. Wiley, New York. ISBN 978–0471505327

Cotton FA, Murillo CA, Bochmann M (1999) Advanced inorganic chemistry, 6th edn. Wiley-Interscience, New York. ISBN 978–0471199571

Figgis BN, Hitchman MA (1999) Ligand field theory and its applications. Wiley-VCH, New York. ISBN 978–0471317760

Ishii T, Tsuboi S, Sakane G, Yamashita M, Breedlove BK (2009) Universal spectrochemical series of six-coordinate octahedral metal complexes for modifying the ligand field splitting. Dalton Trans (4):680–687

Jørgensen CK (1958) The interelectronic repulsion and partly covalent bonding in transition-group complexes. Discuss Faraday Soc 26:110–115

Miyasaka H, Julve M, Yamashita M, Clérac R (2009) Slow dynamics of the magnetization in one-dimensional coordination polymers: single-chain. Magnets Inorg Chem 48(8):3420–3437
Mulliken RS (1955a) Electronic population analysis on LCAO–MO molecular wave functions. II. Overlap populations, bond orders, and covalent bond energies. J Chem Phys 23:1841
Mulliken RS (1955b) Electronic population analysis on LCAO-MO molecular wave functions. III. Effects of hybridization on overlap and gross AO populations. J Chem Phys 23:2338
Mulliken RS (1955c) Electronic population analysis on LCAO-MO molecular wave functions. IV. Bonding and antibonding in LCAO and valence-bond theories. J Chem Phys 23:2343
Mulliken RS (1955d) Electronic population analysis on LCAO–MO molecular wave functions. I. J Chem Phys 23:1833
Ozawa Y, Terashima M, Mitsumi M, Toriumi K, Yasuda N, Uekusa H, Ohashi Y (2003) Photoexcited crystallography of diplatinum complex by multiple-exposure IP method. Chem Lett 32(1):62–63
Ozawa Y, Yoshida S, Kitayama N, Mitsumi M, Toriumi K, Tsuge K, Araki H, Sasaki Y (2008) Single-crystal structure analysis of photo-excited state of halogen-bridged dicopper(I) complexes. Acta Cryst A64:C615–C616
Rosén A, Ellis DE (1975) Relativistic molecular calculations in the Dirac–Slater model. J Chem Phys 62(8):3039
Satoko C, Tsukada M, Adachi HJ (1978) Discrete variational Xα cluster calculations. II. Application to the surface electronic structure of MgO. Phys Soc Jpn 45(4):1333–1340
Schönherr T (2004a) Optical spectra and chemical bonding in inorganic compounds: special volume I dedicated to Professor Jørgensen (structure and bonding). Springer, Berlin/Heidelberg/New York. ISBN 978–3540008538
Schönherr T (2004b) Optical spectra and chemical bonding in transition metal complexes: special volume II, dedicated to Professor Jørgensen (structure and bonding). Springer, Berlin/Heidelberg/New York. ISBN 978–3540008545
Shimura Y, Tsuchida R (1956) Absorption spectra of Co(III) complexes. II. Redetermination of the spectrochemical series. Bull Chem Soc Jpn 29(3):311–316
Tsuchida R (1938a) Absorption spectra of co-ordination compounds. I. Bull Chem Soc Jpn 13 (5):388–400
Tsuchida R (1938b) Absorption spectra of co-ordination compounds. II. Bull Chem Soc Jpn 13(6):436–450

Index

T. Ishii et al. (eds.), *The DV-Xα Molecular-Orbital Calculation Method*,
DOI 10.1007/978-3-319-11185-8

MIX
Papier aus verantwortungsvollen Quellen
Paper from responsible sources
FSC® C105338

If you have any concerns about our products,
you can contact us on
ProductSafety@springernature.com

In case Publisher is established outside the EU,
the EU authorized representative is:
Springer Nature Customer Service Center GmbH
Europaplatz 3, 69115 Heidelberg, Germany

Printed by Libri Plureos GmbH
in Hamburg, Germany